the Transcendental Temptation

the Transcendental Temptation

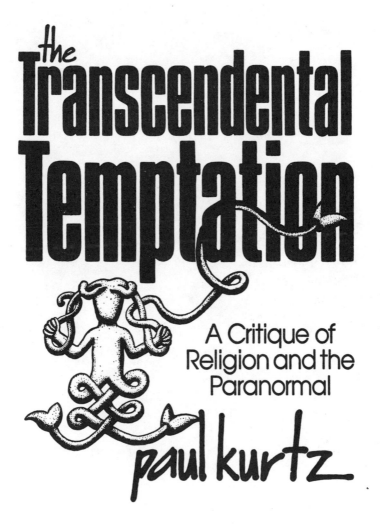

A Critique of
Religion and the
Paranormal

paul kurtz

Prometheus Books
Buffalo, New York

Published 1991 by Prometheus Books

95 94 93 92 91   5 4 3 2 1

Library of Congress Card Catalog No. 86-15082
ISBN: 0-87975-362-5

*Once every people in the world believed that
trees were divine, and could take a human or
grotesque shape and dance among the shadows;
and that deer, and ravens and foxes, and
wolves and bears, and clouds and pools, almost
all things under the sun and moon, and the
sun and moon, were not less divine and
changeable. They saw in the rainbow the
still bent bow of a god thrown down in his
negligence; they heard in the thunder the sound
of his beaten water-jar, or the tumult of his
chariot wheels; and when a sudden flight of wild
ducks, or of crows, passed over their heads,
they thought they were gazing at the dead
hastening to their rest; while they dreamed of
so great a mystery in little things that they
believed the waving of a hand, or of a sacred
bough, enough to trouble far-off hearts, or
hood the moon with darkness.*

W. B. Yeats

Contents

PART TWO: MYSTICISM, REVELATION, AND GOD

PART THREE: SCIENCE AND THE PARANORMAL

PART FOUR: BEYOND RELIGION

Preface to the paperback edition

Is there buried deep within the hearts and minds of men and women a powerful tendency to accept transcendental/paranormal accounts of reality? If so, is this proclivity so strong that it can explain the persistence of orthodox theistic systems of religion—in spite of the overwhelming refutations of their claims—and the recent emergence of new cults of the paranormal to replace ancient doctrines and dogmas?

In the intervening years since the first publication of the hardcover edition of this book, my research has reinforced in my own mind the validity of the transcendental temptation thesis. First, I have been deeply involved with many investigators in the study of "faith healers." We have visited numerous faith-healing services to observe faith-healers at work. And we have been stunned by the degree of self-deception and gullibility displayed by the great numbers of people in quest of miracle healings. Our best efforts have not been able to find any evidence for miraculous cures of people with organic afflictions. Surely psychosomatic illnesses are sometimes helped, but this may be explained by the placebo effect, without involving a hidden or occult cause. Yet the power of suggestion is able to mesmerize otherwise rational people into believing that some transcendental force is at work.

Second, the public has been exposed to a spate of highly questionable reports of abductions aboard UFOs. These "abductees" claim that they have been kidnapped by extraterrestrial beings, been spirited aboard alien spacecraft, and been given messages from semi-divine beings from another dimension. This is strikingly reminiscent of revelations received from afar by prophets throughout the ages. This phenomenon has been popularized in books by Whitley Strieber, Budd Hopkins, and others. Such claims, I submit, testify not to the reality of abductions, but to the easy willingness of some people to believe in another realm and to commune with "entities" from it. And it also testifies to the unguarded willingness of a great many people to accept these claims uncritically based on second- and third-hand testimony—although alternative naturalistic and psychological interpretations can better explain these experiences.

Third, I have had the opportunity to visit China and the Soviet Union to study belief in the paranormal and religion, and since then I have carried on considerable correspondence with Chinese and Russian scholars. Both countries had been under tight totalitarian control, with all of the powers of the state being used to stamp out religious belief and practice. Indeed, official Marxist ideology had dominated both societies for decades, and massive campaigns of education, persuasion, and state terror had been used to inculcate an atheist outlook. Yet, surprisingly, in both societies theistic and paranormal/occult/mystical systems of belief seem to be very strong. This suggests that in spite of large-scale efforts to indoctrinate a materialistic, atheistic outlook, it has been very difficult to suppress the transcendental temptation. Indeed, I found the religiosity of communist countries to be as strong or stronger than that in Western capitalist societies. The reasons for this are no doubt highly complex: the efforts to repress religion may have been counterproductive, and religion may have become a symbol of opposition to all of the negative characteristics of a failing system. But it also points to the power of the transcendental temptation— it is apparently able to thrive underground in spite of extensive persecution.

In further reflection on the nature of this temptation, I am persuaded that it is not genetic; for it is absent among secularists and unbelievers, who make up a significant minority of the population; yet it is apparently a product of human nature and may satisfy some psychobiological needs, though how and in what sense depends on the influence of the culture. To find the moral and psychological equivalents of the transcendental temptation is a difficult challenge. Whether we can succeed in doing so is still—as it was when I first wrote this work—an open question.

Paul Kurtz

January 1, 1991

⸺Preface: The skeptic versus the believer

The question pondered in every period of history is that of meaning. Is there some purpose to human experience, some hidden or divine source for the cosmos? In virtually every age human beings have puzzled about their roots. They have debated the question of whether human life is finite or whether there is a divine promise of eternity. The lines have been drawn between two conflicting approaches to the ultimate nature of human reality in the universe. The proponents of the practical stance call themselves empiricists, or rationalists, or scientific skeptics. They are skeptical about the claim of faith that the universe is divine. They are the atheists and agnostics of old; today they are called secular humanists and are castigated as such by believers. Ranged against them are the disciples of a transcendental and theistic world-view. The former are content to live in the world as it is and to deal with it as best they can; they seek to understand it using the categories of logic and experience, and they are willing to undertake the arduous task of transforming it by their own courageous efforts in the light of their own plans and projects. Opposed to them are all those who are discontented with mundane reality and who seek to escape to an imagined universe based upon faith and credulity. Incompatible with the world they encounter, they find that it is too little; they yearn for deeper mysteries and truths, for the promise or hope that there is an unseen dimension to existence.

The scientific rationalist is a skeptic about the received myths. He believes the dominant religions of revelation to be mythological conjectures full of vain hopes and false illusions. These religious fantasies are kindled by a fearful response to the ambiguities of mortal existence, and they weigh down frail human spirits who are seeking to find in their dreams some refuge from the vicissitudes of fortune. Herein lurk the motives that nourish even the most pretentious religious aspirations. Paradoxically, as Kierkegaard admitted, the more absurd the claim, the more grandiose the dream; the more incredible the mystery, the more committed and devoted the true believers are likely to become.

Today, as of old, we face the same conflict within the human psyche: science versus doctrinaire religion, the empirical world of intersubjective verification versus the world of fantasy, knowledge of fact versus romantic superstition, the courage to become versus the secret longing not to have to do so.

The wounds run deep in human civilization. Some cultures become fixated on the quest for the eternal, and some individuals, overwhelmed by this life, forever seek another. The contrast is between the scientist and the mystic, the philosopher and the theologian, the doer and the follower, the knower and the prophet, the independent person and the dependent soul.

In a sense, the schism between these two parts of ourselves is reflected in our culture and is never completely resolved; for no sooner is an accommodation reached in a historical period than it emerges in the next, often even more pronounced than in the previous, though it may assume different forms. It is as if the species Man has a schizoid nature—his feet implanted on the earth but his imaginative soul soaring toward a heaven of magical unreality. Overwhelmed by the ache of humdrum existence, he seeks an escape to another dimension.

Today, a similar dichotomy exists between scientists and paranormalists, the disciplines of rigorous inquiry and the cults of the occult, the world of modern science and the mythology of fundamentalist religion.

It is often difficult to know which side of our nature will dominate and control us. To l i e and function, one must accept the practical realities of common sense and ordinary life and come to terms with them. One cannot be completely out of touch with the external world of brute existence for long—that is, if one is to endure. We need to bake our bread, build our shelter, forge our plowshares, ward off the threats to our daily existence, and deal with friend and foe alike. Men and women could not long survive if they remained fixated in a dream world of religiosity. Knowledge is an instrument of action. We need to understand the objects and events of the world about us, if we are to cope with them and solve the problems of living that may arise. Yet man is fascinated by questions concerning his origin and destiny. Troubled by disease and death, he often craves something more. Thus he tends to read into his life some divine mystery. He erects cathedrals and monuments, develops creeds and dogmas, engages in ceremonies and celebrations—all to deny his mortal existence and to reinforce and give permanence to that which is absent. *Homo religiosus* invents religious symbols, which he venerates and worships to save him from facing the finality of his death and dissolution. He devises paradise fictions to provide succor and support. Man deceives himself about his ultimate destiny so as not to be tormented by the contemplation of it. Although he must of necessity use his intelligence to cope with the world—at least up to a point—he is forever poised, ready to leap beyond reason in an act of faith. There is, he insists, something more to the human drama, thus straining to resist and deny his existential demise. In acts of supreme self-deception, at various times and in various places he has been willing to profess belief in the most incredible myths because of what they have promised him: Moses on Mt. Sinai delivering the Ten Commandments to God's "chosen people," Christ crucified beckoning man to salvation, Mohammed the true prophet of Allah so appointed by Gabriel, Buddha the Light, Joseph Smith and the New Zion of the Mormons, or the Chariots of the Gods transporting extraterrestrial beings from outer space to observe humankind.

The pathos of the human condition is that many or most human beings cannot easily accept the stark realities of human finitude, the fact that there is no ultimate providence or purpose for our existence. Extraterrestrial beings may exist in remote corners of the universe, but there is no evidence that they have had anything to do with us, can communicate through ESP, or are responsible for our future destiny. One must be skeptical about the very existence of extraterrestrials until it is verified and demonstrated—though it remains an exciting possibility of tremendous import.

Most members of the human race consider skepticism to be disenchanting; since it questions their revered dream-fictions, it is also held to be dangerous. The methods of science and reason leading to technology and industry have contributed enormously to the progress of human civilization. Religious believers cannot deny this. They are willing to use the fruits of science and technology—but only up to a point. They are ever ready to leap beyond science by fashioning a doctrine claiming two truths, insisting they have a right to their beliefs in religious revelations, however fanciful they may be. Skeptics refuse to be lured by the transcendental myths of the day. Unable to accept uncritically the prevailing beliefs, they often stand alone condemned and hated by believers for their "negative" heresies.

One may ponder why skeptical dissent about religion has been held by a relatively small group of intellectuals and naysayers. Are the religious needs of men and women too powerful, the hunger for the divine too pervasive for skepticism ever to prevail? Does religiosity have its roots in biology and genetics? Does it have adaptive value? Is there deep within the human breast a transcendental temptation that reappears in every age and accounts for the ready acceptance of myths about the transcendental? If so, can or should it be overcome, and how?

There was the supposition at one time that, given the availability of education, the increase in literacy, and the elimination of poverty and disease, humankind might someday outgrow the religiosity of its infancy. There was the hope that in its place might develop a mature scientific outlook and a responsible moral and social philosophy grounded in a naturalistic view of the human condition. Under this view the universe possesses no purpose or meaning per se and is indifferent to human achievement and failure. It is not divine in origin or sustenance. The human species is a relatively minor species on a tiny planet in one galaxy among countless others. Hence, the beginning of wisdom is the awareness that there is insufficient evidence that a god or gods have created us and the recognition that we are responsible in part for our own destiny. Human beings can achieve this good life, but it is by the cultivation of the virtues of intelligence and courage, not faith or obedience, that we will most likely be able to do so. We can attain some measure of social harmony and justice, as well as a creative and bountiful life full of potential excitement and exuberance. But what we do depends upon whether we can develop scientific knowledge and moral excellence. The human adventure is not alien to adventure and poetry, romance and beauty. Nor is it insensitive to wonder and awe or the splendid majesty of the cosmos. These are religious qualities, but natural piety need not be devoid of stark honesty about our past and future. Surely we need to avoid a schizoid approach. The method of critical intelligence as used in science, philosophy, and ordinary life is an essential therapy if we are to rid ourselves of false illusions and dogmas.

This was the pagan conviction that prevailed in Hellenic and Roman civilization. It was eventually overtaken by a failure of nerve, and the Dark Age of religiosity descended upon Western civilization. A humanist outlook emerged during the Renaissance and reached fruition during the Enlightenment, with the development of modern science and the ideals of democracy and freedom. But the belief that reason and science would properly emancipate human beings from false mythologies of illusion was mistaken; for the twentieth century has witnessed the growth of virulent new ideological religions (fascism and Marxism-Leninism-Stalinism) and the persistence of orthodox, supernaturalistic religious dogmas. It has also witnessed an outburst of a new set of beliefs in the paranormal and the occult: astrology, UFOlogy, psychic and space-oriented science-fiction religions, and a bizarre magical-spiritual world-view.

Secular humanism provides an outlook on man and the universe, a philosophy of life, an ethic of reason and freedom. It is the story of possibility and outreach. Secular humanism is an alternative to the religions of illusion and salvation. But how it copes with the claims of the transcendent without dogmatically dismissing them and how it deals with life, offering opportunity and power, is a crucial issue. Can secular humanism provide a meaningful substitute for God and the transcendent? Can it deal with the world as it is and yet help us to fulfill our basic yearnings and hopes for what we might become?

Introduction: Not for philosophers only

This book is intended for the general reader interested in the world of ideas. It is not written for academic philosophers. It is very likely that they will not appreciate it; in any case, it is not written in the style of analytic philosophy that is fashionable today.

I am critical of the kind of philosophy that prevails in Anglo-American philosophical circles. I consider much of this philosophy irrelevant. It is like an ingrown toenail, turned within itself, festering with sharp linguistic distinctions and painful sophistries. The trouble with technical philosophy is that it has incapacitated philosophy proper, so that it hobbles about like a mendicant friar cloistered in a distant monastery on a remote island in the midst of an empty sea.

I do not mean to impugn all of my philosophical brethren, because much virtue has been derived historically from the philosophical enterprise: intellectual wisdom, logical clarity, and moral excellence. Yet, there are forms of philosophical disputation that degenerate first into distemper and eventually into rigor mortis—not that philosophers today have much impact upon the contemporary world. Most suffer lives of quiet desperation like celibate monks reading their sacred texts for nuances of inflection and meaning as the world passes them by (though today it is the professional journals that serve as their scriptures).

The tragedy of contemporary philosophy is that it has been castrated; it is too often inconsequential to the larger issues of intellectual concern and social significance. It meditates about formalistic questions, ones without real content or relevance. In the past, philosophers have often castigated and rejected one another. Schopenhauer and Nietzsche railed against the Germanic professors of philosophy. For a long time Sartre lived in isolation from the mainstream of academic philosophy in France, though he exercised a strong impact on French and European society. Charles Peirce, America's greatest philosophical genius, was unable to secure a teaching position; he struggled through life aided by his friend and mentor William James. "Wherever there is a large class of academic professors," Peirce wryly observed, "who are provided with good incomes and looked up to as gentlemen, scientific inquiry must languish."[1]

One of the problems today is that philosophy has become wedded to the academy, with the result that philosophers have taken as their main task either the imparting of knowledge to hordes of diffident college undergraduates or the initiating of candidates for the doctorate by supervising thesis topics of little moment to anyone but themselves. While they do so, genuine intellectual problems concerning meaning and existence are ignored. The fear of scathing peer review censors nonconformist interests; there is always the corroding worry of what one's professional colleagues will say. There is the burden of clannish gossip to weigh a philosopher down and keep him or her in submission.

Not surprisingly, many of the best minds who pursued philosophy have not been university dons. Spinoza ground lenses by day and wrote at night. He refused a position at Heidelberg, lest his freedom of inquiry be compromised. Descartes never held a university position, nor did Locke or Berkeley. Hume's first book fell stillborn from the press; he had to establish a reputation first as a historian. John Stuart Mill was secretary of the East India Company. Marx, unemployed and poverty-stricken, did his research in the British Museum. Kierkegaard wrote in anonymity in Copenhagen. Wittgenstein was an odd figure in the British academic environment, which he eventually captivated.

Of course, there have been university professors who made seminal contributions to philosophy: Plato founded the Academy, and Aristotle, passed over as head of the school, established his own Lyceum—though presumably neither granted Ph.D.'s as we know them today. Kant and Hegel were professors of philosophy, as were Russell, Heidegger, and Dewey. The record, however, has been spotty. The great enemy of philosophy is distraction for alien ends—in this case entanglement in the bureaucracy of education. There are those who have stood outside academic philosophical circles in order to do their own creative work: such was George Santayana, an oracle from afar who lived in Italy in splendid retreat. While teaching at Harvard during it's Golden Age, one morning he reputedly observed, "Gentlemen, it's spring" and left the classroom, never to return.

It is too dangerous to place all of our philosophical eggs in one basket; they are too fragile. If the marriage of philosophy and the church degenerated in the Middle Ages into an official doctrine and if the subservience of Marxist ideology to the party and state in Communist countries meant the death of the creative philosophical impulse, so the

1. *Collected Papers of Charles Sanders Peirce,* vol. 1, ed. C. Hartshorne and P. Weiss (Cambridge: Harvard University Press, 1931), p. 22.

near indissoluble union of philosophy with institutions of higher learning in the present century has tended to stamp an official academic imprimatur on what is considered to be legitimate philosophical inquiry. A similar fate threatens art, literature, poetry, or science, where they are developed in the context of the cloistered university life and filtered through the limited visions of faculty members. Perhaps the most important aspects of academic freedom are intellectual integrity, independence of mind, and freedom from committee work.

No doubt we can learn from both types of philosophy: academic and independent. Perhaps it is necessary for some philosophers to be anomic to the profession, to drink deeply at the well of real life for inspiration and insight, and for academic philosophers, after the fact, plodding and critical, to refine and distill what has been discovered. Philosophical inquiry is kept alive and vigorous by the successive revolts that take radical departures. It remains for the formalists to meticulously analyze the new theories, even though in the process they often trivialize the philosophical enterprise.

A special problem today is that philosophy has become professionalized, like the ministry, with all of the rights and privileges pertaining to its office, or like the medical profession with its analogous high priests. Philosophy is now a specialty like accounting, physiology, or carpentry, and there are tools of the trade that need to be fashioned and honed. Often though, skills are endlessly sharpened, but there is no subject matter. Generally, one has to read only a limited number of books and articles in order to master the technical grammar of a subspecialty. One can be an aesthetician, logician, epistemologist, or ethicist. One does not have to know much else in order to succeed today in the eyes of one's peers. At the professional associations where papers are read and in scholarly journals where they are published one establishes the credentials of one's craft. The union card is the doctorate, and the final sanctification is tenure, which can be achieved at most universities by publishing in a limited number of scholarly journals and perhaps one book. These are generally weighed for quantity, not read for quality or content.

Given the rapid growth of specialization and the compartmentalization of knowledge, one is not required to read outside a narrow literature. Indeed, philosophy today may be defined by what philosophers do; philosophers often read only what other philosophers write and listen only to what their colleagues say. The ingrown toenail is growing deeper. The problem, of course, is not endemic to philosophy alone: economists, sociologists, anthropologists, and psychologists focus on the literature in their fields. Chemists know chemistry, poets poetry, historians history, and engineers engineering, and often little else.

But there are other functions of philosophy and indeed other dimensions to life beyond those delineated by the craft of the linguist or logician. Philosophy should keep alive our sense of wonder. It should be concerned with the larger questions: What is the nature of our knowledge of the universe? What does it all mean—if anything—and how does it fit together, if at all? Does God or Mammon exist? Is there an ultimate purpose to human existence? Does life have meaning per se, or does it only present us with opportunities? Is human nature fixed or open-ended, full of indeterminate peril and adventure? What is the good society, and can it ever be achieved finally and fully? How ought a person to live if he or she wishes to find satisfaction and significance in the few short years of life?

In order to cope with these questions, one has to know something of the natural, biological, behavioral, and social sciences, and also to have studied politics, literature, and the arts—indeed to have reflected on a wide range of lived experiences. Perhaps this is an impossible task; and so philosophers of the academy have eschewed these larger issues and fled in desperation from this quest. When they tackle the philosophy of science, they are concerned more often with its rear action rather than with its avant-garde. It is considered in bad taste to be concerned with the frontiers of knowledge and their implications for man. These matters have been left to historians (such as Toynbee), or scientists (such as Monod, Eccles, or Sagan), or psychologists (such as Skinner, Fromm, or Freud), or theologians (such as Küng or Tillich). The truly significant and basic philosophical questions are left unanalyzed and unanswered. Philosophers today prefer to analyze the meaning of the term *proposition* rather than propose a proposition; they ask what the term *good* means, if anything, rather than suggest what the good life might be; they inquire what the function of the verb *to be* is rather than confront the nature of being. Philosophers who have analyzed and critiqued limited areas have no doubt played a useful role. But in abandoning metaphysical generality and normative recommendations, they have forced ordinary men and educated persons to look elsewhere for guidelines—to scientists and poets, politicians and economists, theologians and soothsayers—anyone who will throw them a lifeline.

I have been a university professor all my adult life. Although I have cherished the ideals of the university and defended its autonomy against those from within and without who have sought to denigrate it, nevertheless I have often found it limiting to my intellectual sojourn. Perhaps I should not bite the hand that has fed me, but in all honesty I must say that I have found other sources of nourishment and enrichment in the marketplace of ideas and action. My philosophical mentors have been Socrates, Marx, John Dewey, William James, and Sidney Hook.

Socrates wandered the streets of Athens questioning the better and worse and debunking both the received doctrines and the latest intellectual fads, and he was condemned to death by his fellow Athenians for so doing. Similarly, Marx did not seek simply to understand the world but to change it radically and he is revered in history for it. It is a pity that Marx was betrayed by his disciples, who sought to build a state party and religion in his name. I identify most closely with the pragmatic approach of James and Dewey, who sought to make philosophy relevant to civilization and who are now reviled by the apostles of fundamentalist doctrine. My most direct inspiration has been Sidney Hook, a controversial philosopher of profound influence, who has dealt with the most perplexing issues of social and political thought. At times he has been a gadfly, seeking by constructive criticism to shed light and provide a rational direction for educating citizens. All of these philosophers were men of both reflection and action. Interested in inquiry, they were not sequestered within narrow boundaries. They considered a central function of philosophy to be practical and ethical, but they did not mean metaethics, the antiseptic treatment of logical niceties neatly separated from the problems of ethical choice. The subject matter of philosophy, as I conceive it, is not simply what other philosophers have said on the metalevel. Philosophers must descend into the concrete world of human practice and belief, sweat and toil like the rest of humanity within the marketplace of ideas and images to find meanings and uncover truths.

Philosophy and the new media

Alas, at the same time that philosophy has become the mistress of the university, the influence of the university has waned in contemporary society. No longer the center of intellectual life nor even the main institution educating leaders, it has become in part a hedonic way-station for half a generation of postadolescents. The real center of intellectual influence and power has, regrettably, become the various media of mass communication.

There are at least three ways of doing philosophy, pursuing the arts and sciences, and furthering research and learning: First, by means of the spoken word as communicated in verbal language, conversation, discourse, and lecture. The oral tradition as a means of communicating and preserving knowledge has deep roots in primitive culture. The first philosophers, Socrates and the Sophists, never wrote anything, but they developed the dialectical skills of talk. Second, by means of the written word. Plato transformed philosophy for successive generations by capturing and embellishing the spoken language in dialogue form. In the process, Plato mistakenly reified language and postulated a realm of ideal essences to explain the meanings of symbols. Essays and treatises became the chief form of inquiry and learning. Written knowledge was enormously facilitated by Gutenberg's invention of movable type. The exponential growth in book, journal, magazine, and newspaper publications was a result. Ideas could be formulated and codified for future generations, while standards of objective analysis and evaluation could be developed. Third, by means of the technological explosion of the twentieth century, which has had an enormous impact that has not been fully appreciated. Since it is so obvious, it is often ignored. The electronic media have radically altered the mode of communication: the telegraph, telephone, movies, radio, television, xerox, microprocessor, and computer—all enable instantaneous communication. They revert back, however, to a prelinguistic age and at the same time project ahead to a new world of imaginative reconstruction. They use signs, metaphors, images, gestures, sounds, and colors, not concepts per se. Radio, phonograph, telephone, and tape recorder focus on tone and sound, music and the spoken word electronically amplified. Photography, film, and television accentuate images and shadow, color, movement, and gesture. The computer and microprocessor produce readouts on electronic tapes and screens.

What do these media do to the principles of syntax found in written language, to the criteria of logical validity, to the standards of objectivity based upon conceptual schemes? What happens to explicit argument, abstract ideas and concepts, when the information communicated is implicit and is visual and auditory? The history of philosophy has been concerned with analyzing the structure of perception and the logical form of written languages. But the new world of knowledge is reducible to artificially induced sound waves, celluloid frames, and picture dots transmitted and unscrambled on a TV or computer screen.

Every age has its sacred institutions, which serve as the fountainhead of influence and power. These are charged with the transmission of important knowledge and values, and they are revered or detested, adored or feared by the general populace. In earlier days, the sacred institutions were the family and the church, the locality or the state, the classroom and the school. Today it is the electronic media: the movie theater, radio, cable, and network television station. The main vehicle for transmitting knowledge and values now assumes the form of dramatic sequences. Suspense and shock, comedy, violence, and tragedy are rendered to imitate and illustrate life and reality. The passive viewer is assaulted and enveloped. Every effort is made to capture his imagination, mold his beliefs and attitudes. He is considered a consumer of ideas and products.

The goal is persuasion. Ideas are not critically analyzed or evaluated, but packaged, and marketed. There is little appeal to critical judgment.

With this rude change comes a profound challenge. The private sanctuary of each individual is invaded by the media; the arts of subtle influence and propaganda intrude into the home. The article or book, which could be absorbed in quiet reflection, is replaced by the cassette, tape, or tube as the main source of information and education.

No wonder there has been a decline in the art of reading, in literacy, and in intelligent comprehension, as well as an erosion of cultivated taste. Hitler used the radio, the newest electronic gadget of his day, to inflame the masses of Germans in favor of his Third Reich. Roosevelt also skillfully employed the radio in his fireside chats to build support for the New Deal. Today, the power of propaganda has been immeasurably increased by television and films. In totalitarian societies the electronic media are used by the party and the state to indoctrinate the masses with ideological dogmas. In Western capitalist democracies the media are used to entertain, advertise, and sell products and candidates. The new electronic technology has had positive as well as negative results. It has extended our senses and our nervous system in a worldwide embrace. Today computers hold forth the promise of instantaneous translation of any language into another language, and the electronic screen and radio encourage the emergence of a general, worldwide consciousness. Consciousness can be transmitted around the globe with little verbalization. At the same time, a banal wasteland of encapsulated values and narrow horizons threatens to impoverish the marketplace of ideas with low quality and little diversity.

The revival of religious fundamentalism by means of the electronic church and the growth of the cults of the paranormal are illustrative of the deleterious power that these newer media can have on public attitudes and beliefs. The scientific outlook is perhaps the most important development of modern civilization. It was bred out of a concern for basic research, and it led eventually to the industrial revolution. But its wondrous technological products are often captured by forces antithetical to the scientific outlook. The power of a myth is measured by its immediate appeal to the emotions: it is the *glance* or the *inflection,* not the sustained argument or analysis, that often decides the merits of a case. Greco-Roman civilization was overwhelmed by barbarians and Christianity, as was the high culture of Germany by Nazism. Are the Western democracies to be overtaken by evangelists on television in plaid suits and toupees? This is the age of the *glimpse,* the *synopsis;* there is little depth of understanding, only passing images. Can philosophical thought play a role in this new situation? Or is its fate to be exile? Is philosophy to be removed from the centers of thought and action?

The need for reflective wisdom

Many technical philosophers will consider any discussion of these issues inappropriate to analytic philosophy. But they betray their own failure to understand: meanings cannot be interpreted independently of the media of communication and the contexts in which they are embedded.

In my view, there is a critical need for philosophy and philosophers to analyze the media by which ideas are transmitted. Two factors are especially relevant here.

First, the amount of technical information and know-how is being increased by leaps and bounds. Specialties within specialties make it difficult or impossible for an individual to encompass and digest everything within his own field of expertise, let alone to be well-informed about other fields. Compound this by the growth of new technologies and research disciplines, and the problem becomes almost insuperable. For example, medical researchers inform us that excessive cholesterol is injurious to the cardiovascular system; other researchers minimize the harmful effects. A leading researcher publishes results that indicate Vitamin E is healthy for the skin and sexual capacity and contributes to longevity (at least in mice); another denies any support for the claim. A geologist defends the shifting-continents theory. At first his professional peers dismiss his theory as outlandish but later accept it as true. Astronomers and physicists postulate black holes and antimatter, defying common-sense views of the universe.

Parapsychologists maintain that precognition (the ability to foresee events before they happen) and psychokinesis (the ability of the mind to alter the state of matter without an intervening physical force as causal agent) have been demonstrated in the laboratory. Skeptical scientists examining the evidence maintain that these claims have not been confirmed experimentally. Faith healers claim that they are able to effect miraculous cures, pointing to the testimony of their clients. Physicians deny such miraculous cures and rely solely on scientific explanations.

Marxists affirm that capitalism is doomed to failure and that the nationalization of the means of production will not only vastly expand productivity and wealth but also provide for a more equitable social system. Some economists maintain that the way to stop inflation is by decreasing taxes (supply-side economics), whereas others disagree, asserting that we must increase taxes (and decrease demand).

How can we evaluate the many truth-claims that are proliferating in the contemporary world? Here philosophy has an important function to perform, for it can help to keep alive the critical intelligence, reflective wisdom, and skeptical disbelief so essential for the development of knowledge. We are surrounded on all sides by people attempting to sell their wares to unsuspecting and gullible consumers. People are being persuaded to accept and to *believe* on the basis of faith. Some encouragement of the *right* not to believe because there is insufficient evidence would seem to be an intelligent response to the information explosion. But this means that the task of the philosophical curriculum, not only in schools of learning but also in the media of communication, is to develop an appreciation for the methods by which we discover and validate reliable knowledge. The best remedy for public gullibility is the cultivation of critical intelligence. To be a scientist or technician in any one field is no guarantee that he will apply ethics or intelligence outside his field. Dr. Mengele experimented with inmates in Nazi concentration camps, and Eichmann used methodological precision in dispatching innocent victims to the crematoria. Philosophical wisdom must have a wider reach than mere technical know-how. The former is an important quality to develop if an individual is to make his way ethically in the world.

A second reason for the need to develop the capacity for philosophical reflection is the rapidity of social change, caused in large measure by the pace of technological innovation. The philosophical spirit is too important to be left to the narrow preserve of the specialist. It ought to be the common possession of every educated person. It is difficult to overestimate the rate of social and cultural transformation that is occurring—and the need of a reflective response to it. Only five hundred years ago Europeans first began exploring the globe, and the United States, only two hundred years ago, had three million inhabitants and part of a continent to conquer. Today, Asia, Africa, South America, and Australia are all involved in a new world civilization that is emerging. The solitary individual and the small social or family group, whether in Nairobi, Florence, Peking, or Vancouver, have been wrenched from the world in which they were formerly rooted.

Technological innovation and social change had been magnified dramatically by the beginning of the space age in 1967 (A.S. 1 = After Space, year 1). For that was the year the human species first hurled vehicles into outer space and was able to overcome the barriers of the earth's gravitational field and to look back in order to view the planet in its majesty and—to some—its unity. In a short time man was able to explore other planets in our solar system. The first voyager was able to leave our sun's gravitational influence, travel beyond it, and transmit radio signals back to earth; space vehicles have now left our solar system. Heroic efforts are being made to contact extraterrestrial civilizations in distant galaxies and to attempt to communicate with intelligent beings from other worlds—if they exist!

Science fiction has far outstripped our present capacities; it has already advanced far into the age of star wars, depicting a struggle for control of the universe. This has helped lead to a basic confusion in the modern world between the ideal and the actual, the possible and the real. Although the realm of the imagination can posit new inventions, these can only be developed by hard research and application. Not all of them will succeed. Science fiction can easily degenerate into a dream world of sheer fantasy and madness, and in the process it can help defeat the impulse of discovery and creation. It is with the realm of the transcendental that it is forever flirting. Nevertheless, a new world emerges out of scientific discovery, and the new scientific technologies always have an enormous impact upon our ideals and values, cosmologies and cosmogonies, ideologies and sociologies—all products of earlier civilizations. The untapped possibilities presented by a new world outlook and new technologies of space travel, microcomputers, biogenetics, and other fields still to be discovered far outdistance in sheer splendor and audacity the ancient myths of religion.

Yet human beings often respond by retreating in fear and trembling. They become fixated on ancient loyalties of ethnicity and territoriality, and they abandon the further promises of scientific discovery. Many assume a prostrate posture, reject modern scientific theories of evolution, the social and behavioral sciences, and the findings of biology and physics because these unsettle their deepest-held prejudices and convictions and shake the foundations of their exalted temples of belief.

The quest for certainty, security, and salvation gnaws at the marrow; many humans chatter before the winds of change, frozen in place by the awesome potentialities of the human species. Frightened by the challenge, they dread an approaching doomsday Armageddon. The images of the future that dominate the imaginations of many in the contemporary world are those that spell disaster: economic destruction, population inundation, resource depletion, and nuclear holocaust. Blinded by myopia, many are unable to synthesize or make meaning out of the information explosion, the technological transformations occurring around them, or the severe social, economic, political, and moral convulsions that have ensued. For some, the only reaction is catatonia, a sense of helplessness and impotence. For others, there is a revulsion and revolt against the modern world and the desire to return to an earlier simplistic age in which Absolutes reigned supreme. Still others become entranced by the paranormal, and they gladly accept anything and everything about it—even the borderlines between the possible and the impossible become blurred. To them, *all* things are possible; hence anything and everything is true.

In this context of the unsettling of our ancient ideas and values on the one hand and the promise of new powers and opportunities on the other, we ask: Do philosophers have a mission to play in the future? Or are they to be passed over by a new parade of computer experts and scientific technicians, and new priests and gurus? Are we to be overrun by illiterate specialists, the metaphysically naive, barbarians of virtue, persons devoid of wisdom? There exist on the present scene few philosophers who are willing to be futuristic missionaries. That is perhaps a bad term, yet there is a genuine need for them. Philosophy can help us make some sense out of the confused present and the onrushing future. Two cardinal philosophical virtues that the pagans have bequeathed to us are wisdom and courage: the *wisdom* to interpret as best we can, in the light of the best knowledge that we have, in terms of evidence and reason, what the world and the cosmos we live in is about; and the *courage* to dream alternative ideal futures of what the good life can become, and then to resolve to bring them about. It is not by looking backward to the early religions of our primeval roots developed in nomadic or rural societies but by daring to look ahead to a postmodern space age and a technological world-civilization that we can best fashion a new outlook and new moral and social values. And it is here that philosophy can, as it has in the past, assist us to regain our future.

But we must first recognize that we need a new general philosophical outlook appropriate to the new age. Whether philosophers wedded to the academy are able to meet the challenge or whether others will assume their role is a profound issue. Will humankind slip into a new anti-intellectual age overcome by a fearful religious quest for salvation, fixated on mysticism and authority, or will it move ahead to accept the dazzling opportunities that may await us as an unparalleled age of secular and humanist promise unfolds?

Part One

SKEPTICISM AND THE MEANING OF LIFE

I: Meaning and transcendence

The value of life: Things left unsaid

All living beings undergo continuous processes of replenishment and renewal. Within each species is the constant striving to persist and to reproduce its own kind, in spite of the surrounding forces in the environment that tend to denude or destroy it. All forms of animal life seek food in order to survive and procreate—though most apparently are unaware of the tentativeness of their existence. In the end nature prevails and every single representative is vanquished: the leaf withers and the lilac dies; the sapling that grows into the magnificent elm eventually rots and decays; the young stallion is reduced to a decrepit horse.

And what of man? He is a creature of intelligence and imagination. He is by now well aware that his life is finite. He sees that the infant, the child, the young person, and the mature adult, all full of possibility and power at one time, eventually grow old and die. Pubescence, adolescence, senescence are all inevitable phases of life. Human beings thus have knowledge of their inexorable demise, and also of the tragic character implicit in the human condition. Life is full of danger: as soon as one is born, one is old enough to die. There is the sudden accident or the incurable disease that can overtake friend and foe alike. These bitter pills are difficult to swallow. Most people will not or cannot in the final summing up accept this ultimate death. Out of their anxieties about it grows the quest for transcendence and religious faith.

That life is or can be good and bountiful, full of significance and pregnant with enjoyment and adventure, is also apparent to those who have the courage to overcome the fear of death and achieve something in the world. Modern civilization, education, science, and technology have helped us to minimize disease and extend the years of a life of rich enjoyment and harvest, without worrying about our ultimate destiny or God's plans for us. Yet lurking in the background of the consciousness of every person is always the potential for despair, the ultimate dread of his own death. No one can escape it: we are all condemned to die at some time, no matter how we may strive to stave it off.

My father was struck down at fifty-nine of a heart attack, in the

17

prime of life, but he told us at his bedside that he knew he was dying. He kissed my mother and said that his life in summation was happy. My grandfather, always ebullient, loquacious, beaming with life, on his deathbed at ninety-two clenched his fist and gritted his teeth and told us that he did not want to die. He refused to accept his approaching end. My grandmother, at eighty-three, gasped her last breath in her daughter's arms and was dead without warning before she knew what happened. Her youngest daughter died of a terrible cancer at the age of fifty. Regardless of how we feel about death, each of us too will someday reach, even after a life of fullness and exuberance, a point of no return.

Let us reflect on the human situation: all of our plans will fail in the long run, if not in the short. The homes we have built and lovingly furnished, the loves we have enjoyed, the careers we have dedicated ourselves to will all disappear in time. The monuments we have erected to memorialize our aspirations and achievements, if we are fortunate, may last a few hundred years, perhaps a millennium or two or three—like the stark and splendid ruins of Rome and Greece, Egypt and Judea, which have been recovered and treasured by later civilizations. But all the works of human beings disappear and are forgotten in short order. In the immediate future the beautiful clothing that we adorn ourselves with, eventually even our cherished children and grandchildren, and all of our possessions will be dissipated. Many of our poems and books, our paintings and statues will be forgotten, buried on some library shelf or in a museum, read or seen by some future scholars curious about the past, and eventually eaten by worms and molds, or perhaps consumed by fire. Even the things that we prize the most, human intelligence and love, democratic values, the quest for truth, will in time be replaced by unknown values and institutions—if the human species survives, and even that is uncertain. Were we to compile a pessimist's handbook, we could easily fill it to overflowing with notations of false hopes and lost dreams, a catalogue of human suffering and pain, of ignominious conflict, betrayal, and defeat throughout the ages.

I am by nature an optimist. Were I to take an inventory of the sum of goods in human life, they would far outweigh the banalities of evil. I would outdo the pessimist by cataloguing laughter and joy, devotion and sympathy, discovery and creativity, excellence and grandeur. The mark made upon the world by every person and by the race in general would be impressive. How wonderful it has all been. The pessimist points to Caligula, Attila, Cesare Borgia, Beria, or Himmler with horror and disgust; but I would counter with Aristotle, Pericles, da Vinci, Einstein, Beethoven, Mark Twain, Margaret Sanger, and Madame Curie. The pessimist points to duplicity and cruelty in the world; I am impressed by the sympathy, honesty, and kindness that are manifested. The pessimist reminds us of ignorance and stupidity; I, of the continued growth of human knowledge and understanding. The pessimist emphasizes the failures and defeats; I, the successes and victories in all their glory.

The question can be raised: How shall we evaluate a human life and its achievements—in the long or the short run? What is the measure of value, the scale of hope? From one's immediate world there may be boundless opportunities. Look at the things that one *can* do, if social conditions permit a measure of freedom and if one has developed the creative verve: one can marry, raise a family, follow an occupation or career, forge a road, cure a malady, innovate a method, form an association, discover a new truth, write a poem, construct a space vehicle. All are within one's ken and scope, and one can see a beginning, a middle, and an end. One can bring to fruition the things one may want, if not in one's lifetime, then in the lifetimes of those who follow and their children's children. France, the United States, the Soviet Union, India, and the New Zion are social entities that human beings have created. The pessimist groans that they have or will disappear *in the end,* in the long run, if not today or tomorrow, then ultimately. He complains about the ultimate injustice of our finitude. One may know that that is true and even come to accept it. But we are alive today, and we have our dreams and hopes, and the immediacies of experience and achievement can be enormously interesting, exciting, satisfying, and fulfilling.

Can we enjoy ourselves today and tomorrow without worrying about the distant future? No, says the pessimist. He is unable to live in this world; he is fixated and troubled by its ultimate disappearance. "How can life have meaning if it will all end?" he complains. Out of this grows a vain belief in immortality and ultimate survival.

"Why not make the most out of this life, if that is all we have?" I respond. Indeed, this life can be full of happiness and meaning. We can make a comprehensive list of all the goods and bads and of all the values and disvalues. For every evil that the pessimist presents, I can counter with a virtue, for every loss, a gain. It all depends on one's focus. A person's world may be full of the immediacy of living in the present, and that is what he may find rich and vital. The life-world involves one's yesteryears. This includes a human being's own small world—the memories of one's parents, relatives, colleagues, friends. It also includes the history of one's society and culture—as memorialized in the great institutions and traditions that have remained. But more, one's life-world shares the recorded histories of past civilizations and the memories of great minds, artists, geniuses, and heroes, as bequeathed to us in their writings and works that remain. There is also the residue of things far past, which, uncovered by science, tell of the evolution of the human species, as the strata of earth and rock dug up reveal more of our history and that of other forms of life. It also encompasses the physical universe, extending back billions of years. The eye of the cosmos unfolds through countless light-years in the telescope of the astronomer: the formation of the rings and moons of Saturn and Uranus, the birth and death of countless suns and galaxies. And what does our present include? Everything here and now on the planet and in the immense cosmos investigated by science. But what does the future hold? One can contemplate his own future: tomorrow, next year, twenty years, or fifty. How far may we go? We can make plans, but we cannot foretell what will be—if anything—one century hence. Of this one may be fairly certain: the universe will continue to exist—though without me or my small influence affecting it very much.

Thus one can argue the case from two vantage points. First, there are the life goals or scene of action that an individual, his family, and culture experiences, its dimensions of space-time, its phenomenological range in immediacy, memory, and anticipation. And this can be full of significance. Does life have ultimate meaning per se? No, not per se. It does, however, present us with innumerable opportunities. Meaning depends on what we give to it; it is identified with our goals and values, our plans and projects, whether or not we achieve them and find them interesting. There is the fountain of joyful existence, the mood of exhilaration and the satisfaction of creative adventure and achievement—within varying degrees, realism no doubt admonishes.

Millions of people, however, do not find life interesting, are in a quandary, and are overwhelmed by the problems and conflicts they encounter. Life apparently leaves them with a bad taste: it is ugly and boring, full of anguish and sorrow. They bemoan a cloudy day, complain about the humidity, rail against the past, are fearful of the future. The tragic dimension of life is no doubt exacerbated by those who suffer a great loss or a severe accident. But nature is indifferent to our cares and longings. A raging fire or cyclone may indeed destroy everything we have built and love; and we are forced to submit to the furies. There is thus the desert of despair, the emptiness of lost zest, the collapse of meaning. In some social systems human beings are their own worst enemies, unable to live or breathe freely, imprisoned in the gulags of their souls.

Yet, within the life-world we occupy, we are capable of intensity, enjoyment, and interest, if we are able to express our proper freedom as independent, autonomous, and resourceful beings, and especially if we live in an open society that encourages free choice. The spark of the good life is creativity—not escape, retreat, or complaint—and the audacious expression of the will to live. *My will be done,* says the free person—not thine. Yes, it is possible—comparatively and reasonably—in spite of the demands and obstacles presented by nature and society, to achieve the good life. I am referring not to a life of quiescent withdrawal or simple self-realization but to the active display of one's talents and powers and the expression of one's creative imagination. One may dig a deep well where none existed before, compose a lullaby, invent an ingenious tool, found a new society, teach a class, pick up and move to another area, succeed on the job or change it. It can be fun; no doubt there may be some sorrow and tears, but basically it may be worthwhile. Thus the pessimist's dilemma, as dramatized by Schopenhauer, is false: life is supposedly empty because we are either faced with desire (here we are anxious, unfulfilled, restless) or satiated (ennui and boredom set in). We *are* goaded by desires and wishes, but these can be a source of high anticipation leading to activity to realize them and genuine satisfaction in doing so. This can mean contentment, pleasure, appreciation, excitement—all positive values.

I am willing to argue the case with the pessimist on the level of the phenomenological, contextual life-world. "What is the matter with you?" I may ask. "Are you sexually repressed? Then satisfy your libido. Are you tired and unhealthy? Then examine your nutrition and exercise. Do you hate your work? Change your career, go back to school. Is someone sick? Try to find a cure. Are you lonely? Then find a lover or companion. Are you threatened by your neighbor? Then form a peace pact or establish a police force. Are you troubled by injustice? Don't just sit there; help mitigate or stamp out evil. Enact new legislation. Come up with a viable solution." Human ills are remediable to some extent, given the opportunity, by the use of intelligence and the application of human power.

But it is the second vantage point that troubles so many: the argument from the *long* run. If I don't survive in the end, if I must die, and if everyone I know must also die—all of my friends and colleagues, even my children and their grandchildren—and if my beloved country or society must perish in the end, then is it *all* pointless? What does my life really mean then—nothing at all? This quandary and the despair that such reflections can generate is no doubt the deepest source of the religious impulse, the transcendental yearning for something more. Can one extend his present life-world and those of his loved ones and community indefinitely in some form throughout eternity? People ask in torment and dread, "Why is there not something more to my existence?"

My wife, who is French, sometimes raises these questions when we are alone in bed late at night. And I have heard others raise similar questions. At some point there is the recognition of one's finitude, as one gradually realizes that he or she is growing old and is not eternal. The lines on one's face and the sagging body point to the fact that one's powers are not eternal. Prayers to an absent deity will not solve the problem or save one's soul from extinction. They will not obviate the inevitable termination of one's life-world. They merely express one's longings. They are private or communal soliloquies. There is no one hearing our prayers who can help us. Expressions of religious piety thus are catharses of the soul, confessing one's fears and symbolizing one's hopes. They are one-sided transactions. There is no one on the other side to hear our pleas and supplications.

Why has secular humanism failed to take hold?

It is at this point that one may ask if the primary reasons why thoroughly secular philosophies that accept human temporality have not succeeded better than they have are psychological and existential. Why do religious institutions, even the most blatantly fraudulent and immoral, persist in spite of the growth of science, near universal education in advanced countries, the possibility of an end to poverty, and the untapping of years more of healthy living? Is there something genetic and sociobiological in the human species? No doubt there are many powerful social causes to explain the persistence of religion: cultural lag, ethnicity and tradition, communications media that present religion as true and that go unchallenged, religious indoctrination and institutionalization, the union in some cultures of religion and state power and in others with economic power.

If one looks back to the Enlightenment and the Age of Reason, we see that men dreamed of using science and education to achieve human progress. They believed that in time superstition would disappear. The philosophers of the eighteenth century deplored the ignorance and dogmatism around them, the sinfulness and hypocrisy of the clergy and the churches; and they believed that reason, if applied, would overcome blind religious faith and superstitious credulity.

That this has not happened is all too apparent. For even highly educated men and women—doctors and lawyers as well as beggarmen and thieves—are wont to believe in salvation by some deity. The end of poverty, the overcoming of disease, and the attainment of affluent standards of living are no guarantee of liberation from their fixation. The churches may wax and wane in influence and appeal, but they do not disappear. And new forms of transcendental religiosity are created overnight to replace those that weaken in their messages. Why does religious tradition have such a profound influence? Why does the weight of the past bear so heavily upon the present that even otherwise rational and skeptical persons find it difficult to resist its allure?

There is no confirming evidence for the existence of a God, and the so-called arguments to prove his reality are inconclusive. We have been through a protracted debate between philosophy and theology, science and religion. Philosophers have won each point of the debate in each battle. As we shall see in future chapters, there is no conclusive evidence for a divine source or purpose for the universe and no proof of God's existence save man's own wish-fulfillment. Moreover, the major monotheistic religions, Christianity, Judaism, and Islam, all rest on myths of revelation. Yet the theologians continue to win the wars. For each succeeding generation appears to be impervious to the victories of skeptics and the defeats of theologians in the past, and it invests its own meaning in the old religious forms and institutions or creates new and even more irrational ones. Why is this so? Why is the message of Christ risen, Mohammed as the prophet of God, Moses on Sinai, or extraterrestrial divinities and demons of the occult still persuasive for each generation anew?

Is it because these messages and prophets promise that there is indeed *more* to this life as lived and endured in the phenomenological life-context and in the universe? And if so, do many or most souls *need* something more? I reiterate my question: Is secular humanism forever condemned to be a minority point of view of a relatively limited number of skeptics, one that is unpalatable to the broad mass of humanity? Or are there other options?

The quest for transcendence

Does man *need* to believe that there is something more, something that transcends this vale of expectant promises and ultimate defeat? This is the great challenge for the secular humanist, who is forever beset with extravagant claims of the supernatural, miraculous, mysterious, noumenal or paranormal realm. Is what we encounter in our perceptions all that there is? Of course not, for we make inferences, develop hypotheses, uncover hidden causes. Knowledge is a product of observation (direct and indirect) and rational inference. Both experience and reason are drawn upon in ordinary life and in the sophisticated sciences to establish reliable knowledge.

We surely should not exclude—antecedent to inquiry—any claims to knowledge about nature or life, however radical they may at first appear. One cannot reject dogmatically new dimensions of experience or reality or refuse to investigate their authenticity. The history of science and philosophy is replete with unwarranted, a priori rejectionism. Yesterday's heretic may become today's hero, and today's martyr, tomorrow's savior.

One must be open to the possibilities of the discovery of new truths and the nuances of fresh experiences.

The transcendent has had many synonyms. Philosophers such as Plato have sometimes used it to refer to "ultimate reality." Plato distinguished reality from the realm of appearances; sensations and observation only denoted the world of objects and particulars in flux. There was a deeper Truth, he said, which only reason and intuition could plumb: the world of universal ideas, in terms of which the contingent world of concrete fact is dependent. The world of Being remains sheltered to the senses, yet it subsists behind it and can be known only by dialectical investigation and reason. Kant referred to the *noumenal* so as to distinguish it from the *phenomenal:* the noumenal lies beyond the range of our phenomena; the phenomenal is knowable, and it obeys regularities and laws as structured by the Forms of Intuition and the Categories of the Understanding. Science is rooted in the phenomenal world, as ordered by the mind. But the "real" world is itself unintelligible to our understanding. We can only get a glimpse of the noumenal, perhaps in our moral life.

We are able to define and stabilize perceptions and ideas by language. Words clothe fleeting sensations, emotions, and thoughts and give them form and structure. We are able to communicate to others the world as we experience and interpret it by relating it to a common world of symbols. But language has many social functions. It enables us to interconnect symbols and to make assertions about the intersubjective world. What lies beyond the rubric of our syntax? Are we imprisoned within its walls? Does the mystic intuition enable us to leap beyond the limits of linguistic discourse?

There is some ground for arguing that the world as we experience it, talk or write about it in language, is not necessarily as it appears, and that there is something beyond, which the parameters of the human mind and language cannot presently comprehend or describe. The growth of human knowledge belies any conceit that pretends that what we know is fixed or has reached its final or ultimate formulations. For no sooner is a theory or hypothesis enunciated and maintained as true than its limitations may be seen, and newer, more comprehensive ones may emerge to replace it. Any effort to limit or fix the body of knowledge has thus far met with failure. On the other hand, we cannot thereby conclude that nothing we know is reliable or that the world as we view it in ordinary life and science is mere illusion and deception.

This apparently is the position of those who use the term *transcendence* with a supernaturalistic and divine meaning, as it has been for many latter-day uses of the term *paranormal.* It applies in opposition to two modes of knowing: the appeal to evidence and the testimony of the senses and the use of rational intelligence to develop theories and evaluate hypotheses. It also suggests that there are other ways of knowing—mystical, intuitive, and revelatory faith, all of which claim to give us a glimmer of the transcendent and allegedly supplement our use of experience and logic. Psychedelic and hallucinogenic drugs suggest that our consciousness is limited by our normal biophysical-chemical brain structures and that we may be able to transcend them and get a hint of another realm of being. Here there is movement toward "mind-expanding" or "consciousness-raising" methods. Transcendental meditation, prayer, and faith-states allegedly open up the possibilities of a new form of awareness. They supposedly bring us to the brink of a nonordinary reality. Similarly, for the alleged world of extrasensory perception or psychic phenomena.

By using such techniques, the psychic or mystic claims to be put in touch with another realm that transcends the categories of everyday life, language, and science. This is termed the higher realm of being. It shows, the mystic claims, that there is indeed "something more" to life and reality and that this gives proper meaning and perspective to this life. It is the source and ultimate destiny of humans, especially in the long run, says the theist.

The questions then are clearly drawn. (1) Is there a transcendent realm of being existing beyond this world? (2) Is it unknowable to ordinary experience, logic, and science? (3) Is it knowable *only* by the intensive use of methods of transcendental and mystical revelation, nonsensory perception and insight?

It is at this point that the scientific humanist and the mystic are at loggerheads; there are not only different theories of reality but also different methods of knowing. The primary question is epistemological: What is truth and how can I discover it?

Epistemology is the most important area of inquiry in modern philosophy. Philosophers have raised fundamental questions about the nature and limits of knowledge.[1] The basic question posed here is: What can be known? Are there limits to human knowledge? Are things that are un-

---

1. Locke spells out the importance of this investigation to himself. He relates in the "Epistle to the Reader" of his *An Essay Concerning Human Understanding* that "five or six friends, meeting at my chamber, and discoursing on a subject very remote from this, found themselves quickly at a stand by the difficulties that rose on every side. After we had awhile puzzled ourselves, without coming any nearer a resolution of those doubts which perplexed us, it came into my thoughts that we took a wrong course; and that before we set ourselves upon inquiries of that nature, it was necessary to examine our own abilities, and see what objects of our understanding were, or were not, fitted to deal with. This I proposed to the company, who all readily assented; and thereupon it was agreed that this should be our first inquiry." (In E. A. Burtt, ed., *The English Philosophers from Bacon to Mill* [New York: Modern Library, 1939], p. 239.)

known indeed unknowable? If this is the case, we have two options. First, one may barter one's life to destiny, affirming that because of the limits of knowledge and the need for something more the transcendental stance is a meaningful response. Is this affirmation of belief a reasonable position? Or is it open to criticism and disapproval? The second option is to adopt the stance of the skeptic, that is, suspend knowledge and deny that such forms of reality are knowable or meaningful. If this is the position taken, then we may ask, "What is skepticism, and what is the range of its application?" Can life here and now be meaningful without a transcendental myth to support and sustain us?

II: Skepticism

Skepticism is a meaningful option for an educated person. It is essential to both the scientific and the secular outlook, and it is especially relevant today. It enables us to live meaningful lives without the alleged benefits and apparent burdens of the myths of transcendence.

The term *skeptikos* was originally used by the Greeks. It meant "thoughtful," "reflective," "inquiry." Various forms of skepticism have developed historically. It has been applied to doubts about the knower and the known, the reality of the external world, the existence of God, the possibility of objective standards of value in ethics and politics, and so forth.

Two main types of skepticism may be distinguished. First, there is total, negative skepticism, in which the very possibility of knowledge or truth is denied, whether it is about the external world or the moral life. In its extreme form this may be reduced to subjectivism, solipsism, or even nihilism. Second, there is selective positive skepticism, in which skepticism is a methodological principle of inquiry. This constructive attitude is essential to all meaningful inquiry and reflection. There are various shades of skepticism in between. No doubt both forms are instructive. We learn from the first as we contemplate nothingness; but we can live and function only with the second as a viable option if we are to pursue our goals in the world that we encounter.

Skepticism as unlimited doubt

The first kind of skepticism has had its proponents from Cratylus, Pyrrho, and Sextus Empiricus in the ancient world to Montaigne, Descartes, and Hume in the modern. Cratylus, an ancient Greek philsopher, apparently agreed with Heraclitus that one cannot step into the same river twice. Since all things are changing, he concluded that communication was impossible: words and speech undergo change and lose their meanings by the time they are expressed and heard. Cratylus supposedly refused to engage in debate and simply wiggled his finger in reply to a question, so as to indicate that there was no possibility of real communication. Pyrrho of Elis lived a skeptical way of life. He attempted to attain peace of mind and happiness, but he refused to assent to any beliefs about the nature of the world. Sextus Empiricus also sought *ataraxia* (unperturbedness), maintaining that one should merely put forth the evidence pro and con without evaluating its adequacy. The suppression of belief, he said, was the only cure for dogmatism.

In one sense, the central theme of modern philosophy is skepticism. Given the intense confrontation between religion and science, it became an important weapon in the hands of philosophers and scientists, who wished to liberate themselves from the dead hand of authoritarian theology.

Cartesian doubt is the supreme illustration of its use. It is the polar opposite of the quest for certainty. Descartes, troubled by the fact that the senses deceive, customs vary, and the so-called authorities of the schools differed among themselves, wished to establish knowledge on a firm foundation. He found that only in mathematics and logic and only by developing clear and distinct ideas and deducing inferences from them could he achieve the certainty that he sought.

In his *Meditations* Descartes used an argument now famous to establish a set of firm truths. Descartes expressed unlimited and total doubt. He imagined that some evil genie had been deceiving him and that everything he had ever believed was untrue—his own existence, the existence of the external world, the existence of God. Descartes manifested a state of indecision: alone in a room, he is puzzled, dismayed, confused about knowledge. What is the difference between the dream state of illusion and the waking state of reality, he pondered. Whether Descartes actually lived through such a total experience has been debated by scholars. One could imagine such a state of universal doubt; many have. Why believe anything is true? What is knowledge? What is it about? Why not doubt the senses and perception?

Descartes thought he had found a way out of the morass. He ingeniously argued that he could not doubt that he is doubting without contradiction, for even to doubt entails that he exists, at least as a conscious doubting being. Since doubting is a form of thinking, he exists as a thinking being (*cogito ergo sum*). But Cartesian doubt could not, in fact, be universal; for Descartes at the very least presupposed the language in which he wrote. Moreover, he assumed his methodological criteria: the principle of contradiction, intuitive understanding and apprehension, clarity and distinctiveness as the tests of truth. No doubt a thoroughgoing skeptic can question even these: Why accept the law of contradiction? Why not entertain contradictions? The answer of course is that without the rule of contradiction (that something could not both be and not be at the same time) all discourse would be impossible. If one denied it, one could not speak, but only shrug the shoulders or make a face. Even to deny it by argument would assume it. Is unlimited universal doubt possible? Hardly. One can doubt things some of the time, but not all things all of the time—else inquiry and life would be impossible.

Descartes also attempted to prove the existence of God. He sought to establish the reality of a Divine Being by using a form of the ontological and cosmological arguments. "There must be as much reality in the cause as in the effect," he affirmed. He has an idea of a Perfect Being, which cannot be caused by him (who is imperfect) but only by a Perfect Being. Why his first premise is justified is unclear; the ontological argument seems to be mere assumption, true by definition. That "the idea of a Perfect Being entails existence" is tautologically true may be held by some people—but only by those who already have that idea and not necessarily by the rest of us. (For a further discussion of the ontological argument see Chapter XI.) The idea is not innate, as Locke observed, but is a product of social convention.

Given his assumption of a mind-body dualism, Descartes is trapped within a world of inner consciousness; he is unable to prove the existence of the external world. He cannot even establish by his conceptual methodology that he has a body. Only by first proving the existence of God, he asserts, who would not deceive his own creatures, could he then say that physical bodies, including his own, exist. Why not? Since Descartes admits that he is deceived about his own existence, then why not about the world of corporeal bodies? Descartes' logic was perhaps convincing to the seventeenth-century mind, but it is hardly convincing to the present. Can one indeed *prove* anything about the physical universe by intuition and deduction? Does not one's concepts need empirical content? A rationalistic test of inner consistency is never sufficient by itself. Rationalism may give some certitude about the deductive relationship between our ideas but never about whether those ideas or pure concepts apply to an external reality. Thus, neither Descartes' feigned skepticism nor his assumed certainty have been demonstrated.

Modern empiricists have also been skeptics. They have dealt with the same kinds of problems that Descartes did, but they were no more able to resolve the problem of reality. They began with two similar assumptions: (1) a mind-body dualism and (2) an inner world of ideas, impressions and sensations. Given these premises, they are confronted with the unresolved problem of how we know whether our ideas are accurate descriptions of objects out there.

Locke considered the mind to be a blank tablet, devoid of innate ideas. He analyzed knowledge by reducing its origin to simple ideas, the basic building blocks derived from experience. From these, he said, we construct complex ideas and draw inferences from them. The starting point of knowledge for the empiricists are phenomena imprinted directly on the senses. How do we know that those things which cause our perceptions—substances in the physical world—are exactly as we experience them? We can never know for sure, said the empiricists; we merely assume that they are. Dualistic materialists such as Locke held that primary qualities (space, solidity, number, etc.) and secondary qualities (taste, color, sound, etc.) are caused ultimately by unknowable substrata. Phenomenalists attack materialism by insisting that we only postulate the existence of independent material substances, but we never observe them directly. "To be is to be perceived," said Berkeley—nothing more than that. When Berkeley attempted to explain causality and the regularities of experience, he postulated a spiritual agency, God, as the ultimate source of our ideas.

It was left to David Hume to take this form of skepticism to its unpalatable conclusion. We have no direct evidence or experience of material substances or mental agencies, nor even of our own personal identity. All that we know are bundles of sensations or impressions. But what causes these, or what they are in themselves, is beyond our comprehension. "How do we come about the idea of causation?" asked Hume. He doesn't necessarily deny causality; he shows only that there are no necessary connections or observed powers—only regularities—in our experience. Impressions appear contiguous in space and successive in time. On the basis of this we infer cause-and-effect relationships, but these are rooted in habit and custom and in our assumption that the future will be like the past. Hume is brought to the brink of solipsism. Locked within his own world of private impressions seemingly without rhyme or reason, he tells us that when he pursues philosophy, he becomes immersed in total skepticism. He is determined, he says, to live, talk, and act like other people in the common affairs of life. Notwithstanding his natural propensities, his animal spirits and passions enable him to accept the world. However, he still feels such remains of his former disposition to philosophical skepticism "that I am ready to throw all of my books and papers into the fire, and resolve never more to renounce the pleasure of life for the sake of reasoning and philosophy."[1] I reiterate, Hume never denied the existence of the external world, nor did he deny causality. It is only when he attempted to trace back the origin of such knowledge that he was led to a skeptical quandary.

Such nihilistic skepticism, though tempting to the rational observer, is patently exaggerated. It cannot be acted upon or lived consistently. The

1. *Treatise of Human Nature,* Book 1, Part 4, Section 7 in *The Philosophy of David Hume,* ed. V. C. Chappell (New York: Modern Library, 1963), p. 189.

solution to it is suggested by Hume himself, when he draws upon animal spirit and custom, thus allowing the habit of expectation to intervene and enabling him to draw inferences and to function in the world.

Kant attempted to respond to Hume by arguing that the mind is not simply passive but that there is an active element that we impute to the known. In perceiving and understanding, we impose order and structure on the given; we interpret, organize, and relate the parts of our experience. Percepts cannot be separated from concepts, experience from understanding; they are interfused in our knowledge of the world.

Solipsistic skepticism is paradoxical. It begins with the common-sense world of objects, seeking to understand how and what we know about them. Tracing our knowledge of them through a causal process of perception, we end up by denying the reality of these objects. But the way out of this paradox is to recognize that there is something puzzling in the causal account of knowing that is offered. Most philosophers maintain that they are engaged in epistemological analysis and that they do not have a psychological theory of experience. On the contrary: the empiricists and phenomenalists have indeed presented an empirical account of how we come to know what we know; they have traced this to simple impressions, sensations, or sense-data as the originative stuff of knowledge. But sense-data are not elementary givens but are highly abstract, theoretical constructs introduced by the epistemologist (after the fact) in his reconstruction of knowledge. Once one begins with phenomenal sense-data or abstract ideas, it is difficult to move beyond them. The egocentric predicament is the end result.

This philosophical theory rests on a questionable psychological assumption. If one examines experience, we find that the starting point is not a two-dimensional qualitative idea but a three-dimensional context of activity, which presupposes that objects in the world are present. The starting point of knowledge is not sense-data reports or cognitive concepts—both abstractions from the life-world—but a common-sense world of action, behavior, and conduct. To employ the language of pragmatism and behaviorism, it is *praxis* or practical activity that is the nexus of experience. In the beginning is the deed, not the word; the touch, not the thought; the act, not the percept. Among the first responses of an infant are sucking and grasping, even before it is able to see and recognize persons and objects in its proximity. Seeing and hearing are forms of behavior, conterminous with exploring and finding out.

A philosophy of action takes us beyond some of the traps of earlier theories of experience, though it is not without its own puzzles. The point is that human experience is not a passive receptacle, nor an inward domain of personal consciousness, but involves an outward expression of thought, sensation, feeling and an active display of our psychological dispositions and motives in a behavioral world. Such behavior presupposes not only an agent dynamically and intimately involved in the scene but objects and events that stimulate one's responses and are in turn modified and changed by them. The external world is a precondition for internal awareness. I am not denying that consciousness exists but only affirming that it is an organic reaction coextensive with the environment; it is a two-way current in which traffic ebbs and flows, influences and changes both our inner and outer world. And this is related to our purposes, ends, goals, and objectives, as we transact in a world that we seek to use, enjoy, and manipulate.

The fallacy of both empiricism and rationalism is to isolate only two elements in human experience: perception and cognition. But there are also feelings, attitudes, and motives, in terms of which we respond and relate to the environment. Added to this are multifarious other forms of motor-affective responses: locomotion, copulation, digestion, retreat, advance, exploration, etc. Action is complex, and there are many levels and types of interactive functions.

In addition to this is linguistic communication, so essential to the growth of mind, science, and knowledge. "Can there be a totally private language?" asks Wittgenstein. No, not really, to make sense. For there is intersubjective communication in terms of which we can use language reflectively, to name and denote objects: the wax candle of Descartes, Berkeley's tree in the forest. Thus, subjectivistic solipsism makes no sense because all the solipsists write and attempt to persuade others of their philosophical insights, and they assume a common world of entities and objects which they designate and name. Language is not used in a solely passive sense but has a variety of functions. It is used in the imperative mood, to command others to do our bidding; and they often do, which verifies our assumption that there are common objective realities and our expectations of how they will react to our wishes and desires. There is the expressive use of language, which conveys our feelings and emotions to others and seeks to arouse in them similar compassionate attitudes. Mind is a form of "minding," the active verb, not the passive noun. Language is understood not by referring to abstract essences or meanings but by examining the actual function of linguistic terms and sentences in contexts of interaction. We calculate sums, order and catalogue things, make promises and contracts, ask questions, tell jokes, supplicate the deity, etc.

This demonstrates at least two points: First, the knower, perceiver, or actor requires objects with which he acts; these stimulate responses within him and they are modified by his range of behavior. Second, the knower, perceiver, or actor uses language to communicate with others, whether by employing a primitive sign language of grunts, groans, and grimaces or by using a highly developed language of symbolism. Again, these presuppose a community of actors. Thus some realistic conclusion seems unarguable. There is a recalcitrant world out there. To deny that would be to divorce our experiences from their roots and make them completely unintelligible.

The philosophy of the act, as I have intimated, is not without its theoretical problems. Perhaps there is an act-centered or praxis-centered predicament, if you will. We may ask: How does action, transaction, interaction or coaction relate to a world of objects in nature? How does active experience relate to that which is known? These are important questions, and there is a skeptical residue. One is not reduced to total skepticism, however, but only to a qualified and selective one. For though we may intersubjectively describe objects and events as encountered and seek to explain how they behave the way they do and why, we never know them entirely independently of our transactions. This entails some form of critical realism. We know things, not simply because we abstract their essences or receive sense-data reports, and then infer what they are, but also because of our active, experiential interactions with them. The known is not equivalent to the knower, however, but retains some independent reality separate from the act of knowing. Yet we cannot say what this is entirely. Kant's unknowable "thing-in-itself" does not seem to have completely disappeared. To say, however, that human praxis is a necessary precondition for knowing, touching, using, or seeing objects in the world does not mean that they are equivalent to them. They are not translatable. The world is not reduced to pure acts: to be is not the act. Action is at least bipolar. The agent and the object both come within one's scene of action. We cannot translate a necessary condition of knowledge into both a necessary and sufficient condition, a hypothetical claim into one of identity.

We may assert, without making a leap of faith, that objects have objective properties, in part separate and distinct from our interactions with them. The whole of science demonstrates the fact that things change and interact independently of our will or fancy. There is a brute facticity to existence, a real, objective independence. Perhaps this is the basic postulate of science and life. But surely it is not irrational, nor is it a mere act of animal faith to assert it. Indeed, we can affirm with some measure of assuredness that if someone would seek to deny it, he would be out of touch with the world. He could not do so consistently. For the existential context is the bedrock precondition for all human activities and purposes succeeding. There is a real world; it resists and opposes our temptations and desires; it can be modified and is malleable, but only up to a point. The child who cuts himself by falling off a fence, the cat who drowns in a pond, the tree destroyed by lightning, the brick wall that blocks my walking through it are all empirical testimonies to the obdurate character of a world distinguishable from our fantasies. If we are to satisfy our desires, fulfill our ambitions, achieve our purposes, we need to find out what it is, what we can and cannot do, and either modify the environment or come to terms with it. What the objects in our world are—their properties and characteristics, how they interact, whether they seem to follow regularities and are explainable in terms of general principles and laws—is another question. There are always puzzles and quandaries that we face. Here, some form of limited skepticism is an essential component of the process of inquiry; for doubt is, as we shall see, an essential element in the process of investigation.

Skepticism as selective doubt

Thus far I have developed a teleonomic theory of knowledge. To say that we know something is to assert a fact about human behavior, as we encounter the external world. Knowledge is a property of human action in a world of objects, but it is not an absolute picture of reality. It is *we* who claim that something is understood and appreciated as true. It is this form of knowledge that we believe in and are prepared to act upon. If what we believe is true, then it can have important consequences for our goals and purposes. That tuberculosis is caused by a bacillus and that certain antibiotics can arrest its development is not only known but acted upon. That the moons of Jupiter encircle the planet in predictable orbits is knowledge that has pragmatic significance as a spacecraft is programmed to pass through Jupiter's gravitational field. We cannot say that *all* knowledge is instrumental, as the pragmatists thought. Although the pragmatic theory of knowledge depicts how we think in ordinary life and in the applied sciences, it does not do as well in accounting for research in the theoretical sciences, except in a general analogous way. It is thus not necessarily a comprehensive theory of knowledge, and any account must allow for the pluralistic dimensions of inquiry.

Nevertheless, the pragmatic theory does vividly illustrate the fact that belief is often formulated in response to a state of doubt. In his famous essay "The Fixation of Belief," Charles Peirce denied the actuality of Descartes' generalized state of doubt.[2] Doubt is specific, said Peirce, not universal. It grows out of a concrete context in which genuine puzzlement is present. This doubt is not feigned, but is real and living. Dewey sought

2. *Collected Papers of Charles Sanders Peirce,* ed. Charles Hartshorne and Paul Weiss (Cambridge, Mass.: Harvard University Press, 1934), vol. 5, pp. 223-47.

to relate doubt to the purposes of the investigation. For the pragmatist it is directly related to an existential problematic situation in which our behavior is blocked and our habits thrown into question. It is in this context that inquiry is initiated; we seek a hypothesis that will resolve the problem and enable action to continue. If successful, our hypotheses are incorporated into the body of belief, published in encyclopedias and textbooks, and accepted as true knowledge.

Dewey thought that even theoretical scientists and mathematicians faced problems and that their theories were introduced to settle their doubt, enabling them to make predictions and draw inferences. This theory no doubt claims too much. But it does at least make it clear that there is a vital role that ambiguity and uncertainty play in the course of inquiry in initiating and resolving problems. It is not subjective doubt that the pragmatists focus upon; they are concerned with how we go about fixing reliable belief in the community. And here the methods of science and critical intelligence are the most effective methods to be employed. For we use objectively related criteria, tested over the years by their demonstrated effectiveness in the course of inquiry.

Skepticism is thus an essential component of inquiry and a necessary aspect of critical intelligence. Without it, science and the development of human knowledge would not be possible. It is both positive and creative.

One is always tempted to raise the question, "What is real?" Such terminology unfortunately does not have a clearly identifiable meaning. One meaning that may be ascribed to it is this: (1) *the real* is a term that describes concrete objects and events encountered in a situation of interaction; (2) it refers to their objective properties and characteristics; and (3) it refers to the uniform regularities that objects and events display independently of our own wishes, fancies, or desires. To know what the objective properties, characteristics, and regularities of objects and events are in any specific case depends upon the context of the investigation. Objects and events can manifest a multiplicity of properties and regularities, depending upon the level of interaction and the focus of observation.

For example, suppose we are in a baseball stadium observing a game. Many modes of description may no doubt apply, if we are to understand it. There are individual players, who demonstrate skills and excellences. There are material components present besides the players: bats, balls, uniforms, and bases. There is a stadium filled with loudly cheering fans, including you. To understand the game of baseball, we need to compute the batting averages and errors of the players for the past seasons. This is published in the program. The game of baseball needs also to be understood analytically in terms of the rules agreed upon by the ball clubs and the interpretations of the umpires. Baseball may also be understood as a sociocultural phenomenon with a relatively recent history and an important role in some societies. Each player may be depicted psychologically in terms of his personality traits and dispositions, and these may be compared with those of his fellow players. Baseball also may be interpreted economically. Each player has a contract; an economic index in terms of his drawing power with the fans and the demand for him by other teams can be made. The players may be described purely biologically and physically in terms of heights, weights, bone structure, physical strength. Moreover, there is physicochemical structure to their behavior that can be understood. Which of these properties refers to the *real* game? In a sense all of these characteristics are pertinent to our description of the players, the team and the game: physical, chemical, biological, psychological, analytic, economic, and sociological. We must thus resist any simple reductive analysis that attempts to break up the observable to a set of entities or principles on a lower level, to its material components or its ideal structures. But rather we need to provide a pluralistic contextual account of what is observed: both the concrete events and their explanations in terms of general principles.

Any number of hypotheses can usefully describe, characterize, and explain behavior, depending upon the context of analysis and the purpose of the inquiry. I have elsewhere labeled this the "principle of coduction," as a methodological device employed to coduce a number of explanations on different levels to account for an object or event.[3] In the preceding illustration, we don't doubt that there is a game, players, and a stadium. The ball may be hit right into the stands and strike you or someone close by you! If this is the first time you have ever been to a game, you may be puzzled by the events about you and try to decipher what is going on. Why did the umpire call a batter out? There were three strikes. Why did he walk him? There were four balls. Even if you are a veteran fan, the game is full of indeterminacy and suspense. Will the outfielder catch the high fly and get it to the third baseman on time? Will the next player strike out? Who will win the game? Or you may ask, was Reggie Jackson safe as he slid into second base? The umpire and the fans can dispute that. What was Pete Rose's batting average? We can look it up. What were the terms of Joe DiMaggio's contract? They may not have been disclosed. Who was the best ballplayer of all time? All of these points may be arguable. In this case, doubt is always specific. For we can always question what we are observing. And we can raise questions about how to interpret what has happened.

Unlimited doubt, as I have said, is self-defeating, contradictory,

3. Paul Kurtz, *Decision and the Condition of Man* (Seattle: University of Washington Press, 1965), chap. 5.

nonsensical—we cannot doubt everything, all of the time, only some things, some of the time. There are, however, forms of doubt that are meaningful. And I wish to distinguish three kinds. First, there are the principles of fallibilism (proposed by Charles Peirce) and probabilism; second, the burden-of-proof argument; and third, the essential role of doubt as a component of the process of inquiry.

Peirce argued that we can never say that something was absolutely certain, had reached its final formulation, or was ultimately unknowable. There is always the possibility that we may be in error. Our observations may need to be corrected. We may have misinterpreted the data. We may be mistaken. No one is infallible. And we must always leave open the possibility that what we today affirm strongly may in time need to be revised. The history of thought is full of such developments. Knowledge needs to be revised continually; at least we should not foreclose inquiry. Even the most firmly held postulates and theories have at some time been questioned by later investigations; and even the most revered truths of mankind, whether in religion, ethics, politics, philosophy, or science, have often been replaced. Only a fool will refuse to see the possibility that he may be in error, or a fanatic, that his first principles and final conclusions are beyond criticism. The Pope may claim infallibility, but surely no one besides certain Catholics are so presumptuous as to think that the Pope has a special monopoly on truth.

There is another reason why fallibilism should be an intrinsic element in human knowledge. That is that the sample on the basis of which we may make a generalization or infer a theory may be limited or partial. New data may be uncovered. These may be bizarre, anomalous, and unexpected. The old theories thus may have to give way to new, more comprehensive explanations. Here *probabilism* is a better term than *fallibilism,* for it points to the fact that a sample is never complete or final. Although on the basis of our observations thus far we may rather confidently assert that all crows are black, we may find a white mutant, however unlikely that may be.

Nor should we be a purist and thereby conclude that no knowledge is possible. Because we can never be absolutely certain that we have not made a mistake, that new facts will not be uncovered, or that our present generalizations and theories will not be modified at some time in the future, can we conclude that no knowledge is reliable? There *is* a well-established body of knowledge: science, history, medicine, the practical arts all belie this form of skepticism. Some measure of doubt, however, is a therapeutic element in the quest for knowledge, whether in the sciences or in ordinary life. There are, however, degrees of doubt and probability. If no inference can be said to be absolutely or finally certain and if we reserve the right to concede that we may be wrong, this does not mean that we do not have relatively strong degrees of reliability in many cases. Where? First in mathematics and logic. If one's axioms and postulates are assumed, we can deduce conclusions and theorems with deductive necessity. That two and three are five, that a straight line is the shortest distance between two points, or that a triangle is a three-sided figure cannot be doubted, if I understand the nature of number, arithmetic, and the principles of Euclidean geometry. Of course, we may question their foundations—as has been done in the history of thought. And given different postulate systems (non-Euclidean geometries for example), one may deduce different conclusions. Still, analytic tautological truths are necessary and certain, if we presuppose certain assumptions.

And there are other hypotheses and beliefs that I can assert without much hesitation: that I see my hand that I hold up before me, that it is raining outside at the present moment, or that the river has crested and is flooding its banks. Given the testimony of the senses, under normal conditions, I need not doubt that what I observe is as I describe it. Of course, I may be mistaken. I may be sleepy or color-blind or drunk, and misperceiving. It may be a mirage or a dream, or I may be hallucinating. The testimony of the senses is often confused and distorted. Thus, we cannot assert with absolute certainty that what is observed is always true—only probable. Two men are standing at a bar drinking heavily. One says to the other: "You better stop drinking, your face is getting bleary!" Where is the bleariness—in the perceiver or perceived? This does not deny that for all practical purposes we may assert that some observed facts are true.

To say that I have observed something to be the case is one thing; however, to interpret or explain it is, of course, another. Similar considerations apply to more complex scientific hypotheses and theories. Often even the most well-established principles have been overthrown by later experiments and theories. Ptolemaic cosmological explanations gave way to Newtonian physics, and this was modified and supplemented by modern Einsteinian physics, which may still be brought into question. Thus no scientist can affirm with complete certainty that his theories have reached their final formulation, that they are perfect, eternal, and beyond question. Only theologians reserve the right to maintain their infallibility. The better part of prudence thus is to adopt the principles of fallibilism and probabilism.

Another consideration in support of selective skepticism is the burden-of-proof argument. We are constantly bombarded on all sides by the claims of those who insist that they have the truth. Whether it is belief in religion, politics, ideology, philosophy, science, the paranormal, ethics, or ordinary life, the question may be asked: Is what they claim true? Was there an eclipse in 322 B.C.E.? Did Christ rise from the dead? Will vitamin C prevent colds? Have space astronauts been visiting the planet earth from another galaxy? Is there life on Mars? Is reincarnation true? May people be possessed by the devil? What is the temperature at the center of the earth? Was Tchaikovsky a homosexual? Did he commit suicide? Will the stock market advance? What is the age of the universe?

For many of these questions, we may not have the answers. We may be unable to ascertain the facts. Some of our knowledge is based upon our own direct perceptions. And we may be rather certain of its accuracy. Some of it is based on the testimony of the senses of other people, second-hand or third-hand; if we think that they are competent observers, we may be willing to accept it. In other cases, the questions at issue are so complex that only the authorities in the field can resolve the matter. Who are the experts? Does every field have one? What do we do when they dispute about the relevant facts and the appropriate theories?

It is clear that where we think the observation or testimony of others or the judgment of experts is reliable, we may defer to them. But in other cases, where someone makes a claim, the burden of proof is upon him to provide the supporting evidence and proofs. If someone maintains that mermaids exist, the burden of proof is not upon me to show that they do not but upon the claimants to show that they do. Old seafaring men reputedly claim to have seen them, especially after weeks and months at sea. The proponents of a claim often ask, "Can you disprove what we have said?" For example, can you prove that God or angels do not exist, or that Sasquatch does not exist, or that UFOs are not manned by extraterrestrial beings from afar? If you cannot, then we have the right to believe, they insist.

Now I suppose that anyone has the right to believe anything that he or she wishes. The questions are whether what they wish to accept is actually true and whether they can distinguish their wishes from objective reality. Here doubt becomes an essential component of rationality. I don't have to prove to my five-year-old granddaughter that Santa Claus does not exist. One may never be able to do that satisfactorily to the inquisitive mind. For if I take the child to the North Pole, she may wonder whether Santa Claus is at the South Pole, or whether he has dematerialized. All beliefs should be considered to be hypotheses; whether or not they are true depends upon the range of evidence and the kinds of reasons that are adduced to support them. If there is insufficient evidence, then the best option may be to suspend judgment pending further inquiry. I am not denying that we can falsify some claims as ridiculous or patently false. Whether tar water can cure the hives or whether the moon is made of green swiss cheese can presumably be tested and confirmed or disconfirmed rather conclusively.

Therefore, some questions can be resolved by reference to a range of evidence; some can be disconfirmed in the same way. Some questions may, however, be beyond present confirmation or proof. And for those, if they are meaningful options, the best course is to suspend judgment, to assume the posture of the skeptic and not postulate a conclusion until the relevant facts are in.

In some cases, of course, we may have some evidence, but this may not be sufficient. We may suspect that something is the case, yet not be able to say that it is so. Here we can frame a hypothesis and say that it is supported to such and such a degree, that it is probable or improbable without being willing to say so conclusively either way.

This is the method that is used in the sciences. And one must grant that it is the sciences that have been the most effective in developing knowledge. Science presupposes some skepticism, but it also contributes to the positive achievement of knowledge.

III: The scientific method

What is science?

Since science presupposes some degree of skepticism in its method, we may ask, "Is there such a thing as *the* scientific method?" Some have denied it. If there is such a method, how does it operate? Scientific methodology is surely not an esoteric road to knowledge available only to a limited sophisticated elite. Rather, it grows out of the interests and purposes of ordinary life, and it is continuous with the use of critical intelligence elsewhere.

Some have interpreted science solely in terms of its findings, that is, as a coherent body of abstract knowledge systematically organized. As such, science is divisible into fields of specialization. Some of these overlap and share common principles. The natural sciences, for example, draw on the principles of physics and chemistry; theories of astronomy reinforce those of geology. Similarly, the life sciences use the principles of genetic and molecular biology. The behavioral and social sciences also interact in sharing in some instances common concepts, hypotheses, and theories.

But science in a deeper sense is a form of human behavior, a way of inquiring into nature. Science, in this sense, refers to an active process of investigation, a procedure for testing hypotheses and laws. The organized body of scientific knowledge accordingly is simply the product of its process or methods of inquiry.

The purposes of science are no doubt multifarious. The aim of basic theoretical research is understanding. Here the task of scientific inquiry is (1) to observe and describe data, (2) to date, record, and classify accurately the subject matter observed, (3) to formulate general hypotheses and laws concerning regularities, (4) to explain and account for what is observed by reference to antecedent conditions and causes, (5) to verify its hypotheses and laws by means of predictions and experiments, and (6) to relate and connect these by means of more comprehensive general theories.

The preceding is a general model, but there are wide deviations from it. And there are developing or proto-sciences, which do not appear to fulfill all of these conditions. Some sciences, particularly the social sciences, have been able thus far to develop only a limited number of lower- or middle-range statistical correlations and tested hypotheses. Some sciences have not achieved the kinds of tested general explanations or high-level theories such as are discovered in physics and the natural sciences.

Scientific inquiry grows out of the practical interests of everyday life. Theoretical research is often stimulated by such concerns. Social, political, and economic needs encourage scientists to find solutions to social problems. The science of mechanics, for example, was engendered in part by the desire of ballistics experts to plot the paths of projectiles fired from the newly invented cannons that were used to batter down fortifications. Atomic-energy research was accelerated by the competitive race for nuclear weapons during World War II and thereafter. Cancer research was encouraged by an increased incidence of the dreaded disease, by the widespread public desire to find a cure for it, and by the expenditure of large sums of money by governmental and private agencies. Here intelligence is pragmatic and utilitarian: the goal is to solve problems, discover remedies, overcome obstacles, find applications.

Some scientific research, however, is largely theoretical, though the quest for explanations fulfills a pervasive human curiosity. The basic question that has been raised is: "Is there a standard methodology of science, a set of procedures that scientists follow in achieving their goals?" Many philosophers of science have attempted to depict what they thought was the process of inquiry and discovery. Thus, empiricists attempted to trace knowledge to its origin in perceptions and sensations, maintaining that the mind is a blank tablet, written upon by experience. Rationalists related general theories to the deductive processes of intuition and inference. Pragmatists connected all hypotheses to problem-solving processes.

Regretfully, these efforts are misdirected. There is no simple logic of discovery that everyone follows. Knowledge may originate in a variety of ways. Penicillin was discovered by an accidental observation by Sir Alexander Fleming. Claude Bernard, wrestling with a problem concerning the cause of sugar in the blood of animals, said he arrived at his explanation by mulling over the experimental data. Einstein developed his relativity theories by questioning the postulates of Newton. The processes of creative discovery thus are complex and diverse. We need to be immersed in data, to know the empirical facts, and to be able to make inferences to solve problems encountered in the process of research.

What is central from the standpoint of science is not so much the genesis of an idea or hypothesis as its validation and verification. Thus it is the criteria of validating or verifying knowledge, how we establish or warrant belief, that should concern us. This does not mean that the scientific inquirer might not possess certain qualities of mind that are essential in the course of his research. He or she must be a keen observer of the facts. His mind must be passive, as it were. He should listen to what nature tells him without allowing his preconceived biases to intrude. He will formulate predictions and draw inferences, if his hypotheses are true, and observe the results meticulously and carefully. The experimental method thus depends directly on the data given. But merely to look at the facts is never enough. The scientist must ask questions of nature. His inquiry may be stimulated by a puzzle or doubt, and his hypothesis, which may be initiated by creative insight or contrived conjecture, is a solution to the problem. Thus the mind of the scientist is both receptive and passive, active and manipulative. The data are not simply given but taken, interpreted, generalized, related to other forms of data. Some abstractive ability is thus involved. Connecting relationships and theories are developed; the coherence of a set of explanations and how they fit together is elaborated. For example, evolutionary principles have theoretical relevance not only in biology, anthropology, and genetics but also in astronomy, geology, and the other natural sciences. Similarly, the germ theory of disease has a wide range of application. Successful scientific theories tend to be comprehensive in the scope of their explanations.

Critical intelligence—reflective, probing, interpretive, analytic, perceptive—operates throughout the process of discovery and validation of scientific hypotheses. But the question I again raise is this: What are the standards of validation? This is particularly pertinent when we ask how to distinguish veridical beliefs from false ones, science from pseudoscience.

Subjectivistic methodology

Some philosophers of science have denied that there are any standards of validation of scientific hypotheses and theories. Paul Feyerabend is perhaps the most extreme exponent of this radical skeptical view. He argues: "The idea of a method that contains firm, unchanging, and absolutely binding principles for conducting the business of science meets considerable difficulty when confronted with the results of historical research. We find, then, that there is not a single rule, however plausible, and however firmly grounded in epistemology, that is not violated at some time or other. . . . There is only *one* principle that can be defended under *all* circumstances and in *all* stages of human development. It is the principle· *anything* goes."[1] He maintains, for example, that the Copernican revolu-

1. Paul Feyerabend, *Against Method* (New York: Schocken Books, 1977). Quoted in Ernest Nagel, "Philosophical Depreciations of Scientific Method," *The Humanist* 36, no. 4 (July/Aug., 1976): 34–35.

tion and the emergence of the wave theory of light occurred because some scientists did not feel bound by the methodological rules and broke them. "When Copernicus introduced a new view of the universe," according to Feyerabend, "he did not consult *scientific* predecessors, he consulted a crazy Pythagorean, such as Philolaos. He adopted his ideas and he maintained them in the face of all sound rules of scientific method."[2]

Now Feyerabend is obviously correct that there are no firm and unchanging, absolute binding principles involved in scientific inquiry. The history of science clearly shows that the establishment of scientific principles is often a rather loose affair and that a whole number of irrelevant considerations enter. Scientists, like everyone else, are human beings, and subjective fancies and predilections may intrude on their work. Many scientists do indeed violate the rules of inference and experimentation. Yet their results remain, and eventually they may be accepted by other scientists and incorporated into the body of scientific principles. The fact that one aspect of a scientist's research may be warranted does not mean that other parts of his world-view may not be flawed. Newton, for example, believed firmly in the Bible and astrology. The point is, scientists do not follow neatly laid out rules of validation in all of their inquiries.

May we then infer that there are *no* rules or criteria for establishing scientific knowledge? Is the only dependable theory epistemological anarchy? Feyerabend believes that although science was liberating during the Enlightenment, it now has become a religion and an ideology and that it is often the enemy of freedom of thought. As a result, he defends astrology and even the teaching of creationism along with the theory of evolution in the public schools. For Feyerabend the theories of science today are no closer to the truth than fairy tales or other myths, and if taxpayers believe in Chinese herbal medicine and acupuncture, voodoo, faith healing, or the Hopi ceremonial rain dance and cosmology, then it should be taught in the schools along with establishment science, which he compares to a state church.

This extreme, subjectivistic theory has puzzling consequences. For if there are no objective standards, then is any belief as true as the next? If there are no rigorous standards for testing claims, then is not anything as true as anything else? But surely there is a difference, for example, between a scientific procedure that can isolate and locate the viruses that cause polio and develop a vaccine against them and one that attributes the disease to the devil. What is meaningful is that there is a relevant range of evidence that decides the question. If we grant that it is difficult to

2. Paul Feyerabend, "How to Defend Society Against Science." Quoted in Klemke, et al., *Introductory Readings in the Philosophy of Science* (Buffalo: Prometheus, 1981), from *Radical Philosophy* 11 (1975): 3–8.

identify firm and fixed rules, this does not mean that there are none. The choice between hypotheses and theories is not arbitrary.

Surely, one of the tests of a scientific claim is its *results:* in technology and practice. However, Feyerabend disputes this point, claiming that other ideologies also have results. Religions build cathedrals and save souls, and revolutionary doctrines overthrow established societies. But the kinds of experimental consequences of scientific inquiry are independent, in one sense, of our own wishes imposed upon the world. When asked why he takes airplanes rather than brooms, Feyerabend replies flippantly, "I know how to use planes but don't know how to use brooms, and can't be bothered to learn." If Feyerabend doesn't believe in objective standards of truth, why doesn't he jump out the window of a fifty-story building?[3] The answer is clear: there are obdurate realities that impose upon, block, or fulfill our desires. Feyerabend's theory is self-defeating, for he can give no objective reasons why we should accept his theory other than taste or caprice. There *are* principles that govern air travel, and one jump out of a high window should lay to rest his views that there are no differentiating standards of inquiry.

According to Feyerabend, science is a social phenomenon, deeply embedded in an entire cultural milieu. It may be related to a religious world-view (as during the days of Aquinas) or stand in opposition to it (during the modern period). It has strong affinities to the state and the economy, which benefit from its findings and applications and either encourage (by support) or impede (by restrictions) its world-view. Lysenko was aided by the Stalinist state because his theory of biological evolution supported the reigning ideology, and nuclear research was massively supported during World War II and after because of its military potential. All of that is granted. But is the conclusion to be drawn that scientific objectivity is a complete myth and all that remains is a form of social relativism?

Thomas S. Kuhn, in a famous book, *The Structure of Scientific Revolutions,* points to the fact that in some historical epochs an entire conceptual framework or paradigm may give way to another.[4] In such revolutionary periods diverse social and cultural factors may contribute to such a shift. Kuhn describes "the transfer of allegiance" from one paradigm as a kind of "conversion experience." At first, there may be strong opposition to a competing theoretical framework, but eventually it may give way and virtually the whole of the scientific community may

3. See Martin Gardner, "Anti-Science: The Strange Case of Paul Feyerabend," *Free Inquiry* 3 (Winter 1982/83): 32-34.
4. Thomas Kuhn, *The Structure of Scientific Revolutions,* 2nd ed. (Chicago: University of Chicago Press, 1971).

adopt it. Often the contributing factors are extrascientific, and the choice of a theory does not resemble a proof in formal logic or mathematics.

Are we likewise to infer from Kuhn's account of what has often happened in the history of science that there are no objective standards of scientific validity? Some have suggested that new paradigms triumph because of some "mystical aesthetic." Some defenders of paranormal world-views today quote Kuhn approvingly to support and sanctify their untested theories. Kuhn himself denies these implications; he thinks that the continued resistance to the new theories (as in the case of Joseph Priestley, for example) may be illogical or unscientific. The man who continues to resist and oppose the new view after the entire profession has been converted to it has ceased to be a scientist. Although Kuhn believes that extralogical forms of persuasion intervene, this does not mean that "there are not many good reasons for suggesting one theory rather than another." He maintains that these are "reasons of exactly the kind standard in [the] philosophy of science."[5]

Testing truth-claims in science

The questions that we need to raise are: "What constitutes good reasons for the choice of a hypothesis or theory and what constitutes good grounds for its acceptance?" Here the focus should be on the *criteria* for testing a claim, not the procedure or method by which it is arrived at.

Here again, I do not think there are easily formulatable criteria or rules one can lay down beforehand. We need to extend our skepticism, to some extent, to science itself and refute any myth that science or the scientific method is unerring and infallible.

Scientists have all the frailties of other human beings; and science, where institutionalized, carries all the weight and drawbacks of its tradition and prejudices. Scientists may be as dogmatic and hostile to new departures in thought as others. Encrusted habits are not the exclusive possession of religion, economic class, or ethnicity but affect all human endeavors.

Still, there are some standards for testing claims. These may be ideal, and there may be wide deviations from them in actual practice. There are at a minimum three important criteria: (1) the need for evidence, (2) logical consistency, and (3) technological and experimental consequences.

5. Quoted in Klemke, et al., *Introductory Readings in the Philosophy of Science,* p. 208, from "Theory Choice," *Criticism and Growth of Knowledge,* ed. I. Lakatos and A. Musgrave (New York: Cambridge University Press, 1970), pp. 260–62.

Evidence

The principle of corroboration. The logical positivists attempted to introduce rigorous criteria that they thought would enable us to determine the adequacy of scientific hypotheses. These criteria have since come under heavy criticism, particularly the principle of verifiability, which was recommended as a theory of meaning and which was supposed to help us distinguish genuine from pseudo-statements. An empirical or synthetic statement, said the positivists, is meaningful if and only if there are conditions under which its truth-claim can be tested. Critics have pointed out that this criterion was much too stringent. Many statements in the sciences are theoretical and have no easily confirmable content. Moreover, one cannot always determine a priori what is testable.

Broadly conceived, the intent of the principle of verifiability was worthwhile: to help us disentangle patently meaningless nonsense in metaphysics, theology, and the pseudosciences from meaningful hypotheses. Unfortunately, the criteria imposed are too narrow, a straightjacket on scientific inquiry; it is an oversimplification to say that one can determine merely by analysis antecedent to inquiry which sentences are genuine and which are nonsensical. Often one is circumscribed by one's preexisting paradigm, which enables one to reject something out of hand.

The principle of *verification,* as distinct from verifiability, is a variation on the empiricist criterion. It is an essential component of scientific methodology. The principle of verification has both a positive and a negative aspect. First, a hypothesis is true (as distinct from meaningful) if it has been verified, directly or indirectly, and if there is a range of evidence that can be brought to bear to support it. As Hume pointed out, however, we can never say with certainty that our inductions based upon observation and experiment are complete. Hence, no hypothesis is definitively confirmed, only probable. Nor can we confirm the principle of induction itself.

Karl Popper has sought to modify this criterion by a negative principle: nonfalsifiability. He argued that the actual procedure in the sciences is to introduce conjectures or hypotheses, often based on creative and intuitive guesswork. This is not necessarily always supported by painstaking data-gathering or experiment. However, the decisive issue is whether the conjectured hypothesis is warranted; and here there are tests that can be brought to bear; at the very least we need to be able to falsify the theory by seeing whether it can be refuted or sustained by the empirical evidence. A theory is meaningless if there are no conditions under which it can be disconfirmed.

I agree with both the positive and negative verification criteria. A more adequate terminology that better conveys what is going on would be the *principle of corroboration*. We first try to establish a claim by reference to clearly established observations and experiments. However, in many cases we do not have sufficient data; some complicated hypotheses can only be established either by a single test (e.g., the Michelson-Morley experiment was thought to confirm Einstein's theory of relativity concerning the speed of light) or by a disconfirming test of alternative explanations. If a theory stands up to the most severe tests that can be formulated, it is corroborated and accepted—though it may, of course, give way when newer explanations and new data are uncovered.

Intersubjective Corroboration. The principle of corroboration refers to *public* evidence. That is, mere private reports of introspection or subjective observations that are seen by only one person are never sufficient. In principle, they must be capable of independent corroboration by other impartial observers. Mystics and seers claim to have some inner road to truth. They refer to a kind of apprehension or illumination that they alone possess. To try to translate this into observable reports, they say, loses its essence, since it is "ineffable." But science cannot accept subjective claims to truth; it needs to replicate the experience or experiment. This means that any scientist in any laboratory under standard conditions could achieve the same chemical reactions or observe the same phenomena with his instruments.

One of the problems of the pseudosciences is the fact that claims so often made by devoted believers resist corroboration by independent or neutral inquirers. One of the basic problems in parapsychological research today is the difficulty that independent researchers have in replicating the results of others. The researchers include not only skeptics, who almost never get results, but also other parapsychologists in other laboratories who find the alleged data elusive and that it is difficult to specify or determine the conditions under which the effect is supposed to occur. A similar problem applies to so-called revelatory knowledge basic to the major religions. How do we know that the experiences of those who allegedly received revelations were "authentic" unless they could be corroborated by others? There must be at least some reliable eyewitness testimony, particularly about the claims that deviate from our normal expectations about the world. I am not talking about hearsay evidence, which is notoriously unreliable, but direct testimony of more than one person.

One should raise a serious word of caution even here about some so-called eyewitness testimony. Although sense perception is the bedrock of any empiricist-experimentalist criterion of evidence, without which science cannot proceed, it may be a snare. In the first case, the untrained or undisciplined observer can often misjudge the "facts" before him. The three blind men who described different shapes of the same elephant illustrate the difficulty. Many people may witness a crime, as was the case in the film *Rashomon,* and each will give a different version of what happened; this is based on one's vantage point. I have personally met people who swear they were abducted aboard a UFO into outer space or have been visited by the spirit of a deceased relative. What is perceived is rarely done so without interpretation in terms of a preconceived idea, hypothesis, or theory, particularly when we are seeking causes. Here a *critical* eye is an essential component of observation. To see something, we must know what to look for. Percepts without interpretive concepts may be unintelligible. We need the creative use of inference and logic to interpret the data, however, not inference that distorts the facts or colors them to fit a preconceived theory but inference conditioned by what is actually there. What is necessary is that there be trained observers who can check the veracity of an observation and then give it some inter-subjective corroboration. We need to ask not only *what* is happening, but *why* or *how;* and only active investigations can answer this by posing questions and unraveling possible responses.

A question often raised is: "What constitutes sufficient evidence for a claim to be accepted?" There are no simple answers. If one is predisposed toward a theory, then only a few instances may be taken as sufficient to establish its credibility. In some cases, a single decisive experiment may be sufficient. In other cases, however, we may need to test and retest the data before we are satisfied that our hypothesis is warranted.

Probability, fallibility, skepticism. Here we come back to our earlier point, namely, that no theory is *ultimately* confirmed decisively, since we may be mistaken and new data may be uncovered. A theory is only as good as the evidence adduced in its support; and this is often only com-parative. All other explanations may either seem less likely or fail entirely; though the theory we have may not be airtight. Here constructive doubt again emerges as the basic ingredient of on-going inquiry. Hypotheses are working ideas or tentative formulations until they are corroborated. Many such hypotheses become well established by repeated corroboration. Given the history of science, however, we must be prepared to allow the possi-bility that they may be in need of revision at some later date.

The question concerning the *degree* of certainty of one's theories is especially relevant to Hume's famous interpretation of miracles: we are more likely to accept a claim if it is not discordant with our own experi-ence or what we already know to be true. The hypothesis may be new, but it need not contradict the body of evidence that has already been amassed. Where a claim promises to overturn a whole body of data and hypotheses that we now accept on the basis of strong grounds, then before we accept it, we must have even stronger grounds to do so. Extraordinary claims thus require extra degrees of evidence. Thus, before we can invoke miraculous or occult explanations that overturn well-established laws and regularities of experience and nature, we would need very strong evidence.

We can see the relevance of Hume's argument in parapsychology, astrology, and the paranormal fields. Often a claim is made that, if true, would supplant well-established principles of both science and ordinary life. C. D. Broad has pointed out a number of principles that parapsychologists apparently wish to overthrow: (1) that future events cannot affect the present *before* they happen (backward causation); (2) that a person's mind cannot effect a change in the material world: there has to be intervention of some physical energy or force; (3) that a person cannot know the content of another person's mind except by the use of inferences based on experience and drawn from observations of speech or behavior; (4) that we cannot directly know what happens at distant points in space without some sensory perception or energy transmitted to us; and (5) that discarnate beings do not exist as persons separable from physical bodies.[6] These general principles have been built up from a mass of observations and should not be abandoned until there is an overabundant degree of evidence that would make their rejection more plausible than their acceptance. Those who use the term *paranormal* believe that they have uncovered a body of empirical facts that call into question precisely those principles. Whether or not they do remains to be seen by future inquiry. These principles are not sacred and may one day need to be modified— but only if empirical evidence makes it necessary.

Now it seems to me that Hume's argument can be pushed too far. We need to be dubious of efforts by a priori reasoning to say that since some hypotheses are incomprehensible to the present conceptual framework they are therefore false. Hume's argument can be used as a club by narrow-minded skeptics to disavow all new revolutionary theories; one must guard against its possible misuses. Nonetheless, we must be prepared to radically alter our conceptual framework, but only if there is sufficient new data that cannot be explained by any other means. All theories in the last analysis must give way to the *evidence*. One needs strong and reliable evidence before one is prepared to overthrow well-established theories. But in the process of scientific inquiry, it has often happened; and we must always be prepared to do so.

6. C. D. Broad, "The Relevance of Psychical Research to Philosophy," *Philosophy* 24 (1949): 291–309.

Logical coherence

Analytic clarity. The beginning test of the adequacy of belief is the precision with which it is stated. Thus before we can evaluate a statement, we need to analyze the meaning of the ideas and concepts that are involved in it. Before we can tell whether the statements that are made are warranted, we must know whether they make sense. Linguistic and logical clarification thus is a prolegomenon to any empirical inquiry. There are many errors that people commit. Ideas may not be clearly stated; they may be vague, even incoherent. Sometimes abstract ideas lack any referent; they may be devoid of clear meaning. It is doubtful that there is a single criterion for clarity. But there are some important guidelines.

First, we should always seek to define our terms, if possible, by the use of synonyms or by pointing to characteristics or properties to which the words refer, particularly if they are descriptive terms. Some terms lack clear connotation. (Ethical terms, for example, such as *good* or *right* are difficult to define by designating observable properties in the world. Yet they are understood.)

Second, we should seek to examine the contexts in which terms are used to see if we can understand what they mean. This means that to understand language we need to be intimately acquainted with the subject matter and how terms and concepts interact in situations.

Third, to do this we need to see how terms are employed and what role or function they perform. We should thus always seek the use of a term and see what job it does in a context in order to comprehend what it might mean.

In the sciences some precision is essential, as far as possible, especially, if we are to develop hypotheses or theories that are testable. In poetry and religion, terms may have only metaphorical or symbolic uses, and they may serve to evoke feelings or moods, to arouse passion and commitment, rather than to stimulate cognitive thought or perceptual imagery. It is important that we do not confuse the meanings of terms in various usages. They may convey information, particularly when found in sentences in the form of affirmative statements or denials; but they may be imperatives (expressing commands), interrogatives (raising questions), or performatives (ceremonial) in function. Contemporary analytic philosophy has made important contributions to our understanding of language. But we are here talking about the meaning of language, not the truth of claims or beliefs. We cannot say that something is true if it is gibberish or nonsensical. But to make it clear still leaves open the question of whether it is true.

We are speaking not only of the clarification of parts of speech, words, terms, and concepts but also of sentences or statements. These may be classified as descriptive, analytic, expressive, prescriptive, imperative, performative, interrogative, etc., in form and function. From the standpoint of knowledge, we are concerned with (1) descriptive, theoretical, or explanatory statements, and whether and how they can be established as true, and (2) prescriptive statements, and whether or not they work and the consequences of their use.

Internal validity. Ideas, hypotheses, beliefs or statements are never held or uttered in isolation. Otherwise, they would be fragmentary bits. We need to see how they are related to other beliefs or sentences in a framework or structure. Our beliefs thus are connected by logical inference, and they may have some internal symmetry. Accordingly, a necessary (but not sufficient) condition for the truth or adequacy of a hypothesis is that it be consistent with other statements that are accepted. Here the rules of deductive logic apply. Some philosophers have thought that this internal consistency was sufficient in itself.

If Rip Van Winkle returned to his home after twenty years, unaware of any of the factual claims made by people around him, he could still check the varacity of their statements by seeing whether or not they contradicted earlier assertions. But he could not fully account for what he encountered unless he corroborated the data empirically. Our premises and theories always need some empirical content, if they are to be taken as true.

Comprehensive coherence. Nonetheless, although we can and do judge specific syllogisms and arguments, or hypotheses and theories in a single science, by criteria internal to that inquiry or discipline, there is always the effort to reach out and generalize. We seek to relate the specific sciences to one another, and these to broader principles.

Some philosophers have thought that if they could develop a comprehensive and elegant enough system, eventually they might incorporate all possible statements and thus comprehend the truth about the universe as a whole. This is known as the doctrine of internal relations. The danger in this is that although one can establish as valid a set of philosophical, theological, or ideological propositions and integrate them into a broader system, whether the entire framework is true is often what is precisely at issue. Thomism and Marxism illustrate such comprehensive systems. Great efforts have been expended ad hoc to account for any discrepancies found in such systems. Unfortunately, the commitment to a total system may distort one's judgment and perception. One may be so swept away by its comprehensive outlook and its totalistic answer to the root questions that factual discrepancies are denied, overlooked, or explained away. Though the arguments used to elaborate and defend the system of belief may be formally valid, they may not be truthful about the material world. Many theories are internally consistent, but since one or more of their premises are false, the system is false. The question to be asked always is: Are the premises factually true or supportable? These should not be tested simply by an appeal to "intuition," or cognition, but by the evidence. Thus, the problem with Thomism is its premise about God, which is unsubstantiated, and with Marxism, the unconfirmed dialectical laws of history.

Here we have gone beyond science to metaphysics, theology, and philosophy proper. It is one thing to seek and develop unifying statements within one science or across the sciences (the principle of evolution illustrates the latter), but it is quite another to develop a unifying theory of history or an ontological theory of ultimate reality. Here we are in the metascientific realm. A comprehensive philosophy of nature is often based upon a generalization from the sciences at one stage of the development of knowledge and upon philosophical reflections on the first principles of science, religion, and the arts. Such enterprises, however, are often perilous. Efforts to construct comprehensive philosophical systems no doubt have some merit, for they give us some perspective and seek to integrate our knowledge. But such systems cannot be tested solely by analytic and/or deductive criteria; they need empirical and consequential tests if they are to have any degree of reliability.

Pragmatic consequences

Experimental test. This criterion is related to empiricism and the requirement that there be corroborative evidence, but it goes one step beyond. For knowledge is not simply a description of what is but of what can be. Belief-claims, in a sense, are plans or rules of action; they make predictions about what is likely to happen if we undertake certain forms of behavior.

To say that "sugar is soluble" means that if we were to put a spoon of sugar into a glass of water, it would most likely dissolve. The statement can be interpreted in operational terms. Many or most sentences in science take a hypothetical or conditional form, such that if we were to do something, then certain consequences would ensue. Every explanatory theory has within it some implied predictions (directly or indirectly); otherwise it cannot be said to be true. By this, I do not mean that every sentence in a theory is equivalent to its operational meaning—this rigorous criterion is much too narrow—but only that a theory could be related, at least indirectly, to sentences that are operationally testable.

Experimentalism is not reducible to empiricism. The latter refers to sense-data and observational reports, passively viewed; the former refers more explicitly to conditions, by which hypotheses are tested. We judge them by their verified consequences.

Problem-solving and technological applications. Another test that is more difficult to generalize about is the fact that our beliefs and hypotheses have a bearing upon our behavior by generating further ideas that can modify the world in which we live. Pure research in the natural sciences has been justified in startling ways in modern times because of the technological application of the principles discovered and the fact that such knowledge has been put to powerful uses.

Ideas are thus prescriptions and directives: they fulfill our purposes and goals and can thus be appraised by whether or not we find the end achieved worthwhile. There is thus an *evaluative* component. Here, the test is not simply whether a statement or hypothesis is descriptively true or analytically sound but whether it is satisfactory or fitting to the situation. We judge ideas in part by their effectiveness—by what they lead to, what they breed, what they bring into existence. Newtonian physics gave us the technology of the steam engine and water pump, and contemporary physics brought us space travel and computer technology. We might not always value the technological applications, but the results of ideas are relevant in some way to their truth. It is the *power* of a belief that can be demonstrated to be true that in some sense provides a test of its adequacy.

Ideas help to assuage doubts, overcome obstacles, and solve problems. They are instruments and tools; but more, they are tools of discovery and creation. Thinking is instrumental—for the ordinary man, whose thoughts can be judged by whether they enable him to cope with life, and for the scientist, whose ideas have observable consequences in the world.

Vindication of the scientific method

In summation, the scientific method involves three main components: (1) evidence, (2) logical validity, and (3) practicality. These criteria are continuous with the way beliefs function in everyday life where critical intelligence is at work. This is relevant to another important issue: the vindication or justification of the scientific method. Two rather fundamental considerations are pertinent. First, the commitment to scientific method is in accord with the use of critical intelligence in ordinary experience, and second, the method has pragmatic convenience.

Many philosophical and theological critics of science have demanded the total justification of science and the scientific method. But this is inappropriate, for science is implicit in our ordinary ways of thinking and in the technological processes of reasoning already involved in practice.

Moreover, the rules governing ordinary thinking are not culturally relative or merely a predilection of Western civilization, as Zen Buddhists and others seem to argue when they claim that the conceptual and empirical scientific method is Western, whereas there is an "intuitive" and "mystical" way of knowing in Asian cultures. Rather, as far as we can tell from an analysis of anthropological data, some teleonomic means-end way of knowing applies to all cultures, even primitive ones. Claude Levi-Strauss' theory of structuralism seems to support the notion of an underlying set of invariant conditions to human mentality.[7] That is, even if the primitive's mind is fixated on his mystical religious tradition, this does not undermine or invalidate his commitment to technological reasoning. Rather, it complements it, and it is always present as a necessary component of living. Indeed, to deny a minimal level of objective thinking would make life in any culture impossible. Thus again, we cannot deny the important uses of common sense and scientific method in certain areas of life, for both the primitive and the modern. To do so would be to fly in the face of the obvious. The question however is: "How far shall we *extend* this method?"

To demand a justification of first principles already deeply embedded and functioning in the life-world of practice is to raise a spurious question. In regard to the justification of deduction, the justification of general rules derives as much from the particular judgments in which we reject or accept deductive inferences as from an examination of the general rules themselves. Aristotle could only formalize principles of deductive inference which were already implicit in the judgments of ordinary men and in the Sophists' skills of argumentation. Thus, rules and particular inferences alike are justified by being brought into agreement with each other. A rule is amended if it yields an inference we are unwilling to accept; an inference is rejected if it violates a rule we are unwilling to amend. Similar considerations apply to the hypothetic-deductive method of scientific inductions. Particular hypotheses are justified if they conform to the valid canons of inductive inquiry; and the canons are valid if they accurately codify accepted inductive practices. Thus, we can stop plaguing ourselves with illegitimate questions about the justification of scientific induction

7. Claude Levi-Strauss, *Structural Anthropology* (New York: Basic Books, 1963).

and examine actual usage instead. The definition of a general rule relates to such usages. This logical point is important. The vindication of scientific methodology is based to a large extent upon an analysis and reflection of what we already do, the rules intrinsic to usages. It is not a metarule imposed on the subject matter of our ordinary processes of inquiry.

Since the scientific method is at the same time a normative proposal, the appeal to existing usage by itself does not suffice—especially if one wishes to justify an extension of the method to other areas. Hence, additional considerations must be brought to bear on the question of justification.

There is another important logical point concerning the kinds of questions being raised. There are no "ultimate" questions, only penultimate ones, and this applies both to the justification of what we already do, the procedures we employ, and what we ought to do. To demand the justification of first principles by deriving them from still more basic first principles would require either that we reduce them to indefinable terms or undemonstrable propositions, which are ultimate, or that we commit a *petitio principi* by assuming the very principle that we wish to prove. Hence, one does not "justify" first principles; one does, however, attempt to make them seem reasonable. Here, a philosophical vindication involves a reasoned argument, wherein we try to make a case for our general position. The procedure is not unlike that employed in a legal context, wherein one attempts to prove a case, not simply by deduction from first principles, but by a drawing together, focusing, or converging of considerations, factors, and arguments, which make the defense seem cogent.

There is a basic contextual consideration here. We never begin such reasoning *de novo* or from scratch. Discussions of vindication are forced upon the case at hand by real situations and concrete problems. In human experience there are no first beginnings and no final endings to the succession of moral principles. Rather, different levels of questions can be raised and appropriate responses given. The defense of the first principle of scientific humanism—that is, the use of scientific knowledge as the ideal—is not unlike the vindication given in everyday life for one set of procedures that we wish to adopt rather than another, whether in gardening, cobbling, or mechanics. Similar considerations apply to reasoning in the ethical domain and the vindication of other normative principles; they are always related to the situation or context before us. Thus, even for those areas in which we do not now use scientific methods and procedures, the kind of vindication that we can give for gaining adoption is the kind of vindication that should appeal to the ordinary man, for it is used throughout life.

Empirical consequences constitute a central consideration: It is not by words but by deeds or fruits that we judge. It is how well a principle works out in practice over the long run that is the basis for our decision to continue to use it, discard it, or modify it. Thus, although the first principles of scientific naturalism are not necessarily true, they are plausible to the ordinary man when he inspects the results of their use; indeed, most human beings tend to judge ideas and ideals by their pragmatic consequences, irrespective of their philosophical persuasions. The consequences of the scientific mode of inquiry are such that: (1) it facilitates the development of new knowledge, since its controlled use has led to steady progress; (2) its use enables us to make the knowledge we already possess more coherent; and (3) no matter what our desires, ends, or values, it enables us to fulfill them most adequately, being the most effective instrument we have for dealing with the world.

Here is the dilemma the scientific naturalist always faces when he attempts to vindicate his position, and it is the charge that Plato raised against the Sophist Protagoras: Why accept what the ordinary man does when he takes into account the standards of consequences or effectiveness? The scientific naturalist begins with the ordinary man and returns to him and to common sense to warrant his position; he does not stay there but considers his principle a normative *proposal*. The problem again is not simply to vindicate existing inductive procedures but to vindicate their extension—say to religion or morality, where they are not now being used—and to make them a model for all knowledge and inquiry. How should we judge proposals, we ask. Proposals imply value judgments. Does not scientific naturalism thus presuppose a more basic value judgment which is controlling?

The only answer that can be given here is to admit, as Mill, Dewey, Hook, and a score of other thinkers have admitted, that there are no discoverable ultimate standards of value by which epistemological principles can be judged. The only argument in appraising key value judgments is a *comparative* argument—in other words and in the last analysis, a pragmatic case based upon a balanced appraisal of conflicting claims.

All other positions face a quandary similar to that hurled at the scientific humanist—though compounded. It is unfair to burden the scientific humanist with the "riddle of induction," for there is a "riddle of intuition" or a "riddle of subjectivism" or a "riddle" for any other method. The intuitionist, mystic, or subjectivist can only justify his position by assuming his method to do so, thus committing a *petitio principi*. The burden of proof rests with these alternative positions. Moreover, insofar as the intuitionist, mystic, or subjectivist attempts to justify his method and employs argument and reason to do so, he is already conceding a point to the scientific naturalist. Even to seek a justification is to presuppose objectivity as an ideal. What does the word *justification* mean? In a sense, it is equivalent to that employed in the objectivistic mode of inquiry and in ordinary life when we demand grounds for belief. To even raise the question of vindication is to suggest a solution. To say that justification is an open question is to open the arena for an objective examination of all claims to knowledge, which is a key methodological criterion of methodological naturalism. The only response of the nonobjectivist is to refuse to justify his method.

We may ask: "Which rules of procedure in the long run appear better or more effective in gaining results?" I deny that the scientific humanist is merely stipulating or that his proposal is an arbitrary, persuasive definition of "knowledge." If all positions involve some question begging and are on the same ground in this regard, we may ask: "Which is the least self-defeating? Which accords best with the facts, with intelligibility as ordinarily understood, or which best enables us to satisfy our diverse desires?" Scientific humanists have made it clear, for example, that they do not exclude on a priori grounds the reports of mystics, intuitionists, psychics, or even of people under the influence of LSD and other psychedelics; they do not exclude, simply by definition, reference to a transcendental realm, God, or immortality. On the contrary, we should examine such reports with care and with caution. But we do insist upon some responsibility in appraising claims made on behalf of subjectivism or transcendentalism. We need to leave the question open and not block inquiry by denying on a priori grounds, as do subjectivists, that such things cannot be known by objective methods of inquiry. On the contrary and as we have seen, all other methods at some point must presuppose the use of objective methods in addition to their own methods. In other words, some objective methods are to some extent unavoidable if we are to live and function in the world, and everyone has to assume them up to a point; the same thing cannot be said of other methods.

We should not maintain a blind faith in scientific methodology or believe that an extension of these rigorous methods to other fields of inquiry must inevitably succeed. We should not be so naïve as to assume that all human problems can be solved and all quandaries settled. If, however, we examine the chief arguments that were used in the past against the scientific study of nature and human behavior, we may question such a priori negativism. One can imagine in pre-Aristotelian or pre-Galilean days someone objecting to the proposal to develop a science of nature as being illegitimate, arbitrary, and contrary to the traditional ways of looking at the world. Yet I readily concede that whether the first principle of scientific humanism—that is, the use of scientific methodology— is the most reliable approach in explaining all the complexities of nature and human behavior is still an open question and can only be judged as inquiry proceeds. Again, that it must succeed is not deductively certain. But in the light of fruitful gains already registered in the natural, biological, and behavioral sciences, we have good reasons to believe that scientific inquiry will continue to progress. Some skepticism, no doubt, should still be present, even about this expectation, especially if it is framed in the form of an overly confident scientism.

IV: Critical intelligence

Rationalism

The question we shall now face is whether the scientific method can, or indeed should be, applied to solve problems in life. This is often difficult to do; moreover we surely do not want scientific technicians telling us what to believe in or how to behave concerning our personal lives. Still, a case can be made for extending the principles of scientific rationality to the decision-making processes of everyday living by cognitively evaluating beliefs and judging truth-claims. But this means that we must also broaden our definition of the scientific method.

The term *rationalism* has been used to describe a distinctive outlook and approach to life. Webster's dictionary defines rationalism as "the practice of guiding one's opinions and actions solely by what is considered reasonable." The term *rationalism* is not without some confusion, however, for it has been identified in the history of philosophical thought with a specialized theory of knowledge that emphasizes rational cognition, intuition, and deductive inference as the primary origin and test of knowledge. Empiricists have objected, claiming that the criteria of analytic truth, logical consistency, and validity are insufficient in themselves to establish truth-claims, for they are without factual content. Empiricists insist that inductive rather than deductive methods should be the test of knowledge. They say that we should examine the evidence before assenting to a claim.

A wider use of the term *rationalism* has come into vogue in the past century. When present-day rationalists call for the use of reason in life, they are opposing appeals to subjectivism, mysticism, authority, faith, or revelation as a basis for knowledge, and they leave room for *both* logic and experience as the test of truth, thus incorporating the best of rationalism and empiricism. To appeal to reason is to appeal to the general principles of objective inquiry, and this is synonymous with the scientific method as it is broadly conceived. The Rationalist Press Association, which was founded in Great Britain in 1899, attracted under its banner a large number of intellectuals committed to these ideals. To quote the brief statement that was issued upon its foundation:

> The prevalence of the "spirit of Rationalism" . . . is one of the chief features distinguishing modern from medieval thought and life. . . . We believe that this spirit of Rationalism is closely associated with the progress of modern science and critical research. . . . Rationalism may be defined as the mental attitude which unreservedly accepts the supremacy of reason and aims at establishing a system of philosophy and ethics verifiable by experience and independent of all arbitrary assumptions or authority.[1]

The use here of the terms *reason, rationality,* and *rationalism* to designate an outlook and a method of approaching questions of belief has much in its favor. Such a use is apt to be confusing however—at least in the mind of its critics—and for yet another reason; it suggests coldness, impersonality, and imperviousness to the psychological content of lived experience. A similar criticism has been leveled against the use of the scientific method as the primary way of knowing. Many naturalistic philosophers who have pointed to the success of the method of science in establishing knowledge and have urged its extension to other areas of life, have been lambasted for placing too much faith in science, and the pejorative term *scientism* has been used to ridicule their efforts.

What is critical intelligence?

The terminology that better connotes the meaning I seek to convey is *critical intelligence.* It incorporates the use of both reason and experience in developing knowledge; and it applies not only to scientific and technical fields of inquiry but also to ordinary life and to normative areas, such as ethics and politics. It also can incorporate under its rubric common sense and practical wisdom. Moreover, skepticism is an intrinsic aspect of critical intelligence.

Fortunately, the capacity for critical intelligence is deeply rooted in human nature. It is the chief instrument that has enabled the human species to adapt and survive throughout the course of evolution. Its development should be the primary aim of education and its presence the chief mark of the educated person. Unfortunately, it often lies dormant in some individuals, who seek to smother their consciousness or deaden their critical sensibilities. Human beings are more than intelligent beings,

1. Reprinted in *An Anthology of Atheism and Rationalism,* ed. Gordon Stein (Buffalo: Prometheus Books, 1980), pp. 313–15.

with emotion and passion contending for control of the human soul and often gaining mastery. Some individuals are so weighed down by habit and custom that they are unable to use critical thought in vital areas of their lives and societies. Critical thought is often more potentiality than actuality. Nevertheless, it must be used by everyone to some extent, if a person is to live and function in the world.

The doctrinaire Christian considers the quest for knowledge the cause of man's downfall. Eve tempted Adam to eat of the tree of knowledge of good and evil, and for that God cast man from the Garden of Eden. For the humanist, on the contrary, the two cardinal sins are (1) gullibility and ignorance and (2) cowardice and fear. These are the original defects of human nature. Gullibility and ignorance are the source of much nonsense in religion, ideology, morality, and politics. Gullibility tempts humans with the lure of wondrous things. The only therapy for childlike faith is the cultivation of a critical intelligence that can examine and evaluate claims to truth and thereby achieve knowledge.

Critical intelligence refers to a capacity or disposition of mind, a method, attitude, or approach that can be used to understand, discern, and assess relationships and to cope with problems. The application of critical intelligence may be thwarted by obstacles encountered by nature, by other persons, or by ourselves. It is sometimes used effectively; at other times it may lie quiescent.

What do we mean by *intelligence?* Does it always refer to a person's so-called I.Q.? Efforts have been made to test the intelligence quotient of groups of individuals statistically and comparatively. Does intelligence measure how many facts a person can recall on a test or how well informed he or she is? Does it indicate verbal ability, mathematical acuity, or how well a person is able to solve problems—abstract, visual, or practical? Intelligence surely involves some ability—native and developed—that individuals possess. Thus it is said that a highly intelligent person is brilliant or insightful, and testers will grade and compare him or her with others.

The power of intelligence, however, is the same as its use. Psychologists estimate that approximately 80 percent of a person's native intelligence is due to inheritance and that 20 percent is acquired. Intelligence can be dormant and untapped, suppressed and squandered, or it can be focused by a highly motivated person. I have applied the adjective *critical.* Here, it is selective in denotation. *Critical* refers to the probing and analytical application of intelligence to a problem or hypothesis; it appraises evidence and evaluates ideas; it sees relationships and infers conclusions. Like a pruning knife it can be a powerful instrument, enabling us to get to the heart of an issue, to repair and correct deficiencies, to discover new truths and discard misconceptions.

A person can possess high intelligence and yet lack the critical component. Such a person's general intellectual ability may be diffuse. The critical element is the focusing power, which enables him or her to grasp the essentials in complex intellectual and practical questions and to resolve them. Aristotle noted in the *Nichomachean Ethics* that there were five intellectual virtues: philosophical wisdom, intuition, scientific demonstration, art, and practical wisdom. He also observed that a person might possess one or more of these virtues but lack others. The truth is, however, that there is a much wider range of intellectual qualities, and a person may possess some of them and not others.

A catalogue of intellectual skills

The following is only a partial list of diverse qualities of mind that individuals may manifest.

1. *Abstract intelligence*—the ability to comprehend abstract concepts, ideas, and their relationships. Logicians, scientists, philosophers, and theologians are adept at this.

2. *Logical agility*—the ability to detect fallacies and deduce inferences validly, to calculate and integrate. Mathematicians and logicians especially demonstrate this skill.

3. *Keen observation*—the capacity to perceive facts clearly and carefully. Reporters, expert witnesses, and scientists constantly use their powers of observation.

4. *Perceptive acuity*—the ability to see what is at issue or to discern underlying causes. Again, scientists and philosophers have developed this capacity.

5. *Intuitive insight*—the powers of the mind to discover and grasp meanings and causes embedded in the facts. Scientists, poets, and philosophers exhibit this.

6. *Interpretive capacity*—the ability to unravel and relate particular events and to interpret them. Scientists, detectives, lawyers, and critics are skilled at this.

7. *Puzzle solving*—the ability to find solutions to puzzles by using cues as circumstantial evidence. A graphic illustration is the solving of crossword puzzles or detective mysteries.

8. *Lucidity*—the ability to express one's thoughts clearly by means of the spoken or written word. Great writers and orators especially demonstrate this talent.

9. *Poetic metaphor*—the skill in expressing imagery, symbols, and metaphors. It is especially used by novelists, poets, and dramatists.

10. *Erudition*—meticulous scholarship and the cultivation of knowledge in the literature, history, and language of a field. Pedants, savants, and college professors are exemplars.

11. *Artistic ability*—the development and exercise of talents in the fine arts. Musicians, composers, sculptors, painters, and dancers are adept in their chosen professions.

12. *Technical skill*—there are any number of specialized proficiencies or techniques requiring training and dexterity. The illustrations are abundant: surgeons, dentists, baseball players, boxers, etc.

13. *Mechanical dexterity*—the ability to figure out physical connections and manipulate, repair, and use devices and machines. Again the list is endless: engineers, mechanics, plumbers, sportsmen, etc.

14. *Inventive and creative ability*—the introduction and discovery of new forms and combinations, new devices and objects. Especially found in inventors, engineers, poets, architects, and artists.

15. *Prophetic ability*—the ability to estimate likely future eventualities based upon knowledge of the past and present. People who possess this are pundits and columnists, and some politicians and historians.

16. *Business sense*—the use of good judgment in financial matters, knowing how to make money, invest wisely, purchase and sell goods and services. People who supposedly possess this skill are economists, accountants, corporate managers, and financiers.

17. *Empathetic insight*—the understanding and interpretation of other people, their attitudes, feelings, and needs. This is especially required of counselors, ministers, psychiatrists, and spouses.

18. *Political sagacity*—the capacity for dealing with people and persuading them and to attain and retain power. This is the craft of politicians, statesmen, corporation heads, and college presidents.

19. *Moral insight*—the ability to discern moral values and principles and to resolve moral dilemmas. We expect this of counselors, psychiatrists, moralists, and ethical philosophers.

20. *Common sense*—practical experience and wisdom enabling one to evaluate and judge situations. Among those who had this gift were Pericles, Marcus Aurelius, Benjamin Franklin, and Abraham Lincoln.

21. *Good judgment*—the ability to evaluate and appraise circumstances and alternatives and to reach wise decisions. This is needed especially by judges, psychiatrists, and marriage counselors.

It is clear that one can display talent and ability in one or more of the preceding skills and not in all of them. Although there is some overlap of these capacities, specialization has become exacerbated and some professions or careers demand concentration. Thus one can be, for example, a skilled architect or engineer, have been educated at the best schools, and be devoted to his field, and yet be totally uninformed about other areas of knowledge or endeavor. One can be a highly intelligent and talented poet yet have a total misconception of the universe of science. One can be a great mathematician and lead a life of disorder and chaos, or a technical philosopher and lack wisdom, philosophical or practical. One may master a body of highly specialized knowledge in a field, but this knowledge and skill need not carry over into other areas. The extreme departmentalization of knowledge is due in part to the fact that there is so much to know that it is impossible for anyone in one lifetime to keep up with the sheer quantity of information. One needs to focus one's talents, if one is to succeed in a field. As a consequence, the great tragedy of contemporary culture, as Emerson observed, is that we have become part, not whole, men and women. This is not due simply to the proliferation of specializations but also to certain psychobiological facts about human beings. Few people can match an Aristotle, Leonardo, Michelangelo, or Buckminster Fuller in the wide range of their abilities. To have talent and demonstrated ability in one area does not mean that one has it in others; nor does being informed in general mean that one has knowledge in the concrete. Being technically proficient in one area does not mean, for example, that one has moral wisdom or sensitivity. One can be a humanitarian who loves humanity in the abstract but despises individuals. One can be shrewd and cunning and an evil knave, or kind and sympathetic and a stupid fool, or know how to solve abstract equations but not drive a car.

Nevertheless, critical intelligence is a general disposition of mind that can be acquired and developed in most people, even while I grant that natural talent for it may be unequally distributed. Indeed, to succeed in a specific field means that one has already disciplined and focused one's critical faculties in that particular field. If a man is to be a good mechanic, he needs skill and dexterity with his fingers and hands, a good eye, and mechanical ability, but he also needs some measure of common sense, critical intelligence, and a measure of skepticism about how things fit together and work. Similarly a competent accountant is precise, has a good mathematical mind, is able to keep a balance sheet, is well-informed about tax legislation, and can also provide good advice to his clients.

A surgeon has skilled know-how, and if he or she is very good, the craft can be taught to others. A doctor understands general principles and knows how to apply them to concrete cases. He or she must exercise good judgment in deciding when to operate or treat and how best to proceed. In all of the preceding examples, the materials are those provided by the subject matter, and the means are those best adapted to the context at hand.

Critical intelligence thus in one way or another is a general quality of mind that is displayed in virtually all fields of endeavor. It is as true of the skilled novelist, who knows his craft and works hard at it, as of the physicist, psychologist, or pharmacologist. Obviously, everyone needs intelligence to gather facts, comprehend principles, and apply skills (sometimes these may be intuitively understood and spelled out). Moreover, he is able to evaluate and judge alternative claims or performances of others. If he demonstrates skill and competence, he will be recognized eventually by his colleagues. The criterion of being an expert within a field is that he is recognized and certified by his peers. Usually (but not always) he has studied at the accredited schools, has a degree, and has been licensed as a professional, if he meets the standards of the credentialing authorities. When a field is in rapid transition or is undergoing radical change, these standards may be uncertain. Professional credentials are never sufficient in themselves, however, for it is the application of objective standards and tests that determine the adequacy of a person's work. In the last analysis, to be competent in a field means that one is able to apply critical intelligence to one's craft.

There are many synonyms for critical intelligence: common sense, practical experience, good judgment, and soundness of mind. And there are others. A person is sensible, balanced, reasonable. He is level-headed, judicious, prudent, sober-minded, clear-headed, discriminating. He is sagacious, astute, and has acumen. He is hard-headed, foresighted, far-sighted. The antonyms of critical intelligence are gullibility, ignorance, irrationality, being impressionable, naive, dense, stupid. Perhaps I have overstated the case pro and con. It is more a question of degree than kind. Yet no human—save an infant or imbecile—can live and function in nature and society without some degree of critical intelligence.

The role of education

The real issue is how widely one's critical intelligence can be extended. The paradox we have seen is that demonstrated ability within a field does not necessarily mean that it will be utilized outside of it, or applied to life in general. This is the most difficult challenge: How can we develop individuals who are capable of applying their critical intelligence throughout life? This surely is the goal of liberal education, which is concerned with developing in the student an appreciation for many fields of human endeavor in the arts and the sciences, as well as rational understanding and the ability to think.

Indeed, this should be, in my judgment, the chief goal of education. What is education if not the process of expanding our awareness and appreciation, increasing our imagination and understanding, providing us with tools that will enable us to adapt and adjust, but most fundamentally teaching us how to think critically? And is not the mark of the educated person not simply the number of facts he can remember but the cultivation of the powers of the mind?

Although the schools are essential to this educative process, they are not the only institutions that should be charged with that mission. Education is not a commodity to be supplied only to the young, from elementary school through college. It is a continuing process for all age groups and at all levels. Perhaps the greatest crisis for education does not concern the young but arises from the necessity of adults to respond intelligently to the new challenges of a rapidly changing world.

Classically, education was the primary task of the schools, whose chief function was to inculcate the beliefs, values, and basic skills of a society and to train for vocations and the professions. It was, as it were, the transmission belt for the traditions of the past. Today, imparting knowledge about the perennial truths of history—though essential—is no longer adequate. With the growth of knowledge, one cannot hope to obtain a degree in a specific field and then rest on one's laurels; there is the ongoing need to keep learning and growing. Virtually all the institutions of society thus have an educative function to perform: the family, labor unions, corporations, business, industry, and the mass media.

The ideal curriculum should involve several components, many of them familiar: training in the basic skills (reading, writing, computation), professional and occupational preparation, understanding of the nature and practice of civic virtues, and an appreciation of history and the arts. Yet, the vital element is to increase, by means of the sciences, philosophy, and the arts, our understanding of the rapidly changing world in which we live and to develop our ability to make reflective judgments.

The latter emphases help to realize the ideals of a general education and of the liberal arts that many schools have abandoned for vocational programs. We need to focus on general education but, in a new sense, by explorations in the liberating arts. The chief goal of this kind of education is to clarify our capacities to formulate judgments so that we can creatively expand their potentialities in a changing world.

Educators have not adequately explained science as a great adventure in learning; nor have they succeeded in developing an appreciation for the scientific method: the appeal to evidence and logical criteria in judging hypotheses, the tentative and hypothetical character of knowing, the skeptical attitude, the use of reflective intelligence as a way of solving problems.

If people are to adapt to the future, then education in the physical, biological, behavioral, and social sciences should be a required part of the course of study, along with other courses in the liberal arts. But the emphasis must not be simply upon science as a static body of knowledge but rather as a method of understanding and modifying the natural and cultural world.

Concomitant with this cultivation of scientific understanding is the need for moral education, a continual process of values examination. By moral education, I do not mean indoctrination and behavioral conditioning, but rather the process of cognitive moral growth.

Moral values, in the best sense, are the product of a process of evaluation that human beings engage in as they respond to the challenges in the environment. An appeal to traditional standards is hardly adequate, even though they be enshrined in religion, law, or custom. Rather, we need to learn how to deliberate about the things that we hold to be good, bad, right, and wrong. An appeal to emotion or passion is never sufficient; we need to learn how to deal with our values critically and intelligently. Values should be treated as hypotheses upon which we are prepared to act, or prepared to modify, if need be. They should grow out of concrete situations and be fashioned in the light of the needs of the situation and an examination of the consequences of our deliberations.

Given the great strains of modern life, we cannot provide young people with ready-made answers—and certainly not with ready-made professions or occupations. No one can anticipate fully the future course of events; the best that we can provide is some resiliency, some help in developing cognitive moral awareness, as a way of life, a means by which a person can respond effectively to life in the light of a deeper understanding of it.

Thus, surely one of the basic goals of education—perhaps the most important—is the development of resourceful people, self-reliant, capable of critical and responsible thinking, able to adjust and adapt, hopefully with some sense of wisdom, some understanding of their own powers, but also aware of constraints and limitations. Individuals especially need to know how to judge truth-claims objectively, how to be skeptical, how to avoid fraudulent and counterfeit promises, and how to live with ambiguities and uncertainties.

V: The justification of belief

The problem of how to establish reliable beliefs is vital for the ordinary person, who may not be au courant with the sciences or the sophisticated use of scientific method but who has developed his critical intelligence. He constantly raises troubling questions: What shall I believe and why? What is true? What is the meaning of life? What may I hope? What should I value?

Deferring to custom

In earlier times, the individual was born into a highly structured society. His beliefs were regulated by social institutions such as the tribe, family, place of worship, or polis. Children were inculcated from birth by parents, teachers, political and religious leaders, and the peer group. They were indoctrinated with the prevailing values and ideals of the community, generally without sufficient questioning about their truth or authenticity. Rewards were meted out for those who accepted the belief system (these were the "virtues"), and sanctions were imposed on those who did not conform (the "vices" or "sins"). The methodology was *custom*. Something was accepted as true because the dominant members of one's peer group believed it to be so.

When asked by Socrates in the *The Republic* "What is justice?" Cephalus, a pillar of the Athenian establishment, merely repeated the traditional verities: to tell the truth, pay one's debts, and take part in the religious sacrifices. However, he suddenly leaves the inquiry when his claims are critically examined by Socrates for their consistency. Although an appeal to custom may provide comfort and security, it does not by itself meet the tests of critical intelligence. Thus one may be a Methodist, Catholic, Hindu, Jew, or Muslim largely because one's parents and grandparents were, and not out of reasoned conviction. To say that one believes something is true because of the mores and traditions of one's society— that grandma's chicken soup will cure the flu; that Christ was crucified and rose from the dead in three days; that one's race or nation is superior to all others—hardly suffices as a justification of those beliefs. It is simply a recognition that one cannot explore every question for oneself or reconstruct all of one's beliefs from scratch. We must rely on others for most of our beliefs, and so deference to social approval is the familiar approach. It is surely not in itself an adequate method, however; for something cannot be said to be true simply because our forebears or countrymen believe it to be so. Many beliefs customarily held *are* indeed true, but this is only because they can be tested and demonstrated on independent grounds. The point is, it is not the customs that make them true; these can only be said to provide added support for a belief after it has been corroborated. There is an important difference here.

If we were to accept custom alone, we would be faced with the problem of cultural relativity, since contradictory belief-systems are held by different groups. For example, Christians believe in monogamy, Muslims in polygamy; some cultures believe that females are inferior to males, others that they are or should be considered equal; some believe in blood sacrifices, others do not. Virtually every kind of strange belief has been accepted by some group of people somewhere at some time. The issue is not what is commonly believed and/or valued, but whether what is commonly held is reasonably held or open to criticism and revision. Does it accord with the facts? Is it consistent with other beliefs that are maintained? What are its consequences? It is the *grounds* of choice and the *methods* by which we select a belief that are important in evaluating whether or not the belief is true, not the fact that it is believed to be true by some culture.

The problem of belief today is especially complicated. The individual in many contemporary societies has been wrenched from his customary framework; he is bombarded with conflicting beliefs and coexisting alternative value-systems. He is challenged on all sides to accept one or another claim to truth. In North American society, for example, a person is confronted with a wide diversity of religious, political, moral, scientific, and philosophical claims. Different missionary cults, such as Jehovah's Witnesses, Seventh-Day Adventists, Jews for Jesus, Hare Krishna, and the Theosophists, implore us to make a donation or join them. We are beseeched to accept a message or devote our lives to a cause—as the only guarantee of health, happiness, prosperity, or eternal salvation. A brief glance at the wide range of schools of psychotherapy that abound today illustrates the problem the individual faces. Which should he adopt (if any) and why: psychoanalysis, primal scream, cognitive therapy, or holistic psychology?

It is the rare individual who can decide everything for himself. Most people do not have the time, ability, or inclination to investigate each and every issue that arises. We need to defer to someone else about most of them. Many are very important, however: Should one be a socialist, liberal, Democrat, Republican, monarchist, libertarian, conservative, or reactionary? Should one believe in free love or follow a strict code of sexual morality? Should one be a vegetarian, nut feeder, or meat eater? There is a tendency to simply follow the example of one's friends, relatives, coreligionists, or ethnic brethren. One lesson to be learned is that although the traditional framework may embody, in part at least, the distilled wisdom of the race at any one time, much of it may be distorted by habit and should be supplanted by new discoveries. Thus one must be prepared to be critical of the customary mores and reform or modify them in the light of educated inquiry.

All things considered, however, the fact that a belief is customarily held in one's community is a prima facie reason to accept it, pending its validation, but it is never a de jure ground. No one can live without conventions: they govern not only our beliefs but also the language we use, our manner and dress, the norms and values to which we adhere. Some rules of the game need to be established. Although these are to some extent arbitrary, a product of chance and of history, they are often necessary. Thus to simply reject all conventions, as the radical reformer insists, and to begin afresh is impossible and impractical. Still, this statement is not entirely true, for we *can* move to a new country or culture and adopt a new language and different manners, beliefs, and values—as when a Russian dissident or Vietnamese refugee emigrates to the United States or France. But here one only moves from one culture to another. One cannot offend or dispense with all social norms regulating conduct. Individuals who attempt to do so are considered anomic, bizarre, even insane. Every social system needs to structure belief and conduct so that we may live and work together harmoniously. Thus deep-seated historic roots develop; and these order our lives. We become habituated to behave, even think, in certain ways. Moral demands are placed upon us, and we are expected to respond in ways deemed acceptable. Individuals who do not conform cannot function very well in the given social system.

Custom thus has a fundamental and pragmatic role. One cannot live entirely without it; for it provides some stability and encourages cooperation. It is only a convenient way station, however, for any one custom may not be true; and it may lose its efficacy.

It is much easier to criticize customs that concern descriptive or causal beliefs about the world than it is to attack beliefs that are expressions of value, though these are often connected. Still, our belief system often seeks to convey age-old truths about reality (particularly our religious beliefs). And these serve as a transmission for old ideas to enable us to cope with the perennial existential problems; birth, death, disease, and tragedy. They seek to provide consolation, succor, hope, and promise.

Belief in the transcendent—in God and immortality—is deeply rooted in custom and tradition. Religious institutions provide the strongest support for tribal superstitions. They celebrate the important rites of passage: baptism of infants, confirmation of adolescents, marriage and burial by means of ceremony and commemoration. People are apt to accept a religious mythology more because of its relation to their ethnic background than because it has demonstrated a truth-claim.

True believers are dedicated and staunch proponents of a belief-value system, defending it against critics and adversaries. They are of two kinds: (1) those who have been so schooled and indoctrinated by their parents and mentors that any questioning of their faith is considered heresy, and (2) those who become committed to a new faith or dogma by conversion. Both may be inflexible and intransigent. It is the first whom I am concerned with now. Religious belief involves a deep commitment, which has its sources in our background and growing years. One is nourished by one's community. It is difficult to break away, for what is at stake is ethnic chauvinism and nationality. These are especially indelible in a pluralistic society since they help define an individual and distinguish him from others.

For an individual to claim that he is a Polish or Irish Catholic, a Southern Baptist or Episcopalian, a Buddhist or Muslim is to say something not only about his beliefs but also about his roots and basic convictions, his loyalties, and his family and friends. Thus to question the basic belief-system of a person is to challenge his or her very being. Custom is rooted in two things: (1) a person's ethnic identity based on blood relations and genetic interbreeding, and (2) the social conditioning and enculturation that have molded his personality and outlook. Some individuals in contemporary society have been able to transcend their roots. Their religious commitments may be only nominal, or they may have married out of their faith, or met people from different ethnic backgrounds. They thus may have been able to abandon their earlier beliefs. But other individuals—particularly from close-knit families and communities—may cling inflexibly to their beliefs and insist that their truths are "eternal" or "sacred" or beyond question. Here, custom is synonymous with prejudice. It is a pity that it is so pervasive in human culture.

The appeal to emotion

What is actually present in much belief is the existence of internalized feelings of commitment. A belief is not simply a cognitive or intellectual state of mind that we hold about the world. It is a psychological state and disposition in which we are emotionally involved and upon which we are prepared to act. When we say that we *believe* something, we are expressing an attitude, either pro or con. Belief is both cognitive and affective; it is in any case clearly motivational. The emotivists thought that interests and attitudes applied only to our ethical convictions; but we can see that attitudes play a role in all of our beliefs—though no doubt it is a matter of degree. Nonetheless, emotional components are particularly recalcitrant, especially when strong beliefs and interests to which we are highly committed are present. It is often difficult to separate feelings from thoughts or feelings from perceptions, as they are intimately fused in conscious awareness. The scientist, after all, is a person, and his feelings and attitudes unavoidably intermingle with his beliefs; there are no easily definable dividing lines within his experience. Nonetheless, his feelings cannot be said to provide an adequate justification for holding a descriptive belief, and they are not relevant to its truth. They most likely are relevant in the area of his normative beliefs, however, for they concern his personal desires and choices.

We must distinguish here between two types of beliefs. Emotions may be present in both cases but are more pervasive in the second: (1) We may believe that *something is the case,* or (2) we may believe *in something*.

Thus, as an example of the first, I may believe that (1) Dallas is one of the fastest-growing cities in the United States; (2) Mercury is the smallest planet in our solar system; or (3) cigar smoking causes cancer. And I am willing to claim that these statements are true. My personal feelings about Dallas, Mercury, or cigars are irrelevant to the truth or falsity of the claims. Yet a belief-state is not equivalent to a statement of fact, which can be abstracted from the context of use and examined analytically and empirically. Beliefs are not neutral statements of fact; they are psychobiological states of behavior. If the belief is vital to me and not trivial, I am prepared to act upon and defend its truth, perhaps even strenuously, against its critics. If I live in Dallas, for example, I may feel especially devoted to the city's future growth and prosperity, and if I am a reformed cigar smoker, I am eager to justify giving up a former pleasure.

Beliefs about the transcendental, i.e., those that go beyond empirical observation or corroboration, are even more charged with emotion. Such beliefs are not dispassionate statements about reality; they are passionate confessions of feeling and emotion. Pascal's famous statement "The heart has its reasons, which reason knows nothing of" clearly expresses what is involved.

Beliefs of the second class more clearly demonstrate the affective-attitudinal-motivational aspect of our convictions. To believe *in* something is to want either to preserve or to defend it, or, if it is not already accepted by those around one, then to try to accomplish that. If I believe in world government, the Second Coming, socialism, vegetarianism, or peace, I confess that I wish to bring these about. And I may seek to persuade you, with all the rhetoric I can marshal, to accept my attitudes and work with me for their implementation.

Clearly, appeals to emotion cannot verify the truth-claims of beliefs of the first class. Although they are relevant to the second class, they are never sufficient in themselves but need to be informed by objective criteria and critical judgment. If we wish to bring about a state of affairs, the feelings and attitudes of people who are involved, their sympathies or antipathies, loves or hatreds are important to know. Knowledge of preexisting de facto values are data to be considered in framing value judgments, but they are hardly confirmatory. The fact that someone or some group approves or disapproves of bringing into being a future state of affairs does not make it ipso facto true; though we surely cannot ignore it.

It is rather important, however, that we see that our emotional states do not and cannot verify a proposition. A scientist cannot allow his attitudes to bias his readings of a meter dial or the conclusions of his research experiments. He may not like the results he uncovers and may even loathe them, but this does not make them true or false. Of course, his feelings may help provide some insight or intuition in forming a conjecture and leading to a viable hypothesis that may turn out to be true. But if it is true, it is because of independent grounds for corroboration. That is the vital point.

The appeal to authority

The basic problem with emotion is its subjectivism, which we shall consider later. Suffice it to point out for now that private roads to truth come into conflict with those of other individuals, and that we must learn how to negotiate a modus vivendi between our convictions and those held by others. Some appeal to social conventions thus appears to be necessary if we are to live and function in a community. It is apparent that we cannot test each and every thing ourselves. Hence we constantly need to defer to others. If custom fails—where our beliefs are simply based on habits of response—then surely there is another source to which we can and do appeal. This source springs from the fact that there exists a division of labor in society and that specialists and experts play a vital role. If it is impossible for any one individual to be able to keep up with and be informed of developments in all the fields of knowledge and inquiry—whether advanced mathematics, physics, microbiology, archaeology, botany, zoology, or epidemiology—he at least knows that others will. The development of the division of labor enormously expands our knowledge and know-how. By specialized concentration, researchers in a field are able to extend their range of knowledge and productivity. We all recognize this; and accordingly many or most of our beliefs depend upon what the authorities discover. If I have an ailment that persists, I consult a doctor, who may send me to still another specialist if he doesn't know the cause or cure. I am willing generally, especially if it is a serious illness, to pay heed to their diagnoses and follow their remedies.

Thus we may ask: Is an appeal to authority legitimate? Does every field have its authorities? Are there even, for example, experts in palmistry, psychic surgery, astrology? Those who believe in palmistry, etc., maintain that certain practitioners of the "art" have talents and insights and that they should be consulted on questions concerning their specialties. One way of evaluating the so-called experts in a field is to submit their claims to peer review. We ask: Are they recognized by other authorities in the field? Are they well educated? Do they have degrees? Are they credentialed? Are they considered skilled and competent by those well versed in the subject matter? We would surely ask these questions of a scientific specialist. But not every field has qualified experts in the same sense. An astrologer proclaims that the horoscopes he casts are predictive. But other astrologers cannot be relied upon in judging his claims in the same way that astronomers test the hypotheses of fellow astronomers. Recognized astrologers, palmists, psychics, black magicians, biblical missionaries, or ayatollahs may all have studied at appropriate schools and be credentialed by their professional associations. They may have acquired skills in a field and know everything relevant, have been trained by its acknowledged leaders and read the basic literature of that field. But not every specialty is a legitimate area of technical know-how and tested claims, and we may not be helped by deferring to the so-called authorities to settle an issue.

For example, what are we to say about *moleosophy,* the alleged "science of moles"? According to its proponents, moles are an indication of a person's character and prognosticate the future. Prediction depends upon the location of the mole, its shape and color. Allegedly, "round moles reveal the good in people; oblong, a modest share of acquired wealth; angular represents both good and bad characteristics . . . Light-colored moles are considered the luckiest." Moles on the ankle show that a man has "a fearful nature," and a woman "a sense of humor, courageous, willing to share love and worldly possessions." Under the left armpit, a mole indicates that "the early years are a struggle, but ample remuneration, even riches will make the later years very happy." On the belly, a mole indicates a "tendency to self-indulgence" and on the buttocks, that a person is "not very ambitious."[1]

How may we evaluate the conflicting truth-claims encountered daily in the world? There are two possible solutions to this difficult question. Some fields are established areas of inquiry, and we may defer to the authorities. But we do so only because we may not have the expertise or time to settle every claim that is made and we believe that there are others who are well-qualified to do so. If we accept their judgment, it is because we believe that it is based upon independent objective criteria—empirical observation, logical inference, predictive experimental consequences—and that it has been tested in part because of someone's demonstrated competence. I am willing to defer to authorities only where I believe that the grounds for their judgments can be reviewed by others (if not myself) and that their results can be carefully scrutinized. I can, in principle, examine the evidence and procedures they used in framing and testing their judgments and review their reasons and replicate their results.

However, not all fields satisfy these rigorous standards of inquiry, and they must remain suspect until they do so. They have no legitimate authorities. This applies to most of the paranormal and occult fields. In other cases, where we do not have sufficient information, where not all of the evidence is in, or where the acknowledged reputable experts differ, then our only sensible option is to suspend judgment and admit that we do not know or know of anyone who does. Skeptical agnosticism is an intelligent mode of response. The salient point is that although it may be convenient and necessary to defer to authorities and draw upon them for their informed judgments, the fact that an authority maintains that something is true does not make it so. An appeal to an authority in itself is never an adequate verification of a truth-claim. What counts is why the authority maintains that something is the case and whether his grounds can be sustained. If he is considered an authority, it is because he is well versed in the rules and conventions, concepts and principles of his craft, and he can apply objective standards to test hypotheses and achieve results. To appeal to authority by itself in order to justify a belief is thus a fallacious mode of reasoning. We may use authorities only if we recognize that they cannot simply define or create the truth or proclaim it from on high. That is one reason we should be especially dubious of so-called religious authorities who claim to have a special road to the truth, which cannot be corroborated by anyone else.

Another consideration is that different types of authorities have different functions. There are the scientific authorities, for example, in

1. Walter B. Gibson and Litzka R. Gibson, *The Complete Illustrated Book of the Psychic Sciences* (New York: Pocket Books, 1966), pp. 275–76.

chemistry, physiology, anthropology, and other fields. They are interested in providing us with basic knowledge about nature. There are also authorities in the applied sciences and technologies: engineers, computer specialists, airplane pilots, dentists, mechanics, etc. They apply knowledge to concrete cases, resolve problems, and employ special techniques to achieve desired ends. There are connoisseurs and critics: individuals with highly cultivated tastes who can help us appreciate and learn what they experience. There are also skilled practitioners, who display and perhaps even teach their arts: artists, writers, ballet dancers, opera singers, musicians, great chefs. And there are individuals of good judgment and common sense who have demonstrated practical wisdom or political sagacity. They provide sage counsel and advice about how to clarify or resolve moral dilemmas or political conflicts.

It is the area of knowledge where people *believe* that something is the case which concerns us here. Here we require that the authority be objective in formulating and corroborating his hypotheses. For these reasons, again, I would reject most "authorities" in the field of religion and the paranormal. For although they may know a great deal about an area and be well-versed in the literature of the field, their claims may lack empirical foundation, be internally inconsistent, or be nonpredictive.

I must confess that I am dubious in the last analysis of all authorities. I have seen how the testing by a team of scientists (STURP—Shroud of Turin Research Project) has led them to conclude that the Shroud of Turin was caused by a supernatural flash of light scorching the image of Jesus Christ on his shroud. A team of skeptics, one that has not received as widespread public attention, has disputed these findings that the shroud was Jesus' and has concluded that it is a forgery. I have also seen how a group of scientists, wishing to be fair and impartial, have claimed that the work of Michel Gauquelin in astrobiology is supported by the evidence. Equally questionable are many of those who wish to deny it. I have been accused of being biased, because after having examined the work of Gauquelin for many years, I am doubtful as to its veracity and particularly its replicability. I may be mistaken, but so may be Gauquelin's champions. Hopefully, time and the course of future research will enable us to resolve the dispute.

One of the most famous cases of the bias of authorities, often quoted by paranormalists, was the refusal, in 1772, of a commission of distinguished French scientists from the Academy of Science to accept the reality of stones raining from the heavens. This commission included Antoine Lavoisier, the father of modern chemistry. They viewed stories of such "thunder stones" with suspicion; yet their judgment was wrong, for meteors do fall to earth. There was also prolonged scientific opposition to the shifting-continents hypothesis and Semmelweis' germ theory.

In some sciences the authorities are so dazzling and impressive—for example, in astrophysics or medical research—that we may be prone to defer to their judgments in all things. But even here we should be careful about going all the way. There is a kind of cult of acceptance that may dominate a group of authorities; this is true in nuclear physics today. Hence, one has to be extremely skeptical of the fads that might dominate the peer group. Yet the authorities themselves may be reluctant to accept new ideas because of the dominant paradigm models that influence them. Or, on the contrary, there are some who are all too willing to swallow what a colleague maintains if he or she has a reputation and status. It only proves the point that it is not the so-called authorities who make something true, but something is true because of objective evidence. Though I grant that even to say this is often an understatement, since what might appear to be compelling evidence or convincing proof to one generation may be totally unpersuasive to another. Hence, the only conclusion is to be wary of all authorities, recognizing however that we need to use them. The proper stance perhaps is ambivalence based upon rational skepticism. This entails skepticism not only about the acknowledged experts but about the skeptics themselves.

Subjectivism and intuition

Science and critical intelligence embrace objectivity; they require corroboration and replication. Are there some forms of truth, however, that can only be known subjectively?

Many individuals, in particular many students I have encountered, are subjectivists. They believe that something is true merely because they have "validated" it personally; they draw upon their subjective feelings and experiences, the rich memories of their private perceptual worlds, their motivations and aspirations. The response to this is similar to the response to the method of emotion. Subjectivism is inadequate as a test of knowledge about the world, though it is undeniably the starting point for knowledge of oneself or empathically, by analogy, of other minds. In many cases, a person's private interpretations are mistaken; thus he needs to correct his private fund of experience by reference to objective criteria. We may be mistaken in what we infer about ourselves or others. Thus knowledge of ourselves is corrigible by interpretation.

In dealing with subjectivism there is another issue at stake, particularly among those who refer to personal feelings as the source of transcendental authority. Undoubtedly, there are many areas in life where there are no authorities at all, where we may not know what is true or false, and where we remain in ignorance. There may be new fields emerging, protosciences on the boundaries of knowledge, and there are aspects of nature totally unknown and thus far uncharted. Where there are no authorities we can consult for guidance, some age-old questions may persist: Is it possible to solve the "riddle of the universe," to find meaning over and beyond this world as we know it, to find something uncatalogable by the principles of logic and unavailable to empirical inquiry, experimental science, and even critical intelligence? Poets and mystics ask, Why be limited by the categories of ordinary life or scientific method? The heart yearns for something more; the quest into the depths of the mysteries of being, they believe, can be universal. At this point emerge poets and mystics, prophets and seers, who claim to have some form of insight into ultimate reality.

Subjectivists claim that all knowledge, in the last analysis, is inner and private. To say that we *know* something means always that there is some subject who experiences or is aware of it. This applies to the perceptions of the world that we observe, the "sense data" reports of tones, colors, shapes, and sizes, and to our cognitive intuitions and deductive inferences. Thus, it is claimed that all knowledge begins and ends with the *subject,* not the object. Intuitionists must grant that we attempt to communicate what we experience and feel and what we believe we have discovered about ourselves or the world, and that we learn from others. This we do by means of language. Signs and symbols, terms and words clothe our fleeting impressions and give them shape and form. We combine particles of spoken and written discourse in sentences, and we organize the discrete parts of our experiences into a broader structure or conceptual framework. But the intuitionist insists that if we eliminate the subject or seek to get rid of him, knowledge will be empty; it is the inner intuitions that are the source of truth. And he affirms that there is a kind of truth that we know within ourselves and cannot translate into language for others without loss of its full meaning. I readily grant that perceptions are qualities that appear in immediate experience; but the mistake of the intuitionist is not to see, as I have already pointed out, that these are not simply inner or isolated but are fused with action and transactions. Immediate intuitions cannot be abstracted; they are connected with behavior, and this is confirmable intersubjectively. Thus, to use subjectivism as intrinsic to empiricism is in error. For the model of the mind as a passive receptacle or tabula rasa is mistaken. We begin with the act, not the intuited fact; with the field, not the subject separated from it. Man is an agent or doer, not simply a passive beholder. Subjective experience makes no sense devoid of the objects of knowledge and transaction.

Similarly, there are cognitive intuitionists (as in the case of Descartes) who think we can apprehend truth by a direct flash of insight. Logical truths allegedly fall into this class, since no proof of their validity is necessary or possible. I would point out that what we are talking about are the operations of the mind itself. And this is not solely private, since other human beings—in this case, mathematicians and logicians—can evaluate the reasoning process to see if it is valid. Objective standards thus explicitly apply.

There is another issue. Are there forms of private knowing that cannot be fathomed empirically or conceptually and are not amenable to the categories of experience or logic and yet give us truths of reality that we cannot get in any other way?

For many, this is the main challenge to science. Has it so defined its subject matter and its standards that it excludes a priori depths of reality and being that are unknowable and unspoken in its own limited terms? Should we not consult the testimony of poets and artists, mystics and seers about the truths which are known directly and cannot be translated into didactic language or conceptual form?

Are there truths conveyed by the arts that one cannot get in the sciences or by ordinary experience? Plato thought that the arts give us a glimmer of absolute forms, although he would have banished the artist from the state as being unreliable or a bit mad. Do the late quartets of Beethoven, the Moog synthesizer, the paintings of Modigliani, or any other work of art render truths in a special way? My response is, first, that if they do, they are surely not about a transcendent realm of absolute beauty or form, whatever that means. Second, it depends upon the specific art. The Platonic view is difficult to defend. Yet the arts surely render meanings, which they are able to communicate, and they can arouse powerful feelings and reactions. Is this a form of knowledge? It all depends upon the context. A painting or a statue may depict a scene or convey an idea of a person—as do the eloquent paintings of Dutch burghers by Rembrandt and the David of Michelangelo. Novels and plays use didactic language and transmit information. They may be highly descriptive of events, objects, personalities. And they may draw in fine detail or present with insight the nature of human character. Songs with lyrics and dramatic opera do the same.

There is thus a descriptive aspect to the arts. Many (but not all) do convey or communicate information and knowledge. However, what is distinctive, I would argue, is the *aesthetic* component and the ability of the work of art to arouse emotion. It is not simply belief that the artist is provoking, but mood. Works of art are dramatic, for they appeal skillfully to the beholder or listener. The difference between a descriptive account and an artistic one is the aesthetic component, which involves the arousal of either pleasure or strong feelings.

A work of art is a product of a process of human production. It can be interpreted in naturalistic terms without any reference to a transcendental source. Science is not a substitute for art; they have different functions. In art, the communication of knowledge is a secondary function; what is primary is the arousal of mood, drama, feeling, and pleasure.

Faith as justification for belief

One kind of appeal often made to support a belief is the existence of faith. In its most popular form this is another version of subjectivism. Many people insist that their beliefs are true because of an inner light, a certainty, or a conviction that they are so. Theologians commend this appeal to faith, extolling it as the highest of the human virtues. They promise that belief in the existence of God based upon the commitment of faith, in spite of paradox, doubt, or a lack of evidence, will be rewarded at the final judgment day. Is an appeal to faith warranted as a method of knowing? Many religionists maintain that we can never abandon it entirely in life; anticipating objections, they insist that even scientists and skeptics presuppose a leap of faith in respect to their own theories. They ask, "Why is one form of faith any less justifiable than the next?"

What is at issue in the dispute is the definition of faith. It is necessary to distinguish at least three kinds: (1) intransigent faith, (2) willful belief (one form is the argument from ignorance), and (2) beliefs framed as hypotheses and based upon probabilities. This distinction concerns the degree of evidence (or the total absence of it) we draw upon to support a belief.

1. In the first form of faith, in spite of strong evidence in opposition to a belief, it is still held tenaciously by some people. We can call this form *intransigent faith*. Many illustrations can be given, especially since it is widespread. In its extreme form, it is totally irrational: a person may not be willing to believe that he or she has been betrayed by another. For example, perhaps even though a husband has committed an act of infidelity, the wife may refuse to accept the facts, despite clear evidence. In this case strong psychological attitudes and habits may block the ability of the individual to recognize the situation. Another case is the persistence of religious beliefs throughout centuries in spite of scientific discoveries. Did Jesus exist? Was his message unique? Was it shared by others in the period? Did Jesus perform the miracles described in the New Testament? Are the virgin birth and the resurrection supportable? According to the best modern biblical scholarship, there is a good deal of evidence requiring

negative answers to these questions, and yet many believers remain intransigent in answering affirmatively.

It is well over a century since Darwin wrote. Yet creationists still stoutly deny the theory of evolution. They reject the wide range of data assembled from biology, geology, paleontology, anthropology, astronomy, etc., which all point to evolutionary processes in the universe. Instead, pointing to the gaps in the fossil record that exist, creationists totally reject evolution. They have introduced the ad hoc creationist thesis, which they call "scientific creationism." But it is a mask for the doctrine of religious creationism outlined in the Book of Genesis: that God created the universe ex nihilo, including all the species, which are fixed in kind. Is this simply a case of balancing the evidence between two competing scientific theories? I doubt it. Scientific evolutionists do not assert that their theories have reached a final form, and they have gone far beyond Darwin. That evolution occurs is well-supported by abundant data. Evolutionists have developed many causal hypotheses about *how* it occurs, but there are still many controversies on that point. For example, is it a gradual process or does it proceed by punctuated equilibria? Uncertainty is part of the normal process of scientific inquiry, where hypotheses are introduced and modified in the course of investigation and in the light of new evidence.

Those who argue for the biblical account of Genesis maintain that the world is between 6,000 and 10,000 years old. (Bishop Ussher said that it was created in 4004 B.C.E.) They also invoke the Flood, claiming that there was a worldwide disaster much as the Bible describes. But any number of questions can be raised: Noah's Ark was too small to contain two (or seven) of each of all the earth's species since, according to the Bible, it was 150 yards long, 25 wide, and 15 high. The known species today are vast in comparison with what was known in biblical times. How could one move about in the Ark given the great quantity of manure that would have piled up? Presumably Noah and his sons would have had to shovel around the clock to keep a passageway clear. If the flood encompassed the entire globe, submerging even the tallest peaks, the sea level would have had to rise 5.5 miles. This would mean that more than one billion cubic miles of water would have been added to the world's oceans. If all of the ice at the poles melted, only 200 feet would be added. Of course, one can affirm that God can do anything, and forget reason and fact. For example, how can we account for the many layers of sediment discovered in geological strata containing different fossils at various depths? God, they affirm, created it in such a manner.

The creationist's response expresses an irrational form of intransigent faith. Should this extreme form of creationism be taken seriously? The fact that it is held by a large number of otherwise educated people indicates that it cannot be ignored. Many adherents of this view have advanced degrees in specialized sciences. Some theologians deride any critical approach to the Bible or any effort to submit revealed theology to scientific investigation. There are two forms of truth, they insist. With Tertullian, some affirm, "I believe, because it is absurd." Is religion based upon a miracle? To obstinately accept what is contrary to the principles of logic and evidence and requires a leap of faith *is* miraculous.

David Hume observed that Christianity is dependent upon a miracle for belief. Mere reason is insufficient to prove the veracity of the Christian religion. "Whoever is moved by *faith* to assent to it, is conscious of a continued miracle in his own person, which subverts all of the principles of his understanding, and gives him a determination to believe what is most contrary to custom and experience."[2] For someone thus to appeal to his faith and intransigently maintain it, though it is demonstrably false, is simply to say, "I believe because I believe."

One should not think that such forms of belief exist only in religion. Intransigent faith is rampant in all walks of life—in politics, morality, the arts, even philosophy and science. Beliefs are deep-seated forms of habit. They are often enormously difficult to modify, especially when they are ingrained in our behavior. To break a habit and its concomitant belief-state may have unsettling results. The problem is especially strong today, given the growth of modern science and the consequent technological and social changes. *Future shock* refers to the inability of humans to assimilate change in behavioral patterns and cherished beliefs. Hence, faith becomes a refuge and subterfuge, enabling someone to withstand challenges to his beliefs.

In some individuals the need for certitude may be intense. They cannot endure the ambiguities in life. To caution prudence or to recommend that one suspend belief or withhold judgment until all the evidence is in falls on deaf ears in many cases. "Fools [read *believers*] rush in where angels [read *skeptics*] fear to tread" eloquently describes the situation. Perhaps there are glandular, even genetic, factors that predispose some people to crave security in belief, whereas others are able to thrive on ambiguity and live with uncertainty. The problem is compounded when we note that intransigent faith is supported by many social institutions and customs. The church, the political party, the state all seek to regulate belief and conduct, and when the sacred cows are questioned, the heretic or dissenter is condemned to the stake or exiled to the Gulag. The true believer will not brook any compromise with his fervent beliefs and values.

2. David Hume, *An Enquiry Concerning Human Understanding,* from E. A. Burtt, ed. *The English Philosophers from Bacon to Mill* (New York: Modern Library, 1930), p. 667.

He will reward those who share his illusions, deify them as gods, sanctify them as saints, or honor them as heroes or patriots. Once someone has built a church or a state on the rock of faith, it becomes difficult to dislodge it.

2. There is a second kind of faith, *willful belief,* which William James labeled "the will to believe." In a famous essay by that name, he claimed that in some cases, notably in religion, we can resolve to believe that something is true. These cases concern beliefs where there is insufficient or no evidence either way to make a rational choice. In the religious sphere, he asserted, we have an option to believe (theism), not to believe (atheism), or to suspend judgment (agnosticism). Many beliefs we entertain may be considered trivial, and we do not have to make a choice. Religious beliefs, according to James, are "momentous," "alive," and "forced." Here, even to suspend judgment is to act; for there are important positive consequences in willing to believe, particularly to our behavioral and passionate life. We can dispute James on that claim, for modern man often finds the Judeo-Christian option dead and boring, not alive or momentous. So unlikely is its truth-claim that the skeptic maintains that he can adopt the stance of the skeptical atheist or agnostic, and that he does not have to affirm a strong position. Nevertheless, James thinks that the believer should have the *right* to believe in such cases and that it is not unreasonable to do so.

Is theism a forced and momentous option? It is not like the following case, which is clearly so. Assume that a man has stepped off a curb and is halfway to the other side of the street when the sees a car rapidly headed in his direction. He wonders if the driver will see him in time and veer away, or is he likely to be hit? Should he beat a hasty retreat or should he run to the other side? If the man is an agnostic, should he stand still to see what happens because he has insufficient evidence to make up his mind either way? The danger, of course, is that the car may careen toward him. Inaction here is clearly a form of action, and we may judge it by its consequences.

Now we must distinguish two cases: first, where there is no evidence either way, and second, where there may be some evidence on both sides, but it is insufficient to allow an informed choice. The case of the onrushing auto most likely fits into the second class, because we know from past experience that pedestrian accidents often occur and hence there are probabilities at stake. A rational choice would be to either retreat or advance; this is based upon some calculation of likely probabilities. We need to act upon the best available evidence even though we do not have the time to chart the probabilities fully. We can make a prediction and wait to confirm it. But a reasonable person will recognize that it is senseless to take risks and adopt the option, for the stakes are too great if there is a chance that the auto will hit him.

James argues that the wish is often father to the fact, that ideas breed consequences and that we judge those ideas pragmatically by how well they work out in practice. Thus the door is left open for religious faith. To which I reply: religious beliefs, especially when false, can have negative and deleterious effects. No doubt they provide hope and meaning for some persons; for others, however, they narrow horizons and limit choices. One might will not to believe in religion, and do so on pragmatic, psychological, and ethical grounds.

The skeptic has no illusions about life, nor a vain belief in the promise of immortality. Since this life here and now is all we can know, our most reasonable option is to live it fully. In not being deluded about the human condition we recognize that we are responsible in large part for what will happen to us. "No deity will save us, we must save ourselves," says *Humanist Manifesto II;* out of this recognition can emerge important ethical, psychological, and social consequences.

Now I readily grant that for some people some religious questions may be considered so momentous that they feel they may have to choose to believe and act. They believe that their eternal destiny is at stake. I doubt that, and will address these questions in Chapter XIV. I submit that the most viable option in cases where there is no objective evidence and we *have* to choose would be to flip a coin and hope for the best. It then becomes a question of one's prejudices and values intervening. But if we don't have *any* evidence either way, then why not admit it and leave it at that? Again, we run the risk, retorts the believer, that some vindictive deity will punish us for failure to believe. Sidney Hook, when asked what he would do if he were confronted by his maker at Judgment Day for not believing, retorted, "I would say 'You didn't give me any evidence,' and trust that if God were a rational being he would respect rationality in his creatures!"

One area where willful faith may be important, even decisive, is the area of value. For example, we may wish to bring something new into the world or we may have ideal ends that we think are worthy of achievement; whether they are achieved depends in large part upon what we believe and what we do. Perhaps we resolve to support a candidate or work for a cause; and we say that we have faith *in* him or it, in the hope that in realizing our normative goals we can lead better lives or contribute to a better world. Whether one's faith in these prescriptive recommendations is wise can only be determined by the results. However, we may succeed or fail precisely because our believing it made us act, which brought it into existence. However, we should be clear that such beliefs about the world are *not* empirically true. Thus, whether or not God exists is one question; faith does not make it true. Whether religious beliefs and practices are morally worthy is another.

A variation on James' argument is often encountered; it is sometimes hurled at the skeptic in defense of faith. It runs like this: "You ask me, do I believe in *X* (substitute whatever you like: God, ghosts, angels)? You ask me for confirming evidence. I can't prove it. But can you prove that *X* does *not* exist? You say you can't. Well, if you cannot, then why am I not entitled to believe in it until such time as you show me it is false?"

This argument is patently fallacious; it has been labeled in classical logic *argumentum ad ignorantiam* (argument from ignorance). It may be difficult, however, to convince a person who is persuaded by it to the contrary. I would respond as follows: Some questions for all practical purposes can be proven to be true, and some can be disproven; and so we can settle many or most disputes that concern facts. That it is 44° Fahrenheit outside at present and not 34°, as you allege, we can resolve by taking several thermometer readings (assuming that our thermometers are properly functioning). That the Mt. St. Helens volcano erupted spilling ash throughout the state of Washington can be corroborated by the factual data recorded.

Some questions are difficult to resolve either way. I may not be able to disprove a hypothesis decisively yet consider it to be highly unlikely. Here the burden of proof is on the person who introduces it. The burden of proof is not to be placed primarily on the person who rejects the claim. If a person offers a hypothesis, the critical question to be raised is: "What are the grounds in support of the hypothesis?" If there are none, then this does not mean therefore that one belief is as good as the next and that the affirmative is equal in reasonability to the denial. The response in such cases should be either in the negative or an agnostic one. We are not justified to take a leap of faith and affirm it as true.

If someone asks me to prove that elves and fairies do not exist, what can I do to convince him? We can travel everywhere in our search and find no such creatures. But he may be unconvinced and insist that they may be somewhere we have not looked. If there is no adequate confirming evidence that Bigfoot exists, the believer may insist that he is lurking in some rain forest or mountain fastness and that we have not discovered him yet. To which I may say, yes, it is surely possible that other species are still to be discovered, but until there is some scientific evidence for a new form of life, it is improbable that Bigfoot exists. The argument from ignorance is especially appealing for some minds in cases that are not simply improbable but are also beyond the range of possibility. For example, to claim, as some disciples of the paranormal have, that there are indiscernible beings from another space-time dimension in outer space and that they intervene in our destinies, makes it difficult to know how to deal with the claim, especially since these beings are "indiscernible." Since they are paranormal, it is claimed, they may block all the normal means of perception. But then how do I know they are there? If I can't disprove that they are—by definition I cannot—are you justified in affirming their reality? The two options are not equivalent: the burden of proof as I argued earlier is upon the fideist, not the skeptic.

3. There is still a third form of faith that merits discussion, though the term *faith* is perhaps inappropriate here. I am referring to "hypotheses based upon evidence," i.e., beliefs held with varying degrees of probability, based upon the evidence, but none with complete certainty. Does faith intervene here? Does it involve a leap beyond the corroborating grounds? It all depends upon the relationship between the degree of evidence and the strength of belief. As I have argued in Chapter II: Skepticism, hypotheses should not be taken as inflexible or absolute beliefs. The principle of fallibilism should always be held in reserve. Truth values are simply equivalent to or a function of their supporting grounds, If faith is that portion of our convictions that goes beyond the evidential base, then faith in itself is inadmissible. Skepticism refers precisely to that portion that goes beyond and which we are not justified in believing. I readily grant that few people, if any, can so structure their beliefs that they are always equivalent to their grounds. In principle, they are; in practice probabilities convert to certainties and doubt gives way to commitment. A kind of animal instinct operates, making it possible for us to live and function.

Still, it is clear that the hypotheses and theories of the sciences and the reasonably formed judgments of critical intelligence do not involve faith in the sense that the element of faith is confirmatory. The fact that a faith-state is present does not justify the belief, as it does in the case of intransigent or willful belief. Faith never proves a point; it only registers our convictions. The belief, if it is warranted, is justified by reference to evidence and reason.

It is necessary to raise one further question again; it concerns the status of first principles. Are all first principles based upon faith and hence essentially irrational? Fideists who believe in God, the transcendent, or immortality have argued that to accept the epistemological principles of inductive and deductive logic is a supreme act of faith, analogous to the belief in God. But these cases are dissimilar. John Stuart Mill observed that first principles are not amenable to proof as other hypotheses are. Nevertheless the criteria of inductive and deductive logic, as I have argued in Chapter III, are not based simply upon irrational leaps of faith. They are corrigible; they are tested by the whole fabric of rationality; they are presupposed by ordinary common sense and imbedded in the processes by which we characterize the everyday world and make choices. The consequences of denying them would invalidate the entire structure of behavior and would make any form of knowledge impossible. We do develop hypotheses that we say are true, and we act upon this knowledge; and in that sense every logico-empirical confirmation of our beliefs adds further justification for our root principles. The secular and scientific humanist in the last analysis, however, remains a skeptic; and this applies even to his first principles, which are not held on religious grounds but are simply convenient rules of inquiry, vindicated by their consequences.

Part Two

MYSTICISM, REVELATION, AND GOD

VI: The appeal to mysticism

Is there a form of knowledge that cannot be derived from the usual sources of human experience, and that need not be justified by reference to them? According to the classical religious tradition, there is; it has been called the "mystical experience" or "mystical consciousness." Mysticism is supposed to untap an esoteric form of knowledge of a transcendent realm of being that lies deep within the soul. This knowledge is incapable of being translated into ordinary sensory empirical terms; it defies the categories of cognitive logic; it is *extra*-ordinary in its essential nature. Yet it is supposed to be the most important kind of knowledge we can discover; and it points to a reality beyond our normal consciousness of the world.

I wish to deal with three issues. First, what is mystical experience? How have the leading mystics characterized it? Second, how have they interpreted their mystical experiences? Is mysticism evidence of another kind of ultimate reality? Is the existence of a transcendent source the sole explanation that may be inferred from it; or are there alternative, naturalistic and causal explanations of such experiences? Third, it is important that we distinguish between direct and secondhand accounts of mystical experience. A firsthand report is from the standpoint of someone who claims to have had an extraordinary experience of ecstasy and insight. Secondhand interpretations are from those who have not had such an experience but are asked to evaluate a claim to determine whether it should be taken as trustworthy. What should be their response to such a claim?

What is mysticism?

A definition of mysticism is not easy, since there are so many varieties; it is often difficult to find a property that underlies them all. There is the mysticism of the Eastern religions: the Upanishads is a source, and Shankara is the great teacher and interpreter of Advaita. There is also an extensive Yoga system of mental concentration. The Buddhist mystical tradition begins with the Buddha himself and extends to the Satori experience of contemporary Zen masters. In Asian religions, the self becomes

91

infused with divine Being, while the idea of a personal God is foreign. Monotheistic religions also express mysticism. In the Muslim tradition, there is Sufism. There are Jewish Chassidic mystics. Christianity is infused with the philosophy of neo-Platonism and Plotinus, and it abounds with great mystics and saints including Dionysius the Areopagite, Meister Eckhart, St. Gregory, St. Bernard, Heinrich Suzo, Catherine of Genoa, St. Theresa of Avila, Mme. Guyon, St. Marguerite Marie, St. John of the Cross, Jan van Ruysbroeck, and Jakob Boehme.

Students of mysticism have sought to distinguish mystical experiences from other forms of religious experience or divine encounter. Paul's alleged encounter with God on the road to Damascus, as recorded in the New Testament (Acts 9) is not mystical per se; nor should mysticism include "the hearing of voices," "the seeing of visions," or "the receiving of revelations" by means of verbal commandments. Hence, mysticism would exclude Moses' instructions from on high in the form of the Ten Commandments, Jesus' Sermon on the Mount, or Mohammed's reception of the divine messages recorded in the Koran. This special class of revelatory claims we shall deal with in subsequent chapters.

Nor should mysticism be used synonymously with the occult or the paranormal, with apparitions, demons, ghosts, telepathy, precognition, clairvoyance, or extrasensory perception. Mysticism should not be used so ambiguously as to include aesthetic experiences: the intense moods aroused by poetry, music, the arts, the inspiring appreciation for the beauty of nature. These are unwarranted, metaphorical extensions of the term. More precisely, mysticism is applied to that specialized class of experiences in which mystics claim to have some direct encounter with a spiritual reality; this is usually discovered in an overwhelming and unique encounter overflowing with intensity and rapture.

William James, in *The Varieties of Religious Experience,* suggests that our normal consciousness, which includes rational-empirical components, is but one special type of consciousness and that there are "potential forms of consciousness entirely different." We sense objects in the spatial-temporal world by means of colors, tones, shapes, and sizes. This world is malleable to our purposes; objects in our behavioral field are affected by our actions and vice versa. These we interpret in terms of relational concepts. Mystics claim that over and beyond these normal forms of awareness there is another range of consciousness in which we come into contact with a reality much deeper, more authentic and enduring than anything we confront in everyday life. They are unable to translate this reality into meaningful perceptual or conceptual equivalents. Nor can they communicate by means of linguistic discourse precisely what it is. It is "ineffable," they insist, unutterable and indefinable.

Plotinus denies that discursive reason can adequately describe a mystic encounter with the One:

> The discursive reason, if it wishes to say anything, must seize first one element of the Truth and then another; such are the conditions of discursive thought. But how can discursive thought apprehend the absolutely simple? It is enough to apprehend it by a kind of spiritual intuition. But in this act of apprehension we have neither the power nor the time to say anything about it; afterwards we can reason about it. We may believe that we have really seen, when a sudden light illumines the Soul; for this light comes from the One and is the One.[1]

Some species can sense smells or hear sounds well beyond the human perceptual range. Is it not possible that there is an extended awareness which the mystics can untap, opening up dimensions unknown to ordinary consciousness? This is a special challenge for the scientific empiricist and skeptic. If knowledge about the universe has its roots in some experience (whether potential or actual, direct or indirect), then why not admit the claims of the mystics, who may be so attuned to reality that they can penetrate to the essence of Being. There is a fairy tale about a stranger visiting a kingdom where everyone is blind; it is impossible for the stranger to convey to the blind inhabitants any sense of the richness of the visible world. No doubt it is difficult to explain to a frigid woman or a celibate monk the meaning of an orgasm, if they have never enjoyed a sexual climax. Why not allow the possibility of an extended sensibility, which opens up splendors undreamed of by the prosaic imagination?

Classical mysticism in its specialized form has been relatively rare. W. T. Stace estimates, for example, that there were only one hundred well-known Christian mystics. Many of the great scholars who have devoted an enormous amount of time to studying the phenomena admit to never having had such an experience. Cuthbert Butler, a Catholic theologian, and author of an important study on mysticism, admits that he has never had any such experience himself, "never anything that could be called an experiential perception of God or His Presence."[2] William James confesses to being an outsider. He says that most of the writing on mysticism that he has seen "has treated the subject from the outside, for I know of no one who has spoken as having the direct authority of experience in favor of his views. I am also an outsider. . . ." James suggests "that states of mystical intuition may be only very sudden and great

1. Reprinted in W. T. Stace, *The Teachings of the Mystics* (New York: Mentor, 1960), p. 115.
2. Cuthbert Butler, *Western Mysticism,* 2nd ed. (New York: Harper and Row, 1966), p. 136.

extensions of the the ordinary 'field of consciousness' . . . an immense spreading of the margin of the field."[3]

The problem, of course, is that mysticism, by its very nature, violates the standards of scientific objectivity. If we were to accept its data, it would need to be amenable to intersubjective analysis and corroboration. Unfortunately, this is precluded almost by its very nature. Until recently it has been virtually impossible to replicate it, that is, until the introduction of psychedelic drugs, which seem to arouse similar psychological experiences. Should we reject mysticism a priori? This appears to theists dogmatic by definition.

The problem for skepticism is that mysticism, at least in its classical sense, is an unusual and bizarre experience. Although mystics claim it to be ineffable and untranslatable, some of them do try to explain and interpret it in language, and this we can submit to careful scrutiny. The first thing we discover is that there are significant differences in the accounts of the mystics, particularly between Eastern and Western mysticism, and there are radically dissonant interpretations. Nevertheless, according to some commentators, there is a common characteristic. The most important feature of mysticism, Stace maintains, is that it involves the apprehension of an ultimate "nonsensuous unity" to all things, a powerful sense of the "oneness to reality." In this undifferentiated unity, divisions and multiplicities disappear. This pure unitary consciousness is allegedly devoid of any sensory-intellectual content.

Rudolf Otto describes the Hindu mystic tradition of Advaita as follows:

> True Being is Sat alone, Being itself, the internal Brahman, unchanging and unchanged, undivided and without parts." That is (a) the multiplicity of things exist only through "Maya" (which is usually translated as "mere appearance"). Sat itself is the One only, (b) in itself Brahman or Being absolutely and immutably "One only," without parts, without any multiplicity, and therefore without the multiplicity of differences and delimitations. . . . This Eternal One in its uniform nature is wholly and purely Atman or spirit, pure consciousness, pure knowledge. . . . Thus it is at once "anantam" without end, and beyond space and time. . . . The soul of man, the "inward Atman," is nothing less than this one eternal, unchangeable, homogeneous Brahman itself.[4]

Mysticism thus involves a path for the soul to take, a ladder by

---

3. *The Works of William James: Essays in Philosophy,* ed. F. Burkhardt (Cambridge, Mass.: Harvard University Press, 1937), from "A Suggestion About Mysticism," *Journal of Philosophy, Psychology and Scientific Methods,* Feb. 7, 1910, p. 157.

4. Rudolf Otto, *Mysticism East and West: A Comparative Analysis of the Nature of Mysticism* (New York: Macmillan, 1932), p. 19.

which it can ascend into a mystical intuition of the One. This involves, in the words of Plotinus, a union with the divine:

> When the soul turns away from visible things and makes itself as beautiful as possible and becomes like the One ... And seeing the One, suddenly appearing in itself, for there is nothing between, nor are they any longer two, but one, for you cannot distinguish between them, while the vision lasts.[5]

St. Augustine expressed the meaning of Christian mysticism with the descriptive phraseology: "My mind in the flash of a trembling glance came to Absolute Being—That Which Is."[6] Mysticism thus implies some direct contact of the soul with transcendental reality, an intense attachment to and identification with it.

A major strand in mysticism is the belief that the way to achieve the mystical experience and discover the unity of existence is to turn within. The path to God is by means of introspection and introversion; God is immanent in the soul.

> God is hidden within the soul, and the true contemplative will seek him there in love. (St. John of the Cross)

> The soul finds God in its own depth. (Jan van Ruysbroeck)

> Where the soul is, there is God. (Meister Eckhart)

> We should seek God in our interior. (St. Theresa)[7]

The mystical experience culminates in a tremendous burst of passionate and emotional fulfillment. This involves the opening of the heart, an arousal of love. There is supposedly a "union" in the sense that the soul of the mystic, as it were, melts away or becomes fused with the divine, flowing into God as God flows into the soul. There is often an allusion to an intense and shimmering light, which suffuses the mind and being of the person and dominates everything. Out of this experience emerges an indescribable joy, a sense of bliss, ecstatic happiness, a buoyant experience of peace, an inner sweetness.

Louis de Blois (a Benedictine abbot who died in 1566) described his experience as follows:

> It is a great thing, an exceeding great thing, ... to be joined to God in the divine light by a mystical and denuded union. This takes place where a

5. Stace, *Teachings of the Mystics,* p. 115.
6. St. Augustine, *Confessions,* VII, 23. Quoted in Butler, *Western Mysticism,* p. 4.
7. Quoted in Douglas Clyde Macintosh, *The Problem of Religious Knowledge* (New York: Harper, 1940), chap. 2.

pure, humble, and resigned soul, burning with ardent love, is carried above itself by the grace of God, and through the brilliancy of the divine light shining on the mind, it loses all consideration and distinction of things, and lays aside all, even the most excellent images, and all liquified by love, and, as it were, reduced to nothing, it melts away into God. It is then united to God without any medium, and becomes one spirit with Him, and is transformed and changed into Him. . . . It becomes one with God, yet not so as to be of the same substance and nature of God.[8]

St. John of the Cross (a Carmelite who died in 1591) describes his experience:

The end I have in view is the divine Embracing, the union of the soul with the divine Substance. In this loving, obscure knowledge God unites Himself with the soul eminently and divinely. . . . This knowledge consists in a certain contact of the soul with the Divinity, and it is God Himself Who is then felt and tasted, though not manifestly and distinctly, as it will be in glory. But this touch of knowledge and of sweetness is so deep and so profound that it penetrates into the inmost substance of the soul. This knowledge savours in some measure of the divine Essence and of everlasting life.[9]

St. Francis of Sales (Bishop of Geneva, who died in 1622) depicts the soul's union with God. He says:

As melted balm that no longer has firmness or solidity, the soul lets herself pass or flow into What she loves: she does not spring out of herself as by a sudden leap, nor does she cling as by a joining or union, but gently glides, as a fluid and liquid thing, into the Divinity Whom she loves. She goes out by that sacred outflowing and holy liquefaction, and quits herself, not only to be united to the well-Beloved, but to be entirely mingled with and steeped in Him. The outflowing of a soul into her God is a true ecstasy, by which the soul quite transcends the limits of her natural way of existence, being wholly mingled with, absorbed and engulfed in, her God.[10]

The first questions that need to be raised are: What does the experience as recounted mean? What does it point to? What interpretation shall be placed upon it?

One of the problems is that interpretation has been colored by the social and cultural context in which a mystical experience occurred. Christian mystics tend to read Christ into the message, but Muslim, Hindu, Buddhist, or Jewish mystics differ in their cultural interpretations. Buddhists are not theists and Hindus are not monotheists; and so the

---

8. *Spiritual Mirror,* c. II, quoted in Butler, *Western Mysticism,* p. 9.
9. *Ascent of Carmel,* II, c. 24, 26, quoted in Butler, *Western Mysticism,* p. 10.
10. *Treatise of the Love of God,* VI, 12, quoted in Butler, *Western Mysticism,* p. 11.

notion of a Supreme Being is not present, nor is there expectation of personal immortality. Hindus refer to contact with Atman or the world-soul, Buddhists to Nirvana or nothingness. Their experiences are not evidential for the monotheistic interpretations of the Western religions. Stace suggests that perhaps no religious significance should be read into mystic experience: it is an experience in itself without any special theological trappings.

Can we cull from such experiences, at the very least, a transcendental inference? Does it point to a realm beyond, of which only some individuals by special efforts can get a brief glimmer? Should we pay close attention to what their experience tells us about ultimate reality? Or, on the contrary, is the mystic's experience an altered, pathological state of consciousness? Is it a distortion of normal consciousness and perception? Bertrand Russell observes: "From a scientific point of view we can make no distinction between the man who eats little and sees heaven and the man who drinks much and sees snakes. Each is an abnormal physical condition, and therefore has abnormal perception."[11]

William James says that "mystical experiences are, and have the right to be, authoritative for those that have had them." But, on the other hand, he agrees that those who have not had the experience are not called upon to accept their validity. Yet he concludes: "It must always remain an open question whether mystical states may not possibly be superior points of view, windows through which the mind looks out upon a more extensive and inclusive world."[12]

Some naturalistic explanations

The religionist who accepts mystical experiences as veridical is often unschooled in comparative scientific analysis or unaware of alternative explanations for the mystic state. Committed to a theistic world view, he finds theistic interpretations plausible. Yet a number of alternative naturalistic explanations are possible. Classical religious interpretations were made in the prevailing cultural milieu, particularly during the medieval period. With a shift from a religious to a scientific culture, naturalistic causal interpretations are meaningful.

An interesting question to be raised is this: What are the preconditions necessary to achieve the mystic state? It is clear that it is a relatively rare occurrence and that it requires an enormous effort. This does not deny

11. Bertrand Russell, *Mysticism.* Quoted in Walter Kaufmann, *Critique of Philosophy and Religion* (Garden City, New York: Doubleday, 1961), p. 315.
12. William James, *Varieties of Religious Experience,* pp. 422-24, 428. Quoted in Butler, *Western Mysticism,* p. 137.

that many ordinary human beings have reported mystical-like experiences, a sense of oneness with nature and of peace; but these are not the kind of experiences that great mystics and saints report.

Reading the accounts of their lives, we find that their experiences usually involve a long and arduous period of preparation before the mystic culmination. There is first an intense longing for the mystical experience, a desire to discover God or achieve Nirvana. Second, there is an intense effort at a spiritual life, including constant prayers and devotions. Third, the mystic must repress all physical desire, worldly goods, sexual thoughts, or carnal lust. Fourth, he or she must lead a life of asceticism, self-denial, and self-control. This may include fasting, lack of sleep, hard work, even self-flagellation and mortification. Fifth, the mystic generally lives an isolated and secluded life, whether in darkened caves or isolated cells, and he or she withdraws from the world. The ground is thus laid for such an experience by sensory deprivation, sexual repression, and enforced withdrawal; in short, the preconditions for a psychotic-like reaction are present.

Now, the direct way, according to the mystic, to achieve the sought-for state is to turn within by engaging in contemplation and meditation. This may begin with the chanting of a mantra, the repetition of a prayer, the focusing of one's concentration upon one object or experience. According to yoga, the technique is repetition and concentration. Only then does the desired detachment occur. Recent research shows that a "relaxation response"—which is far short of a mystic experience—can occur by focusing attention on any object, without the need for reciting a prayer or mantra. A preliminary state of dissociation seems to emerge. It is only after a long and arduous path that the mystic may achieve his desired release. When it occurs, he enters a kind of trancelike state, in which the awareness is disengaged from bodily sensations. An out-of-body experience occurs.

St. Augustine described this state as "midway between sleep and death." He says that "the soul is rapt in such wise as to be withdrawn from the bodily senses more than in sleep, but less than in death."[13]

And St. Gregory affirms that "the mind disengages itself from things of this world and fixes its attention wholly in spiritual things." He says that "by dint of a great effort," it mounts up to a momentary perception of the "unencompassed light." Then, "exhausted by the effort" and blinded by the vision of the light, he continues, "it sinks back wearied to its normal state, to recuperate."[14]

13. St. Augustine, *De Genesi ad litterum,* 26, 53. Quoted in Butler, *Western Mysticism,* p. 50.

14. St. Gregory, *Benedictine Monachism.* Quoted in Butler, *Western Mysticism,* p. 67.

Other mystics seem to have an impression of floating in mid-air, or levitating. St. Theresa recounts that "often my body would become so light that it lost all weight." She was aware of being "lifted up." Heinrich Suzo also reports an impression of "floating."[15]

All of these are common forms of out-of-body experiences that people report today—somewhere in the realm between hypnagogic and hypnopompic sleep. The real question is whether an exacerbated focus on these experiences is a symptom of mental illness. These experiences resemble psychotic states. The schizophrenic often is out of cognitive touch with reality. He may not be able to clearly distinguish objects in the spatial and temporal world. His normal perceptions are distorted. His own sense of selfhood may be in doubt or lost. He may seem to stand outside of himself, and he views himself as a spectator, or he finds himself swallowed up by reality. Schizophrenics are often enmeshed in hallucinations and fantasies. Time and speech may be emptied of their usual meanings. In a well-known case, Freud describes the psychotic reactions of Daniel Paul Schreber, a paranoid, who believed that he was in contact with God. Schreber thought that the nerve filaments of his brain were connected with God's. He heard voices and saw these filaments in the form of rays:

> With my mind's eye I see the rays which are both the carriers of the voices and the poison of corpses to be unloaded on my body, as long drawn out filaments approaching my head from some vast distant spot on the horizon. . . . The radiant picture of my rays became visible to my inner eye, while I was lying in bed not sleeping but awake. . . .[16]

A student whom I know recounted a similar "mysticlike" experience. An older man, he had lost his wife to cancer and was burdened with raising three children. He lost his job because of "emotional difficulties." One day, he recounted to me, his mind suddenly went blank. The world was bathed in white light and he had no sense of time—he felt buoyant and at peace. Three days later, he found himself in a psychiatric ward. Apparently he had been wandering nude on the streets, but had no remembrance of it. Since then, he has been in and out of psychiatric wards as he tries to make sense out of his life.

One aspect of the life of the mystic that is particularly intriguing is the role that sexual repression and symbolism plays in this experience. Two striking ingredients are present. First, there is repression and frustration, a sense that pleasures of the flesh, especially sexual enjoyments, are

15. Quoted in James H. Leuba, *The Psychology of Religious Mysticism,* rev. ed. (London: Routledge, Kegan Paul, 1929), p. 250-58.
16. Quoted in Ben-Ami Scharfstein, *Mystical Experience* (Indianapolis: Bobbs-Merrill, 1973), p. 135.

evil and sinful. Second, there is an apparent form of sublimation and some sexual gratification in the mystical encounter. According to James Leuba, there are deep-seated psychological and organic needs that help explain the behavior of mystics.[17] There is a tendency for "self-affirmation" and "self-esteem," a desire "to devote oneself to something or somebody," a need "for affection and moral support," a need "for peace, for single-mindedness or unity." Last, there are "organic needs," for "sensuous satisfaction," especially in connection with one's sex life. Christian mystics fall under the influence of two ideals of monastic Christianity: "self-surrender to God's will and chastity." On the one hand there is the ideal of "self-renunciation," in which physical sex is forbidden; on the other, the demand to surrender obediently to a loving and righteous God. Very few, if any, of the prominent mystics led a normal married or sexual life: the mystical life in a sense grew out of perturbation and perversion. It expressed a pathological response to the need for sexual fulfillment, by means of a deflected discharge of the libido.

The most notable illustration of this is St. Theresa. Her mystical experience has unmistakable phallic symbols and orgasmic overtones. She says:

> I saw an angel close by me, on my left side, in bodily form . . . I saw in his hand a long spear of gold, and at the iron's point there seemed to be a little fire. He appeared to me to be thrusting it at times into my heart, and to pierce my very entrails; when he drew it out, he seemed to draw them out also and to leave me all on fire with a great love of God. The pain was so great that it made me moan; and yet so surpassing was the sweetness of this excessive pain that I could not wish to be rid of it. The soul is satisfied now with nothing less than God.[18]

A similar sexual element is found in a description by Mme. Guyon, a French mystic (1648–1717). "I crave," she cries, "the love that thrills and burns and leaves one fainting in an inexpressible joy and pain." God, she says, answers her cries, sets her aflame with passion, and after the gratification, still trembling in every limb, she says to him, "O God, if you would permit sensual people to feel what I feel, very soon they would leave their false pleasures in order to enjoy so real a blessing."[19]

Pierre Janet, the French psychiatrist, treated a patient, a lonely woman named Madeleine, who was noted for her mystical experience. She reported:

17. Leuba, *Psychology of Religious Mysticism*, pp. 116–17.
18. Quoted in Evelyn Underhill, *Mysticism* (New York: World/Meridian, 1972).
19. Quoted in Leuba, *Psychology of Religious Mysticism*, p. 76.

I feel myself under the spell of a pure and sweet hug which ravishes the whole of my being, an inexpressible heat burns me to the marrow of my bones. . . . The flesh, which is dead to the [sense] perceptions is very much alive to the pure and divine enjoyments. I go to sleep sweetly cradled in God's embrace. God presses me so hard to Himself that He causes me suffering in all my body, but these are pains that I cannot but love. . . .

Madeleine also reveals her desires for children in her fantasies about her relationship with God: "On one occasion she remarks that her nipples are inflamed, 'because He suckles so much.' She is God's mother and wet-nurse, she lets God play, she scolds Him, and so forth."[20]

Elements of sexuality are also present in male mystics, which tends to suggest a latent homosexuality. The statues and paintings of an almost nude Jesus, which appear in the great cathedrals of Europe, perhaps helped to arouse prurient feelings.

St. John of the Cross graphically describes his union:

O gentle subtile touch, the Word, the Son of God, Who dost penetrate subtilely the very substance of my soul, and touching it gently, absorbest it wholly in divine ways of sweetness. . . . What the soul tastes now in this touch of God, is in truth, though not perfectly, a certain foretaste of ever-lasting life. It is not incredible that it should be so, when we believe, as we do believe, that this touch is most substantial, and that the Substance of God touches the substance of the soul. Many saints have experienced it in this life. The sweetness of delight which this touch occasions baffles all description.[21]

Jan van Ruysbroeck says:

In this storm of love two spirits strive together: The Spirit of God and our own spirit. God, through the Holy Ghost, inclines Himself towards us; and thereby we are touched in love. . . . This makes the lovers melt into each other, God's touch and His gifts, our loving craving and our giving back: these fulfill love. This flux and reflux causes the fountain of love to brim over: and thus the touch of God and our loving craving become one simple love. . . . For that abysmal Good which we taste and possess, we can neither grasp nor understand; neither can we enter into it by ourselves or by means of our exercises. And so we are poor in ourselves, but rich in God.[22]

St. Bernard describes his experience as the "mystic kiss" of Christ. If fundamentalists were to read Catholic devotional literature, they would probably regard the following as pornographic:

20. Quoted in Andrew Neher, *The Psychology of Transcendence* (Englewood Cliffs, N.J.: Prentice-Hall, 1980), p. 113.
21. Quoted in Butler, *Western Mysticism*, p. 10.
22. Ibid., pp. 8–9.

[The Kiss of His Mouth] signifies nothing else than to receive the inpouring of the Holy Spirit. . . . The Bride has the boldness to ask trustingly that the inpouring of the Holy Spirit may be granted to her under the name of a Kiss. When the Bride is praying that the Kiss may be given her, her entreaty is for the inpouring of the grace of this threefold knowledge [i.e. of Father, Son and Holy Ghost], as far as it can be experienced in this mortal body. . . . This gift conveys both the light of knowledge and the unction of piety.[23]

Another form of sexuality is masochism. In many cases, the mystic is able to achieve his intense rapture only after self-torture and flagellation. The church extolled physical suffering; it found grateful devotees who willingly and lovingly submitted to its demands.

Heinrich Suzo (1300–1366), a Dominican monk, recounted his experiences in his *Autobiography*. Like others who were committed to a monastic life, he was plagued by temptations; and he struggled to overcome them by inflicting physical pain and torment upon himself. He thought that only by practicing extreme asceticism and torturing his body could he find expiation for his sins. During a long period of his life, he recounts that he was able to achieve frequent, even daily ecstasies. However, a time finally arrived when the bleeding saint realized that he could not continue in his path and still live. "He was so wasted that the only choice was between dying and giving up these practices." At this point, he tells us, he threw all of his instruments of torture into a stream.[24]

Interestingly, the Marquis de Sade and others have found similar pleasures in sadomasochistic pain, though they have not called it "divine." The church has exacerbated pathology and psychosis and considered such a state divinely sanctioned and virtuous. It has done so by establishing the monastic life, enforcing celibacy, demanding strict vows of chastity, rewarding penance and suffering, and emphasizing passivity and obedience.

Hallucinogens and mysticism

Of special interest in explaining mystical experiences is the fact that drugs can induce similar experiences. There are many verified accounts of the use by primitive tribes of certain plants to bring on hallucinations. Mexican Indian tribes used peyote, Native Americans smoked stramonium, and the Samoyed shamans of Siberia ate poisonous toadstools to induce ecstatic visions. The Pythia at Delphi in ancient Greece fasted for three days, chewed laurel leaves, and afterwards breathed noxious vapors. The worship of deified wine was found in the Dionysian rites. Some Muslims,

23. Ibid., p. 98.
24. Leuba, *Psychology of Religious Mysticism*, p. 62.

all of whom are forbidden liquor, attempted to gain intoxication by extravagant dancing. The whirling dervishes were Sufi mystics, who could induce exalted states. Apparently there is a deep tendency within the human species to want to alter consciousness by various stimulants and intoxicants; marijuana, cocaine, heroin, alcohol, coffee, tea, and tranquilizers are only some of our modern forms.

In recent years psychedelic and hallucinogenic drugs have been tested under controlled conditions. William James reported that the use of nitrous oxide can induce "ineffable feelings." Sir Humphry Davy wrote that nitrous oxide caused emotions that were "enthusiastic and sublime." In the 1950s Aldous Huxley described a kind of "transcendental religious experience" stimulated by mescaline.[25] More recently, the widespread use of psychedelic drugs such as LSD (lysergic acid diethylamide-25) has enabled its study under scientific conditions. Some researchers have reported that LSD can produce something similar to the ecstasy and mystery of a mystical experience. Psychedelic and mystical experiences have common elements. First, there is a slowing down of time perception, and one loses a sense of past or future, being immersed in the present moment. Moreover, there is for some users a sense of an "undifferentiated unity." There is a loss of self-identity. Objects melt and are fused, colors are vividly enhanced, shapes and sizes may be distorted. Many LSD users, like mystics, say that it is impossible to describe the experience, yet claim that they have had a vision of a heightened reality. Some users claim that psychedelic experiences extend the awareness and perception of beauty and complexity, as well as of the self as a person. Unfortunately, many users, particularly those who repeat its use, have had bad trips, in which fearful and loathsome experiences may ensue. Some have suffered psychotic episodes. Involuntary flashbacks are not uncommon. Are the persons most likely to become drug users predisposed to mental illness, or does the drug provoke it? In any case, a change in body chemistry—which mystics of old could stimulate only by long and arduous processes of sensory deprivation, fasting, and prayer—can apparently be triggered immediately by psychedelic drugs. Is God to be found by taking pills or drinking the contents of a bottle?

Are drug and mystical experiences the same or do they differ? Is one genuine and the other artificial? There is a rather famous experiment that suggests that LSD does produce something like a mystical experience. The experiment, known as the Good Friday Experiment, was conducted by Dr. Walter Pahnke in 1963. The double-blind study involved twenty volunteer divinity students. Half of the students were given a psychedelic

25. Aldous Huxley, *The Doors of Perception* (New York: Harper, 1954).

drug and half a placebo. The experimental group reported "overwhelming mystical experiences," including "encounters with an ultimate reality" or "with God." The control group reported few experiences that could match in depth and intensity those of the experimental group. Moreover, in a follow-up study six months later, it was found that the lives of many of those who took the drug had been "transformed."[26]

Timothy Leary, during the heyday of the counterculture, reported having deep religious experiences after eating "sacred mushrooms" in Mexico. After his "illumination" in August 1960, which he described as a profound "transcendent experience," he reported that his life was changed and that he has since devoted his time to trying "to understand revelation potentialities of the human nervous system and to make these insights available to others."[27] Leary maintains that the LSD-induced experience is religious and that it provides "the ecstatic, incontrovertibly certain, subjective discovery of answers" to basic spiritual questions. These include: What is the ultimate power that moves the universe? What is life? What is man? What am I? and What is my place in the plan? Leary also reports that various kinds of psychotic (psychedelic) experiences are similar to those that are described by the mystics.

Evaluating the mystical-psychedelic experience

How shall we evaluate such experiences? They can be examined from the standpoint (1) of someone who has had the experience, and (2) of someone who has not and wonders whether it is evidential. The chief reservations of someone who falls in the second class—and that is most people—is that a classical mystical experience, taken by itself is private and subjective, unanswerable to controlled examination or corroboration, and therefore cannot be easily accepted as evidential. If psychedelic devotees are correct that we can now experimentally induce similar experiences, then these can be studied in the laboratory. What is the moral to be drawn—that we all should rush out and take mescaline, LSD, and other drugs to bring about new forms of insight and truth? Or, do these new powers undermine the credibility of mysticism by pointing to a chemical basis of behavioral perception and consciousness and the possibility that such altered consciousness, rather than being extended, is

26. Walter Pahnke, "LSD and Religious Experience," in Richard C. De Bold and Russell C. Leaf, eds., *LSD, Man and Society* (Middletown, Conn.: Wesleyan University Press, 1967); also Edward F. Heenan, *Mystery, Magic and Miracle Religion in a Post-Aquarian Age* (Englewood Cliffs, N.J.: Prentice-Hall, 1973), p. 32.
27. Timothy Leary, "The Religious Experience: Its Production and Interpretation," in Heenan, *Mystery, Magic and Miracle Religion,* pp. 36–37.

distorted? Are psychedelics the key to a new frontier of knowledge, or do they merely induce self-delusion? There are dangers here, given the fact that all too many drug users have provoked what seem to be psychotic reactions: enthusiastic students, wishing for transport and release, ended up by blowing their minds; they were burned out by the temptation and quest for a new reality that never arrived. Instead they often became out of touch with the world, and were unable to function adequately in it. One suggestion that has been made is that madness and psychosis take us to another world, and this reveals new "truths" that would otherwise be left unspoken. R. D. Laing, for example, argues that "madness may not be a breakdown, it may also be a breakthrough. It is potentially liberation."[28]

Is mysticism, at least in its exacerbated form, a kind of madness? And is this a form of truth? Is this religious virtue—madness—justifiable? There is a great danger that the skeptic will simply dismiss his religious opposite on ad hominem grounds. It is all too simple to reject a point of view by questioning the character, motives, and in this case the sanity of those with whom one disagrees. Perhaps it is the skeptic who is blind and ignorant. I have wondered at times: Is it I who lacks a religious sense, and is this due to a defect of character? The tone-deaf are unable to fully appreciate the intensity of music, and the color-blind live in a world denuded of brightness and hue. Is mysticism, indeed psychosis—if these are parallels—a special kind of experience that enables a person to break out of a limited perceptual and conceptual world? Perhaps. One leaves the possibility open.

On the other hand, from the standpoint of one who has not experienced another dimension, there are practical problems that can be raised by common sense and critical intelligence. And one can hardly wish to exchange places with the unfortunate patients in psychiatric institutions, even if we grant them certain kinds of "truths" and visions. We have no doubt mislabeled many people as mad when they were only unconventional nonconformists, as Thomas Szasz has provocatively demonstrated. Nonetheless, to sanctify the world of the schizophrenic as another form of "higher truth" is to abandon all categories of rationality that the human species has evolved over the millennia. It would mean a retreat to frenzy, superstition, and fear, a retreat from reason and sanity into a world of confusion and disorder. For these reasons, one has the right to question the claims of mystics about the "truths" they allegedly have uncovered.

28. Ronald D. Laing, *The Politics of Experience* (London: Penguin Books, 1967), pp. 31–45.

VII: The Jesus myth

The appeal to revelation and miracles

The sacred texts of the three major monotheistic religions of the world, Judaism, Christianity, and Islam, are based upon alleged miraculous revelations received from on high. These documents affirm that three great figures, Moses, Jesus, and Mohammed, have had contact with God and brought messages to earth as his prophets or, in the case of Jesus, as his son. God supposedly revealed himself in a special way to Moses and Mohammed, and Jesus proclaimed a divine message, which he claimed was transmitted from his father; and he was said to be able to perform miracles. The immediate disciples of Moses, Jesus, and Mohammed were so impressed by the claims made by and for these prophets that they became apostles of new religions.

The question we need to examine is whether such historic revelations should be taken as veridical. Are they special sources of truth? Can we authenticate the declarations by their followers about the divine origins of the prophets' message? Although it is extremely difficult for scientific skeptics to take these revelations seriously, given the strength and persistence of biblical and koranic belief, such "evidences" cannot be dismissed out of hand.

Appeals to revelation have been made by others: Paul of Tarsus, Joseph Smith, Baha'u'llah, Ellen White, Mary Baker Eddy, Oral Roberts, and the Rev. Sun Myung Moon. How should we deal today with an individual who claims to have prophetic powers or special spiritual or magical gifts? If they do not talk to God, today they at least claim to have unique powers. We will deal with these phenomena in due course. For the present, I wish to focus on historical claims that a few exceptional, God-intoxicated individuals were designated or selected in some sense to proclaim divine truths to humanity. These truths are not simply expressed in ineffable mystical symbols; nor are they held to be true of transcendent realities over and beyond all human experience. They have been communicated to mankind, either by verbal means or in the form of parables and allegories—often in connection with miraculous works and healing. Since they have a historical dimension, they can be submitted to historical analysis and investigation. I will not treat the three great founders of Judaism, Christianity, and Islam in chronological order, but will first deal with Jesus, perhaps the most remarkable religious prophet whose reputation has survived from ancient times into the present.

Biblical criticism

So much has been written and proclaimed about Jesus that one wonders if there is anything new to be added. Unfortunately, since the lion's share of the commentaries about Jesus are by those who have professed faith that Christ is the son of God, it is difficult to get an unbiased judgment about the historical Jesus and the claims made about his divinity. Powerful churches have been founded in his name and persistent warfare has been waged to suppress criticism and heresy. Rather minor differences in textual interpretations have been the basis for the foundation of new sects, and they have engendered intense rivalries and hatreds, as the disputes between Roman Catholicism, Greek Orthodoxy and the numerous variations of Protestantism demonstrate. The story of Jesus has so fascinated every generation that each has reread the Gospels in its own terms, interpreting Jesus' life and message in the light of its own needs, interests, and language. The mythology of Jesus is still potent enough to transform behavior and subvert all principles of logic and common sense.

This does not mean that there is not a considerable body of critical texts. Many of the great philosophers, such as Spinoza and Hume, read the Bible carefully and published skeptical critiques. In the eighteenth century, biblical religion was submitted to devastating criticism by Voltaire, D'Holbach, Diderot, Thomas Paine, and others. Hermann Samuel Reimarus, in *The Goal of Jesus and His Disciples,* interpreted Jesus as a political revolutionist who worked to deliver the Jews from the Roman yoke. He did not succeed as king of the Jews, and after his failure his disciples misinterpreted his teachings. In the nineteenth century, D. F. Strauss, in his influential *Life of Jesus,* distinguished the supernatural and mythological interpretations of Jesus from Jesus the man; he focused on the latter. Ludwig Feuerbach provided a materialistic interpretation of Christianity and religion. Bruno Bauer tried to show that Jesus was a myth, the personficiation of a second-century cult. In France, Ernest Renan, in *La Vie de Jesus,* applied rationalistic criteria to the New Testament. The Abbé Loisy also published rigorous textual criticism and was excommunicated for denying the existence of Jesus. John M. Robertson and others published similar works in England. Albert Schweitzer attempted to naturalize and humanize the teaching of Jesus in his book *The Quest of the Historical Jesus.*

As a result of this biblical research a number of unconventional, skeptical conclusions were drawn that were highly damaging to the Christian faith. Scholars have argued that there is no strong evidence that Jesus existed. Even if we assume that he did, we know very little about him that can be corroborated. The main Gospel documents and the New Testament are replete with contradictions and inaccuracies, and there is little, if any, independent confirmation of their claims. The Jesus that appears in them is primarily the creation of a newly emerging church of the first and second centuries that needed to justify its views to itself and the outside world. Our knowledge of the birth, ministry, crucifixion, and resurrection of Jesus is therefore fraught with uncertainty. If the Christian religion rests upon a historical claim, and if that claim is full of mythological additions, then it is difficult to accept Christianity as reliable.

In the twentieth century many theologians, from Barth to Bultmann, have reacted to this impasse by adopting a paradoxical and essentially irrational response. They reject the historical method entirely, or (as Bultmann) accept it only as a strategy to evoke faith in the Gospel. We cannot know anything, they admit, reliably and definitely about the historical personality of Jesus or his life. Once the Christ figure is demythologized, Christianity is viewed essentially as a form of faith in the *kerygma* (literally, proclamation). It involves belief in the essential message of Christ and in his incarnation and resurrection. This curious posture abandons all standards of objective critical intelligence. It has not refuted the results of skeptical inquiry, only ignored them and instead proclaims an absurdity as true. "I believe because it is absurd," affirms Tertullian, Kierkegaard, and later-day apologists for Christianity. There is a sublime contradiction here that enables some minds to willingly accept belief in the absurd, insisting that it is justifiable, even heroic. The reason for this, of course, is that the quest for the historical Jesus leads to the inference that faith in Christ as God incarnate has no rational foundation. The only hope for Christianity is to insist this is true in spite of all the evidence to the contrary.

Today it is important that we take a fresh look at the New Testament and the legends built around its central figure. We should not abandon the rich tradition of biblical criticism, even though some Christian scholars have given up the inquiry, maintaining that it is illegitimate because the true Jesus of history is unknowable. We should continue to pursue biblical criticism as an important field of inquiry and insist that the biblical record be examined in a rigorous manner, using the best methods of textual analysis, linguistic and philological criticism, the findings of the sciences, history, comparative anthropology, and archeology. New documents have been discovered in recent decades that shed further light on first- and second-century Palestine. The Dead Sea Scrolls and the Nag Hammadi codices present a clearer picture of the culture from and in which Christianity emerged. Future discoveries may add to our knowledge and demonstrate that Christianity was not unique.

My own point of view in examining the Bible comes from the empirical study of new religious sects as they have flourished on the contemporary scene. In rereading the Bible, I have been struck by the parallels with the paranormal and cultic belief-systems today. We can see the emergence of various Christian or quasi-Christian sects in the nineteenth century, such as the Mormons, Seventh-Day Adventists, Jehovah's Witnesses, and Christian Scientists. All were based on the Bible, but they added new materials drawn from the latter-day revelations of their founders. These sects have grown rapidly, yet their foundations are weak, even fraudulent. Similarly for the claims of psychics and faith healers, astrologers and UFOlogists, holding new paranormal faith systems. The analogies with classical religions are remarkable, and they tell us something about a pervasive trait of human nature: a willingness to accept the incredible without sufficient evidence. In addition, the fact that so many of the contemporary religious sects draw inspiration from the New Testament urges us to raise questions about its historical veracity.

What, in particular, shall we say about the historical Jesus? The traditional story is well known: he was born in Bethlehem of a virgin, he grew to manhood and began to claim that he was the Christ (a messiah) as prophesied in the Old Testament, and that he came to deliver mankind. He manifested miraculous powers, was able to cure the sick, raise the dead, exorcise demons, and perform wondrous acts that violated the laws of nature. He also proclaimed an ethic of obedience to God and love of all mankind, an ethic of humility, acquiescence, simple sincerity, and trust. He had many enemies, was seized and brought to trial before Pontius Pilate in Jerusalem, and was crucified. He rose from the dead in three days and appeared to his disciples. His resurrection has been taken as evidence for his divinity and also as a promise to the early Christians that "whoever shall believe in him shall have eternal life."

What is the evidence for these claims? It is largely an assemblage of twenty-seven documents known as the New Testament, including the four Gospels (Matthew, Mark, Luke, and John) and the letters of Paul of Tarsus and others. There are a number of questions to be raised. First, should these writings be taken as definitive eyewitness testimony for the existence of Jesus and/or proof of his divinity? Second, is there independent confirmation of his existence and divinity?

   Christianity as a religion rests on a historical case and any number of typical empirical claims: (1) that there was an individual named Jesus; (2) that he was born of a virgin; (3) that he performed miracles; (4) that he was crucified and then was resurrected; (5) that he was the son of God; (6) that he proclaimed the end of the world and the kingdom of God; (7) that those who believe in him will be rewarded with eternal salvation. It is possible for one to accept some of these propositions, for example, that Jesus existed, and to reject some or all of the others.

Did Jesus exist?

There is now a strong tradition in biblical scholarship—though today it is only a minority view—that the historical Jesus did not exist. How do we go about settling this question? We can ask the same question about any other famous name in antiquity. Did Socrates exist, Isis, Osiris, Julius Caesar, Alexander?

   How do we resolve the question of the existence of the historical Jesus? One can read the biblical scriptures themselves. The four Gospels, as we shall see however, can hardly be taken as reliable or objective historical documents for they were testaments written by committed missionaries to justify and extend the Christian faith. In order to evaluate their historical accuracy, it is important that we find independent corroboration. Thus we need to examine nonbiblical literature of the first few centuries C.E. I am referring here (1) to any parallels between Christianity and pagan or Jewish precursors; (2) to any secular references by pagan and Jewish authors; (3) to what the early critics of Christianity had to say; and (4) to the extensive apocryphal literature, gospels, and epistles that were subsequently rejected by the church.

   Is the story of Christ a myth comparable to the legends of Osiris, Attis, Krishna, Dionysus, Mithras, and other deities who attracted large followings in the ancient world? In reading the classical Graeco-Roman literature one is struck by the similarities between the story of Christ and other venerated figures. Such stories often involved a divine being who becomes incarnate as a man, enters into our world in order to perform miraculous or atoning deeds, and then returns to heaven. It was common to ascribe to the Roman emperor divine attributes. For example, writing in 30 B.C.E*, Horace composed an ode to Augustus hailing him as the incarnation of the god Mercury. Sons of God and resurrected deities were numerous in ancient times.

---

   *I shall use B.C.E. to designate "before the common era," and C.E. to refer to "of the common era," instead of B.C. (before Christ) and A.D. (*anno Domini*).

Apollonius, who lived in Asia Minor (Tyana) in the first century and died about 78 C.E., had a career strikingly similar to that ascribed to Jesus. As a young man, he was sent to the Hellenistic city of Tarsus to study. Apollonius eventually became a Pythagorean and adopted the strict ascetic rules of the brotherhood, renouncing meat, wine, and marriage. He distributed his inheritance among his relatives and lived as a penniless monk, wandering through the cities of the Mediterranean world. He taught that behind the world was a supreme unknowable deity. Apollonius' followers maintained that he was the son of God and credited him with many miracles. He was able to walk through closed doors, heal the sick, cast out demons, and was even said to have brought a dead girl back to life. He bade his disciples injure no living creature and to shun hatred, jealousy, and enmity. "We cannot hate our fellowmen," he preached, supporting a very simple brand of communism. Accused of witchcraft and sedition, he went to Rome to respond to charges before Domitian, was jailed but escaped. His followers claimed that Apollonius appeared to them after he died and that he ascended bodily into heaven.[1]

There is considerable evidence that Apollonius existed as a real person, as there are numerous references to him in ancient literature. Whether he had divine attributes, as his followers insisted, is another matter. Similar questions can be raised about Jesus: Did he exist? Are the legends built around him and transmitted through his followers accurate accounts of his life and his powers? Was he indeed a messiah uniquely sent by God to save Israel and suffering humanity?

The historical investigation of the origins of the Christian message has been enormously enriched by the discovery of the Dead Sea Scrolls at Qumran in Jordanian Palestine. Documents of parchment and papyrus first uncovered in 1947 have had a bearing on Christian origins. They were apparently written by a Jewish cult known as the Essenes. What is striking for the religious historian are the many similarities between this literature and the New Testament, between the Essenes and the early Christian communities. The Essenes practiced a form of ascetic exclusiveness and maintained a rigid discipline. This close-knit community expressed mutual love and respect, and it practiced primitive communism (like the early Christians), disagreed with established Judaism, participated in a ritual meal or eucharist banquet of some kind, paid special heed to biblical prophets, and baptized its initiates. Of special interest was their teaching concerning a future messiah, to be born of the lineage of David, who would bring into being a divine world order. This kingdom of God would emerge only after a period of wars and revolutions. The

1. See Philostratus, *The Life of Apollonius of Tyana,* 2 vols. Loeb Classical Library (Cambridge, Mass., Harvard U. Press, 1912).

messiah would lead the forces of light against the powers of darkness through an apocalyptic battle. When the kingdom of God was fully established, the messiah would help liberate Israel and would play an intermediary role between God and men. There are no doubt differences between the scrolls and the New Testament, the Christian savior and the Essene messiah, which Christian scholars are wont to point out. Yet the scrolls undermine the idea that the Christian message was unique. On the contrary, they demonstrate that Christianity, like all other cults, grew out of and manifested beliefs that were common in the period. Rather than being God's revelation in human history, human desire influenced the form and character that the revelation was to take.

The quest for the historical Jesus and the nature of his message are important. But it is notoriously difficult to ascertain the facts, because, outside of the Gospels, there is little reliable confirmation, either of his existence or the actual details of his life.

We will ask: When were the Gospels written? Were they contemporaneous with Jesus? Are they to be taken as reliable? Are the details contained in them supported by independent corroboration? Could Christianity have been developed by Paul, Peter, and others, without a historical figure?

Among the early references to Christ are two passages in the work of Josephus, the Jewish historian (who died about 100 C.E.). But the longer passage has fairly conclusively been shown, even by Christian theologians, to have been an interpolation, probably written by a Christian. It is a highly complimentary description, and it is unlikely that an orthodox Jew such as Josephus would have written it.[2] Moreover, it appears in the middle of a section dealing with other matters. Josephus' comment that "the tribe of Christians . . . are not extinct at this day," is rather odd if written during Josephus' life, but not odd if interpolated much later. Indeed, references to Josephus' quotation do not appear until the fourth century. The second passage consists of only a phrase referring to James, "the brother of Jesus, him called Christ." Whether this is authentic is also open to dispute.[3]

---

2. "Now there was about this time, Jesus, a wise man, if it be lawful to call him a man; for he was a doer of wonderful works, a teacher of such men as receive the truth with pleasure. He drew over to him both many of the Jews, and many of the Gentiles. He was (the) Christ. And when Pilate, at the suggestion of the principal men among us, had condemned him to the cross, those that loved him at the first did not forsake him; for he appeared to them alive again the third day; as the divine prophets had foretold these and ten thousand other wonderful things concerning him. And the tribe of Christians, so named from him, are not extinct at this day." (Flavius Josephus, *Antiquities,* Book XVIII, Chap. 3. Trans. William Whiston)

3. See Gordon Stein, *An Anthology of Atheism and Rationalism* (Buffalo, N.Y.: Prometheus Books, 1980), p. 179.

Another reference often cited is by the historian Tacitus, who in his *Annals,* written about 120 C.E., says: "He from whom the name (Christianus) was derived, Christus, was put to death by the procurator Pontius Pilatus in the reign of Tiberius." In this reference he does not use the name *Jesus.* Moreover, in another reference he is repeating what was commonly said in his time by the Christians. Tacitus mentions that "an immense multitude of Christiani" were present in Rome during Nero's time. This refers to Christians or those who believed in a messiah (of which there were numerous pretenders), not necessarly to the followers of Jesus, who were fairly small in number. In any case, Tacitus does not refer to Jesus, but only to his followers.

Another reference in non-Christian literature is in a letter of the younger Pliny (c. 110), asking Trajan for his advice on treatment of the Christians. Suetonious (c. 125) mentions similar persecution and says that Claudius banished "Jews who, stirred up by Christ, were causing public disturbances." But all of this demonstrates only the existence of Christians, which is not doubted, and not of the historical reality of Jesus.

What should we make of this? The nonbiblical evidence for Jesus is fairly uncertain. Gordon Stein estimates that some sixty historians and chroniclers lived in the first century in the Roman world. If Jesus had had such an effect on the multitudes as the New Testament writers proclaimed, then one would think that his impact would have been noted by more historians.[4] The passages cited are fragmentary at best. Should we deny that Jesus existed? Could Christianity, like other ancient mythological religions, have come into being without a historical figure behind it?[5]

Christianity bears some similarities to the ancient mystery religion of Egypt, which involved two key figures: Isis, goddess of motherhood and fertility, and Osiris, her brother and husband, god of the underworld and judge of the dead. According to myth, Osiris lived in a body, and he had suffered, been mutilated, and died; then his body was reconstituted and raised from the dead and into heaven by the god incarnate in him. Both Osiris and Isis appeared on earth as human beings; they were gods who became incarnate in mortal bodies, as was claimed about Jesus.

Mithraism also has many striking similarities to Christianity. The religion had deep roots in Persian and Hindu history and began to attract widespread attention in Rome at the end of the first century C.E. By the end of the second century, the cult had spread rapidly through various

4. Ibid., p. 178.
5. For a recent viewpoint that strongly maintains that Christianity could have developed without the actual existence of Jesus, see G. A. Wells, *Jesus of the Early Christians* (London: Pemberton Books, 1971); *Did Jesus Exist?* (London: Pemberton Books, 1975); and *The Historical Evidence for Jesus* (Buffalo, N.Y.: Prometheus Books, 1982). See also the Dutch writer Van den Bergh van Eysinga.

classes of the Roman world, including the mercantile class, the army, and the slaves. Mithraism thrived under the auspices of the emperors because of its support of the divine right to rule; well into the second and third centuries it competed with Christianity for allegiance, but Constantine destroyed it when he established Christianity as the official religion of the empire at the end of the fourth century, though it survived through the fifth century on a much reduced scale.

Extensive temple remains and statuary from the period attest to its great influence. The religion was built around the mythical Mithras, who was regarded as a mediator between humanity and the unknowable God who reigned in the universe. Mithras was allegedly born of rock, a marvel witnessed by certain shepherds, who brought gifts and adored him. Mithras was able to perform great miracles. For example, when a terrible drought was inflicted upon the earth, Mithras discharged an arrow against a rock miraculously bringing water from it. Analogies with early Christianity are unmistakable. The Mithraic communities, composed of people of humble origin, expressed a fraternal and communal spirit. Moreover, the Mithraic priests used bells, candles, and holy water in their communion services, where the worshippers were offered consecrated bread and water; they sanctified Sunday and the 25th of December. Their creed insisted on moral conduct, demanded abstinence and self-control, encouraged celibates and virgins to serve God, and postulated heaven and hell, a primitive revelation, a mediator of the Logos emanating from the divine, an atoning sacrifice, a constant struggle between the forces of good and evil (with the ultimate triumph of the former), immortality of the soul, the last judgment, the resurrection of the flesh, and the fiery destruction of the universe. After death, all men would appear before the judgment seat of Mithras; unclean souls would suffer eternal torment, while the pure would be received into the full radiance of heaven. Mithras was clearly a mythological figure, and eventually the religion gave way under Christian assault. Our knowledge of Mithraism provides another reason to raise the question of whether Christianity could have developed without Jesus.

Nonetheless, it is interesting that few of the bitterest pagan or Jewish opponents of Christianity denied Jesus' existence. These include Celsus, Lucian, and Porphyry. The latter questioned the historical accuracy of the New Testament, and ancient critics recognized the mythic elements in Christianity. Nevertheless, they did not dispute the historicity of Jesus. Although it may be difficult to assert with any degree of certainty that Jesus existed, it seems likely to me that some such man lived, most likely in Palestine in the first half of the first century, that he was crucified or hanged (though he may not have died), and that a sect of Christians developed proclaiming his divinity. (Though I will not venture the same claim about a historical Mithras). We know very few authentic facts, however, about Jesus beyond this bare outline.

What of the Christian case in his behalf? We need to go directly to the ancient Christian literature, but when we do, we encounter several problems. First, it is difficult to know what is the authentic text of the New Testament and whether it has been accurately transcribed. The oldest extant fragment we have (consisting of only a few verses from John) is dated no earlier than 150 C.E. Second, there are any number of testaments that have survived. These include the so-called apocryphal literature, containing the writings of Clement, Barabbas, Thomas, Nicodemus, and others, some of which are uncomplimentary to Jesus. At least thirty gospels were known to have existed in the first few centuries of the Christian era. It was only in the fourth century, following the bitter battle at the Council of Nicaea in 325, that a decision was made on what to include and exclude from canonical literature; only the four Gospels we now have were deemed "official." And it was not until 367, more than three centuries after the deaths of the original followers of Jesus, that the first official listing of the twenty-seven books of the New Testament was recorded in a letter from Athanasius, patriarch of Alexandria.

The problem for historical evidence then is that if several pieces of literature about a period survive, can we accept some and reject others? For example, Socrates was surely a historical figure; Plato wrote many dialogues in which he was idolized. Xenophon and Aristotle also wrote about him, as well as the dramatist Aristophanes, who portrays him rather unsympathetically. But we don't refuse to read Aristophanes. The apocryphal literature deals with Jesus as a young boy, who is able to work miracles and has amazing powers. But in one instance, at least, he kills a young playmate.[6] What is fact and what is fiction? How do we decide?

In any case, it seems clear that none of the New Testament Gospels was written by eyewitnesses who knew Jesus. Indeed, the true authorship of Mark, Matthew, Luke, and John is uncertain. For probably many decades after Jesus lived, the Christian testimony about him was part of an oral tradition. It is most likely that stories about his words and deeds were circulated by his followers, including alleged eyewitnesses. These

6. In the gospel about Jesus' infancy, we read: 22. "Another time, when the Lord Jesus was coming home in the evening with Joseph, he met a boy, who ran so hard against him, that he threw him down; 23. To whom the Lord Jesus said, As thou hast thrown me down, shalt thou fall, nor even rise. 24. And at that moment the boy fell down and died." The First Gospel of the Infancy of Jesus Christ, *The Apocryphal New Testament* (London: Wm. Hone, 1820), p. 57.

Also we read in Thomas's Gospel of the Infancy of Jesus Christ, chap. 2: 7. "Another time Jesus went forth into the street, and a boy running by, rushed upon his shoulder; 8. At which Jesus being angry, said to him, thou shalt go no further. 9. And he instantly fell down dead." Ibid., p. 61.

accounts were preached in the earliest Christian communities in Jerusalem, Ephesus, Antioch, and Caesarea. These separated communities developed different traditions. The oral witness was reinterpreted as the people
who allegedly knew or had seen Jesus began to die off. Since the average
life span was fairly short at that time, this happened quickly. Fearing that
the record would be lost, efforts were made to preserve accounts of Jesus'
life and teachings. It is generally accepted that the earliest written account
was the Gospel of Mark, which was probably written between 65 and 70.
According to Eusebius, author of a church history in the fourth century,
Mark did not know Jesus, but was a disciple of Peter in Rome. [7] According to legend, Mark was induced by the Christians to write his memoirs,
based on Peter's oral recollections, at best a second-hand account. Many
scholars now doubt whether Mark actually received his material from
Peter; what is thought likely is that he took it from a general oral tradition
then existing in his Christian community. Indeed, who Mark was is unknown. Thus the author of the Gospel according to Mark remains anonymous. Oral traditions are notoriously inaccurate. Scholars believe that
there did exist another document, one called *Q* (from the German *Quelle,*
meaning "source"), now lost, which may have contained some 250 sayings
of Jesus. If so, this was probably written about 50 C.E.. The Gospel of
Matthew, probably written between 80 and 85 in Antioch, incorporated
most of Mark's account and repeated at least 610 out of 661 passages, but
greatly expanded the content and also drew from the *Q* document. Matthew added a story of Jesus' birth and sought to find prophecies in the
Old Testament that he thought Jesus fulfilled.

It was only in 85-90 that Luke, a Gentile, wrote his Gospel repro-
ducing a good portion of Mark and adding other material. Luke is also
the author of Acts, which is an account of the early Christian church.
Luke is a third-generation Christian, who tells us in his Gospel: "Many
writers have undertaken to draw up an account of the events that have
happened among us, following the traditions handed down to us by the
original eyewitnesses and servants of the gospel" (1:1-2). The interesting
point is that neither Matthew nor Luke knew Jesus. The first three Gospels are called the synoptics. It is widely accepted that Luke and Matthew
knew the Gospel of Mark and drew upon it, which makes them twice
removed from direct testimony.

7. "When Mark became Peter's interpreter, he wrote down accurately, although not in
order, all that he remembered of what was said or done by the Lord. For he had not heard
the Lord nor followed Him, but later, as I have said, he did Peter, who made his teaching
fit his needs without, as it were, making any arrangement of the Lord's oracles, so that
Mark made no mistake in thus writing some things down as he remembered them. For to
one thing he gave careful attention, to omit nothing of what he heard and to falsify nothing
in this." Eusebius, *Ecclesiastical History,* Book Three, chap. 39. He is quoting Papias, an
early churchman (about 140).

The fourth Gospel was apparently the last one written, and is different from the others. Its authorship is unknown. There have been all sorts of speculation about who the author of the fourth Gospel was. The author himself indicates that he was "the beloved disciple." But who is that? Was it John, an early disciple, Lazarus, whom Jesus allegedly raised from the dead, or someone else? There are elements in the Gospel of John not contained in the other Gospels, and the early church fathers acknowledged the differences. Clement of Alexandria notes, in about 200, that, although the other evangelists preserved the "facts of history," John wrote a "spiritual Gospel." The earlier tradition of the church considered John to have been one of the early disciples, who escaped the fate of his brother James, which was execution. However, most scholars now believe that this was not the case. Some have suggested that the author lived about 100 in Ephesus. But there is no consensus, and most commentators take the text as anonymous. The author was evidently familiar with the Gospel of Mark, and he may have consulted Luke. Even so, the Gospel of John leaves out a large portion of the synoptic story and muddles the sequence of events in many important respects. For example, its account of the Last Supper differs from that related in the synoptics. Moreover, the author reflects the influence of Greek philosophy, especially in his use of the divine Logos in the prologue to the narrative. It seems apparent that John was written later than the other Gospels, and was not an eyewitness account. The author would have had to live a very long time to have known Jesus, and the average life span of people in the Middle East then was fairly short.

Thus far we have not referred to the epistles of Paul, which played an important role in the building of the church. Indeed, many commentators have attributed to Paul the most important role in creating Christianity as a religion. The letters of Paul, which appear in the New Testament, most likely were written *before* the Gospels. Paul clearly did not know Jesus, though he claims to have seen and heard him in a vision on the road to Damascus. Paul, originally named Saul, was a devout Jew, a Pharisee, in Tarsus, capital of Cilicia, a Roman province in Asia Minor. A Roman citizen, Saul had persecuted Christians but was converted to their beliefs after he had his vision of Jesus.

After his conversion, Paul set himself to spread the Gospel of the dead and risen Christ to the Gentiles. Almost half of Paul's letters are written to various churches in Europe and Asia. What is interesting is that many of the stories that appear in the four Gospels do not appear in the letters of Paul, which were most likely written between 49 and 62. Paul undertook three missionary tours through parts of the Roman empire, attempting to build the church, and according to legend he was

eventually imprisoned and beheaded in Rome, about 64. Some accounts
of his travels appear in Acts.

Who was the historical Jesus?

The questions I would like to raise are: Was the account of Jesus' life in
the New Testament (assuming that he existed) accurate? Should the Gospels and Epistles be taken as evidence?

Now we have three problems: First, there are serious contradictions
in the various accounts. Second, the writers were not eyewitnesses to the
events but propagandists who were interested in attracting followers to the
faith. Third, even if their stories and second- and third-hand accounts
came from eyewitnesses, who transmitted these accounts by oral tradition
and then a written account (in the *Q* document), still these eyewitness accounts must be held suspect. My general objections to the latter are based
on Hume's argument against miracles and on the general unreliability of
eyewitness accounts from primitive, uneducated, or illiterate people.

How accurate are the accounts of Jesus' life recorded in the New
Testament? Not very, I submit, because of the conflicting, unsubstantiated
stories presented in the Gospels.

The Jesus of the Gospels appears as a powerful charismatic figure,
but most of his personal history is shadowy, and most of his life is not
even accounted for. The Gospels give us legend and myth rather than
historical fact. The New Testament is full of fictions introduced in order
to justify his messiahship and divinity. Virtually every fact stated in the
New Testament is open to historical question. For example, Christians
today celebrate December 25 as his birthday. But December 25 was also
taken in ancient times as the birthday of *Sol Invictus* (Unconquered Sun)
and, as I noted, was the chief celebration of the Mithraic cult. The last
week of December was a pagan holiday. It also approximated the Jewish
celebration of Chanukah, which occurs near the end of December. Scholars speculate that Jesus was probably born four years earlier, because of
the dates of Herod's reign.

There is no certainty about where Jesus was born. Was he born in
Bethlehem? We cannot corroborate this place by independent historical
sources. What do the Gospels have to say? Here we also encounter difficulties. Luke and Matthew say that he was born in Bethlehem, while
Mark and John imply that his origin was Nazareth in Galilee; Paul does
not mention his birthplace. Yet this was supposed to have been a
remarkable event. John indicates that there was some dispute in his day
about whether Jesus was the messiah, because the messiah was supposed

to originate from the seed of David in Bethlehem, but he says that Jesus came instead from Galilee (John 7:41-43; also 7:52). In order for Jesus to be the messiah of the lineage of David, he would have to have been born in Bethlehem. Was this birthplace invented by Luke and Matthew to justify him as the messiah to the Jews? If so, even Matthew and Luke disagree about the circumstances surrounding his birth. It is implied in Matthew that Joseph and Mary lived in Bethlehem when Jesus was born, stayed there about two years, and then fled to Egypt. They returned to Palestine after the death of Herod, but moved to Nazareth. Luke, on the contrary, maintains that Joseph and Mary lived in Nazareth, but traveled to Bethlehem while Mary was pregnant to register for a tax census. (There is no historical confirmation that the Romans had a census that required people to return to their original domiciles, a massive undertaking.) They left Bethlehem some forty days later and returned to Nazareth.

It is clear that Jesus *had* to be born in Bethlehem, if he was descended from King David. But was this so? Both Matthew and Luke attempt to prove genealogical origins. Matthew traces Jesus to Abraham, while Luke provides the generations all the way back to Adam. According to Luke, there were forty-one generations between David and Jesus; according to Matthew, only twenty-eight (actually, he lists only twenty-seven). The names in these genealogies differ. For example, Luke (3:23-26) had Jesus as the son of Joseph, who was the son of Heli, whereas Matthew (1:15-16) has the father of Joseph as Jacob.

> Luke: "And Jesus himself began to be about thirty years of age, being (as was supposed) the son of Joseph, which was the son of Heli." (3:23-24—KJV*)

> Matthew: "And Jacob begat Joseph, the husband of Mary, of whom was born Jesus, who is called Christ." (1:16—KJV).

But if we trace Jesus' ancestry further, we get discordant genealogies all the way. Let us look at only seven generations.

Luke	*Matthew*
Joseph	Joseph
Heli	Jacob
Matthat	Matthan
Levi	Eleazer
Melchi	Eliud
Janna	Achim
Joseph	Sadoc

*I will be using both the King James version (KJV) and the *New English Bible* (NEB) translation (Cambridge, 1961-70).

The divergence of these genealogies is surprising, as if they were made up by the first Jewish Christian communities independently of each other. Some Christian apologists have suggested that one of the genealogies is that of Mary. But the text clearly denies this, and if Mary is kinswoman to Elizabeth (Luke 1:36), then her background is not Davidic but Levitical, of the house of Judah. One of the books of the apocryphal New Testament, which was not declared canonical by the Council of Nicaea in 325, the Gospel of the Birth of Mary, says that Mary was not of Davidic descent. It identifies her father as Joachim, which does not appear in either Luke's or Matthew's list of Jesus' genealogy.

It is very difficult, even today of course, for anyone to trace his or her genealogy back for many generations, but Luke and Matthew attempt to do it, and they differ even about the grandfather and great-grandfather of Jesus, which again suggests that these passages are pure fiction. In any case, both cannot be accurate.

A further discrepancy is that while Matthew and Luke maintain that Jesus was born of a virgin, Mark, John, and Paul do not mention this remarkable fact at all. Again, this miraculous event is referred to in order to demonstrate that Jesus was the Jewish messiah. An unresolved textual problem concerns the virgin birth. According to Matthew 1:18, Mary was "found with child of the Holy Ghost." (KJV) Joseph has a dream which reiterates the virgin conception of Jesus. Luke 1:35 also has an angel appear to tell her: "The Holy Ghost shall come upon thee, and the power of the Highest shall overshadow thee: therefore also that holy thing which shall be born of thee shall be called the Son of God." (KJV)

If this is the case, then in what sense was Joseph the father of Jesus? And if Jesus was *not* the son of Joseph, then he could not be descended from David! The Gospels are impaled on an irreconcilable contradiction.

Many scholars have pointed out that the authors of the Gospels were endeavoring to prove that Jesus was the messiah as prophesied by Isaiah in the Old Testament. Now I think that the whole notion of prophecy is questionable. Whether precognitive prophecies are veridical or have ever been demonstrated is open to doubt. Often what is claimed to be a psychic prediction is either an inference based upon empirical or rational expectations (which may turn out to be right) or sheer guesswork! More often, prophecies are so general that they can be made to match any number of future events. I think that this is what happened in regard to Isaiah's prophecies—yet so much of the faith in Jesus Christ is predicated on them. The passage in question is as follows: "Therefore the Lord himself shall give you a sign; Behold, a virgin shall conceive, and bear a son, and shall call his name Immanuel." (Isa. 7:14—KJV)

In the first place, the meaning of the term *virgin* is open to dispute. The Hebrew word used by Isaiah is *almah,* which means "young, unmarried female," not *bethulah,* which literally means "virgin." Indeed, the New English Bible translation has helped clear this up: "Therefore the Lord himself shall give you a sign: A young woman is with child, and she will bear a son, and will call him Immanuel."

All of which means that Luke (who was most likely Greek, not Hebrew) has confused the terms *virgin* and *young woman* in the Septuagint, the pre-Christian, Greek translation of the Hebrew Old Testament. Jesus had to be "born of a virgin" in order to fulfill the prophecy of Isaiah, as Luke read it, after the fact. It is highly unlikely, however, that Isaiah was prophesizing something scheduled to happen seven centuries later. The passages in Isaiah probably concern the Syro-Israelite conflict of 734 B.C., with the prophet seeking to assure King Ahaz that the invasion of Judah would not succeed. But his prophetic words were stretched by Christian propagandists to prove the messiahship of Jesus by showing its basis in ancient scripture.

Matthew, no doubt, was a Hebrew and thus was in a better position to know the meaning of the term *almah,* "young, unmarried woman," although he probably had also read the Septuagint translation. But serious questions were raised about Mary's morality (as we shall see shortly), particularly the fact that she was pregnant before she married Joseph. Was the virgin birth an invention introduced, in part, to overcome charges surrounding Jesus' illegitimacy? Being born out of wedlock was considered a heinous sin by the Jews. Hence, Mary's betrothal or marriage to Joseph (which it is, is unclear) and her impregnation by the Holy Spirit is supposed to overcome any suspicions about Jesus' illegitmacy. Indeed, the Jews of the day, according to John 8:41, taunted and ridiculed Jesus because of questions concerning his real father and the fact that he was "born of fornication" (KJV). We are not "baseborn," they say (NEB), perhaps implying that Jesus was considered illegitimate.

Aside from all these caveats, the most important issue is the evidence put forth for the claim that Jesus was born of a virgin and hence was the son of God. Neither Paul, whose epistles preceded the writing of the Gospels, nor Mark mention the virgin birth. Jesus' conception surely was perceived as a miraculous event by Matthew and Luke. Why the silence by the others? Was it because it was invented by the subsequent tradition and so accepted by Luke and Matthew? There was no one to testify that Mary conceived in a supernatural manner; nor was anyone in the room with her when the Holy Ghost appeared. Hence, there is no corroboration for the claim. According to Matthew:

> Now the birth of Jesus Christ was on this wise: When as his mother Mary
> was espoused to Joseph, before they came together, she was found with
> child of the Holy Ghost. (1:18—KJV)

Joseph was disturbed by this, but as he was not willing to make her a
public example, he thought to put her away privately. However, he had a
dream in which the angel of the Lord visited him and told him that
Mary's pregnancy was the work of the Holy Ghost:

> But while he thought on these things, behold, the angel of the Lord ap-
> peared unto him in a dream, saying, Joseph, thou son of David, fear not to
> take unto thee Mary thy wife: for that which is conceived in her is of the
> Holy Ghost. (Matt. 1:20—KJV)

Whether this dream actually did occur, we, of course, have no way of
knowing. Dreams are notoriously untrustworthy, and to rest a religion on
a dream is dubious, to say the least. According to Matthew, Joseph did
what the angel told him: he took Mary home to be his wife but had no
intercourse with her until her son was born.

Concerning Mary's case, all that we have is her testimony that she
was a virgin and that the Holy Ghost visited her. According to Luke, an
angel greeted her and told her:

> Fear not, Mary, for thou hast found favour with God. And, behold, thou
> shalt conceive in thy womb and bring forth a son, and shalt call his name
> Jesus. (1:30-31—KJV)

> The Holy Ghost shall come upon thee; and the power of the Highest shall
> overshadow thee. (1:35—KJV)

There are no other witnesses to this event. Luke himself was not a witness.
Indeed, writing around C.E. 90, he makes it clear that he is only writing a
narrative following the tradition handed down to him by the original
eyewitnesses and servants of the Gospel (Luke 1:1-4). To claim that Luke
was "inspired" by God is to beg the question, for we need to know on
what basis we can validate that he was. To say that it is a matter of faith
does not counter intelligent skepticism about the grounds for the authenticity of the Gospel.

In Luke's narrative there is an account of Mary's relationship to
Zechariah, a priest, and his wife, Elizabeth, a kinswoman of Mary, who
were childless. According to Luke's legendary embroidery, the angel Gabriel also appeared to Zechariah and informed him that his wife would
bear a son, apparently John the Baptist, which she did. Mary lived with

Zechariah and Elizabeth for about three months. Mary did not remain a virgin throughout her life, since Jesus had several brothers and sisters and nowhere is it stated that they were also conceived by divine interception.

If we think of the context of the next century, when Christianity was developing as a cult, we find an entirely different story about Jesus' parentage and birth. The early Christians claimed that Jesus was the son of God. Many similar legends of divine parentage were familiar to the Roman world. It allegedly applied to the emperors, who were called sons of God, and earlier a Greek myth had it that Danaë, mother of Perseus, was impregnated by Zeus, appearing as a shower of gold. Why not Mary for the poor and uneducated Christians?

The critics of Jesus

How did the educated citizens of Rome and the non-Christian Jews view the claims for Jesus' divine parentage? Many apparently found the story incredible, and they provided other versions of his background. It is always a source of amazement that people are willing to believe the most fanciful stories; the more fanciful the exaggeration, the more likely are the convictions to grow in irrationality and intensity. The early Christian claims about Christ were not without their critics and detractors. Indeed, Christianity had to wage a continual battle of apologetics. Many of the early fathers of the church were forever on the defensive. This suggests that even then refutations by critical intelligence were employed against the emerging mystery religion. The philosophers and intellectuals, as well as the dominant political and social forces of the day, sought to refute Christianity, and some used violent means to suppress it. Yet, in the end, Christianity succeeded in overwhelming Greco-Roman culture, one of the great tragedies of human history. Why that happened has mystified thinkers ever since. Why Christianity persists today after two millennia is equally mystifying. The answers do not point to its being true—far from that—but rather to the frailty of critical human intelligence in civilization at large and the willingness of large sections of the human species to devour myths, however weak or fraudulent they may be.

Among the many early critics of Christianity, perhaps the best known today are Celsus, a pagan philosopher who lived in the later part of the second century, and Porphyry, in the third century. There were undoubtedly many more critics; but virtually all the works critical of Christian myth have been censored, destroyed, or lost during the subsequent era in which Christianity dominated Europe and western Asia. Virtually all of Porphyry's books against the Christians, save a few scattered fragments, have been lost. We are fortunate, however, that tradition has preserved some of Celsus' views. Ironically, we owe a debt to Origen, an early father of the church who quoted extensively from Celsus' treatise *A True Discourse* (written about 161-180) or paraphrased his views in order to refute him. This we find in the lengthy work *Contra Celsus (Against Celsus)*, written in the first part of the third century.[8]

According to Origen, Celsus maintained that certain Christian believers were "like persons who in a fit of drunkenness lay violent hands upon themselves, have corrupted the Gospels from its original integrity . . . to a many-fold degree, and have remodelled it, so that they might be able to answer objections." This supports the contention that Christian propagandists wrote the Gospels to meet objections to their religion and not as a record of historical fact.

Celsus is a skeptic who wishes to use reason. According to Origen, he recommends "that in adopting opinions we should follow reason and a rational guide, since he who asserts to opinion without following this course is very liable to be deceived." And he "compares inconsiderate believers . . . to soothsayers, . . . or any other demon or demons." "Amongst such persons are frequently to be found wicked men, who, taking advantage of the ignorance of those who are easily deceived, lead them whither they will, so also . . . is the case with Christians." Certain persons "who do not wish either to give or receive a reason for their belief, keep repeating 'Do not examine, but believe!' and, 'Your faith will save you!'"[9] No doubt Celsus found Christianity a threat to pagan civilization, which he wished to defend.

Origen objects to Celsus' approach and that of other philosophers. He recommends the importance of faith in Jesus Christ, particularly for the vast bulk of mankind who, says Origen, are unable to follow the niceties of philosophical logic.

Celsus goes on to attack the idea of a virgin birth, which he considers ludicrous. He offers an alternate account of Mary's pregnancy:

> When she was pregnant she was turned out of doors by the carpenter to whom she had been betrothed, as having been guilty of adultery, and that she bore a child to a certain soldier named Panthera.[10]

He follows with an account of Jesus' birth:

8. *The Writings of Origen,* vol. 1, trans. Rev. Frederick Crombie (Edinburgh: T. and T. Clark, 1879), pp. 33 ff.
   9. Ibid., p. 405.
   10. Ibid., p. 431.

Born in a certain Jewish village of a poor woman of the country, who gained her subsistence by spinning, and who was turned out of doors by her husband, a carpenter by trade, because she was convicted of adultery; that after being away from her husband, and wandering about for a time, she disgracefully gave birth to Jesus, an illegitimate child.[11]

This view was apparently also maintained by Jews of the time. The main body of Judaism rejected Jesus' claim that he was the messiah. In the centuries following the death of Jesus there was considerable proselytizing by Christians and there was a Jewish response, parts of which are expressed in the Talmudic writings. The Talmud stories maintain, along with Celsus, that Jesus was born without a legitimate father, as the product of an illicit relationship between Mary and a Roman soldier named Pandera, Panthera, or Pantera. Still another scenario inferred from the apocryphal books of the New Testament suggests that Mary may have been impregnated by one of the high priests of the Temple (perhaps Zechariah), who then selects Joseph as her husband as a cover-up.

It is difficult today to ascertain the truth. My main point here is that Mark is silent about Jesus' birth and the other Gospels provide contradictory statements, so that the notions that Jesus was born of a virgin in Bethlehem, is of Davidic descent, and is the son of God are not corroborated by empirical data. Indeed, alternative accounts appeared in the century following Jesus' death; these would suggest plausible, prosaic explanations of his life and career. There is no way to corroborate other parts of the legend: that the infant was born in a manger, under a special astronomical event over Bethlehem, or was visited by three wise men or Babylonian astrologers, or that Herod massacred innocent male infants because of fear of the birth of Christ. All efforts to verify these claims have failed, which points to the mythic aspects of the Christ legend.

The ministry of Jesus

Aside from these questions of his origin, one may ask: What about Jesus' ministry? What did he achieve in his own lifetime? What was his distinctive message? What about the presumptuous claim that his word had divine authority?

In regard to these issues, biblical sources are contradictory, and the meaning of the life of Jesus is open to a variety of explanations. Actually we can say very little with certainty about the life, personality, or sayings

11. Ibid., p. 427.

of Jesus. Many different interpretations of Jesus' life can and have been given. It seems possible that he was a first-century Galilean Jew, who practiced Judaism and did not depart from the Torah. Accordingly, his role must be seen in the context of Jewish society and the times in which he lived. Israel was then occupied by Roman legions, and Jesus—or at least his first Jewish followers—hoped that the Jews might be delivered from their yoke. Much of this interpretation of Jesus is no doubt colored by the fact that many of those who believed in his divine ministry in the next generation and wrote the early Gospels had witnessed the Jewish revolt against Rome, the destruction of the Temple of Jerusalem in 66-70 (approximately thirty-five years after the death of Jesus), and the great slaughter and dispersion of the Jews thereafter. Thus Jesus, as seen retrospectively through the eyes of the Gospel writers, was a messiah, "the King of the Jews" who came to deliver Israel from bondage as prophesied in the Old Testament. Both the Gospel of Matthew, which is appealing to the Jews, and Luke, appealing to the Gentiles, are aware of these latter-day events.

Every person who lives and breathes is influenced by and expresses the needs and aspirations of the time and milieu in which he exists. What was distinctive about Jesus was, first, his moral-political message, preached and practiced with fervor and devotion—this was, in part, an ethics of love and humility. Second, there was the miraculous and paranormal basis of his messianic ministry, which attempted to clothe his words and deeds with special divine authority. This was largely due to the transformation by Paul of the message into one of salvation through faith in Jesus Christ as the savior of mankind.

The ethical teachings of Jesus

Jesus' moral code as it appears in the Gospels was somewhat universalistic in scope and emphasis. In many respects, however, it was not unique but was anticipated by the Old Testament and the Essenes. It was also characteristic of much first-century rabbinical teaching, especially its praise of the pure of heart, the merciful, and the peacemakers. Efforts to build a universal morality are found in pagan sources, such as the writings of the philosophers, especially Plato (of which Jesus was probably unaware), and in the ethical focus of Stoicism, which emphasized the brotherhood of men and was widely held by the educated classes of the Roman Empire. Nevertheless, there is something powerful, dramatic, and poignant in Jesus' Sermon on the Mount and his many parables designed to illustrate a morality that would speak to all of mankind. It apparently had an overwhelming effect upon many of those who later received his message. According to the Gospels, Jesus preached that it is not the letter of the law but the spirit that counts. He was concerned not for the rich and powerful of society but for the poor, weak, downtrodden, and helpless, even sinners and fallen women. Instead of the familiar ethic of retribution, "an eye for an eye, a tooth for a tooth," Jesus preaches genuine sympathy and empathy, forgiveness and love.

The morality of Jesus is very simple in its eloquence. It is virtually impossible, however, to follow it consistently in practice. There are passages which, when quoted out of context, seem extremely condemnatory and harsh, as when he says:

> Ye have heard that it was said by them of old time, Thou shalt not commit adultery. But I say unto you, That whosoever looketh on a woman to lust after her hath committed adultery with her already in his heart. (Matt. 5:27-28—KJV)

That surely seems like an extreme dictum to abide by, for it would make sexual desire evil. Yet sexual passions are deep-seated in human nature. Jesus' admonitions furnished a basis for repression and consequent human suffering. Much that followed in the ascetic tradition of Christianity can be traced to passages such as these. This, unfortunately, led to warfare against the body, which was considered by Paul and Augustine to be sinful and corrupt. Similarly for Jesus' statement:

> Whosoever shall put away his wife, saving for the cause of fornication, causeth her to commit adultery: and whosoever shall marry her that is divorced committeth adultery. (Matt. 5:32—KJV)

This, no doubt, applies to a patriarchal society in which the place of women is demeaned. Surely there is nothing wrong with marrying a woman who is divorced (the Roman Catholic church notwithstanding); and to prohibit remarriage is harsh and legalistic. Or take the key statement attributed to Jesus:

> I say unto you, That ye resist not evil: but whosoever shall smite thee on thy right cheek, turn to him the other also. And if any man will sue thee at the law, and take away thy coat, let him have thy cloak also. (Matt. 5:39-40—KJV)

In regard to the first, if we were not to resist tyranny and evil, protecting ourselves from aggression or an infamous crime, we would ourselves be evil. How are we to cope with Attila, Cesare Borgia, Hitler, or Stalin? Surely not by becoming defenseless pacifists? In response to Jesus' second recommendation, if one is unjustly tried in a court of law, then one should defend oneself as best one can through the legal system and not collapse by giving away a cloak if someone has already taken our coat. Thus, the literal interpretation of many of his prescriptions can lead to more harm than good.

I readily grant, however, that Jesus' overall dictum, "Love your enemies, bless them that curse you, do good to them that hate you" (Matt. 5:44—KJV), has a ring of moral excellence and nobility to it. It provides some guide for us in much of life, where we should seek to emphasize tolerance, cooperation, and charity. "Always treat others as you would like them to treat you," said Jesus (Matt. 7:12—NEB). This general principle has much to commend it, and it has shone through in every generation. This moral insight and hope of Jesus thus speaks to every man. Universal love and beneficence, however, is only one principle among others. There are several moral principles and values that we may equally cherish, not the least of which is the need to use reason and critical intelligence as a guide to life. Still others commend themselves especially today: principles of human rights, freedom, equality, and justice. It is the *application* of such general principles and values to the concrete situations of life that raises moral problems. We cannot simply apply an absolute set of commandments as religious believers would have us do; it is how we interpret these and what to do when they conflict that pose the dilemma. In moral inquiry, we need to be sensitive to the complexities of moral dilemmas, and the simplified ethic of Jesus is never sufficient to aid us in making wise choices. This is all too apparent to anyone concerned with philosophical ethics, as distinct from religious morality.

That Jesus himself followed his own principles is questionable. There seems to have been a double standard—one that applied to his followers, another one that applied to himself. If there was a historical Jesus, was he a follower of the principles ascribed to him by the more zealous advocates of the moral faith of those who called themselves Christians? One can make the case in reading the Gospels that Jesus himself was not an absolutist; he was hardly an ascetic. He ate and drank with tax-gatherers and sinners and did not condemn the pleasures of life. (Perhaps this only describes the social composition of the movement.) Indeed, the Gospels report that he and his disciples were rebuked for not fasting like others and that he was called a "glutton" and "drinker." It is the spirit of the law, not its legalistic letter, that we should follow.

Surprisingly, there are many cases in which Jesus abandons his ethic of love and curses those with whom he disagrees. He rebukes those who do not follow his ways: "I never knew you; out of my sight, you and your wicked ways!" (Matt. 7:23—NEB). He attacks the scribes and Pharisees, saying "Woe unto you, scribes and Pharisees" (Matt. 23:14 ff.—KJV), calling them hypocrites. All too human, finding it difficult to maintain forbearance, he lashes out against those he deplores, saying: "Ye serpents, ye generation of vipers, how can ye escape the damnation of hell?" (Matt. 23:33—KJV). These are hardly the words of one unreservedly given to the ethic of love. In Luke, he rebukes not only Pharisees but scribes and lawyers, exclaiming "Woe unto you also, ye lawyers!" (Luke 11:46–52—KJV). When he is accused of being impure, he retorts: "I tell you this: no sin, no slander, is beyond forgiveness for men; but whoever slanders the Holy Spirit [including Jesus] can never be forgiven; he is guilty of eternal sin." (Mark 3:28–30—NEB). The God of love is consumed by wrath when it suits his purpose.

The Jesus of the Gospels was far from being humble; in fact, he was arrogant at times. The Gospels report that he spoke with great authority and confidence until his trial and crucifixion, when he was facing imminent defeat and death. Otherwise, he demands that others follow him, even give up their parents, wives, and children in allegiance to him. He insults his mother and his relatives. After he is accused of being mad, his mother and brothers stand outside a house with a crowd and he is told: "Your mother and brothers are outside asking for you." And he replies, "Who is my mother? Who are my brothers?" Looking at those sitting around him in a circle, he said, "Here are my mother and my brothers. Whoever does the will of God is my brother, my sister, my mother." (Mark 3:32-35—NEB). On another occasion, he addresses Mary in an insulting way, "Woman, what have I to do with thee?" (John 2:4—KJV).

Worst of all, the Jesus of the Gospels had a superiority complex. Did he really believe that he had a divine mission and that all others should submit to him or was it simply an act—or a combination of both? "You must not think that I have come to bring peace to the earth," says Jesus. "I have not come to bring peace, but a sword. I have come to set a man against his father, a daughter against her mother, a son's wife against her mother-in-law; and a man will find his enemies under his own roof." (Matt. 10:34-36—NEB). "No man is worthy of me who cares more for his father or mother than for me . . . [or his] son or daughter," says Jesus. (Matt. 10:37-38—NEB). If we assume that Jesus were a mere mortal, how should we interpret these statements? At one point, he says to Peter, "Away with you, Satan. You are a stumbling block to me. You think as men think, not as God thinks." Surely, this is extreme self-glorification. When asked what a person should do to get into heaven, Jesus replied that he should give up everything, wealth and family, and follow him. He even promised his twelve disciples great power. They would sit with him at Judgment Day to judge the critics of Israel. When we read the Gospels carefully, Jesus appears as an extreme egotist. A woman comes to him and pours a small bottle of fragrant oil over his head. The disciples were indignant at the waste and suggested that it should have been sold and the money given to the poor. Jesus differs with them and applauds the woman. "It is a fine thing she has done for me." (Matt. 26:10—NEB). He expresses indignation, threatening those who disobey his words with eternal punishment. He threatens Judas, who is going to betray him, saying: "Alas for that man. . . . It would better for that man if he had never been born." (Matt. 26:25—NEB).

Was Jesus disturbed?

All of this suggests that Jesus was a disturbed personality. It is difficult to be certain, since we have no way of submitting him to intensive psychiatric diagnosis. He usually spoke in parables to both his disciples and the populace, often uttering vague statements, not unlike schizophrenic personalities who are out of touch with reality. He seemed to be especially worried about his cloudy parentage. He called himself the "Son of Man" and said his "Father was God," which suggests a sensitivity exacerbated by possible gossip about his mother's transgression and his own illegitimate birth. Had he not suffered this psychological confusion or psychotic disorder, posterity might never have mistaken him for the son of God, and Christianity would never have been born. Did Jesus really *believe himself* to be the messiah or the son of God? If he ever had such pretentions of divinity (his disciples thought he did), then he was deranged. He kept preaching that doomsday or the last days were at hand and that only *he* provided the truth and the light and could save Israel and humanity from impending destruction. "I tell you this: there are some of those standing here who will not taste death before they have seen the Son of Man coming in his kingdom." (Matt. 16:28—NEB). "I tell you this: the present generation will live to see it all." (Matt. 24:34-35—NEB).

Jesus' disciples took all this seriously and indeed expected him to return imminently and to bring the promise of salvation. Much of this was, no doubt, rooted in the false hopes and the fears of his followers, but perhaps he helped to feed their expectations by his psychotic delusions of grandeur, whether as the messiah, the King of the Jews, or the son of God. Several passages in the Gospels tell us that many people in Israel, including his relatives and neighbors, thought that he was mad, "possessed" of a demon. In John, we read: "And many of them said, He hath a devil, and is mad; why hear ye him?" (John 10:20—KJV). "He is possessed, he is raving" (NEB). (See also John 8:52). In Mark, we read:

"And when his friends heard of it, they went out to lay hold on him: for they said, He is beside himself." (Mark 3:21—KJV). The scribes who had come down from Jerusalem said, "He hath Beelzebub,* and by the prince of the devils casteth he out devils." (Mark 3:22—KJV). They accused him, saying, "He hath an unclean spirit." In the New English Bible, we read the following interesting account of his visit to his hometown, Nazareth. Jesus and his disciples had entered a house and apparently had such a crowd collected around them that they had no chance to eat: "When his family heard of this, they set out to take charge of him, for people were saying that he was out of his mind." (Mark 3:20-21—NEB).

Surely, if anyone were to come forth today claiming to be a messiah and preaching the message of Jesus, he would be considered crazy. There are thousands of similar bedraggled souls who have ended up in mental asylums. Jesus was so exasperating to the established authorities that they crucified him, a much worse fate. But was one of the reasons that he was psychotic? I do not mean to be unfair to Jesus, and no doubt devout Christians will be offended at the suggestion.

One might also ask the hypothetical question: Was Jesus a homosexual or bisexual? There is, for example, no evidence that Jesus married. There are indications that there were mysterious rites of initiation into the mysteries of the "Kingdom of God" that his closest disciples underwent. One passage relates that a youth was with Jesus the night of his arrest. Perhaps he had participated in such rites. In Mark, we read that following the seizure of Jesus all of his disciples deserted him and ran away: "Among those following was a young man with nothing on but a linen cloth. They tried to seize him; but he slipped out of the linen cloth and ran away naked." (Mark 14:51-52—NEB).

The recent discovery by Morton Smith of a letter apparently written in the second century by Clement of Alexandria gives some credence to the possibility of erotic practices. This letter suggests that there was a secret Gospel of Mark that was subsequently censored by the church fathers. The author of the letter attacks an early Christian sect, the Carpocratians, for engaging in libertine practices and finding some justification for these in Jesus and his disciples. According to the Gospel, Jesus came to Bethany and entered a tomb to raise a youth from the dead. The story is similar to the account given in John about the raising of Lazarus from the dead. The youth, the story goes, "loved" Jesus and "began to beseech him that he might be with him." The Gospel continues: "In the evening the youth comes to him, wearing a linen cloth over (his) naked (body). And he remained with him that night, for Jesus taught him the mystery of the

---

*Beelzebub was considered to be the prince of the devils. In Milton's *Paradise Lost,* Beelzebub was said to be a fallen angel, ranking next to Satan.

Kingdom of God."[12] According to Smith, there were various rites that Jesus performed, the communion or eucharist meal, in which he commanded his disciples to "eat my body" and "drink my blood" and also the baptismal rites reserved primarily for his inner circle of disciples. Although these may have had some erotic components, they also imply that Jesus engaged in a form of magic and that his religious celebrations were similar to those performed by other magicians of the day.

Was Jesus a magician?

Indeed, an important interpretation of Jesus' ministry, which should be taken seriously, is the view that Jesus was a conjurer. From what we have learned about how magicians, gurus, and psychics have performed down through the centuries, this hypothesis is plausible.[13] Indeed, the very fact that many of the pagan and Jewish critics of Jesus (such as Celsus) in the first centuries after his death accused him of being a magician should mean that the possibility bears some scrutiny. Moreover, the Gospel itself reports that Jesus was criticized by his detractors in his own day for practicing sorcery and magic. It was precisely because the disciples of Jesus believed that Jesus could perform miracles that his divinity was proclaimed. In other words, the authority of Jesus' divinity was attributed first to the fact that he was able to do wondrous things and, second, to his resurrection. Indeed, these claims are used to support the basic article of faith of the Christian religion: that the Incarnation had occurred and that God had assumed human form and revealed his plan for salvation for suffering humanity through Jesus Christ, his son.

We know that magic, superstition, and fanaticism were widespread in the first and second centuries. Lucian, a Roman author born in Syria, testifies to how extensive they were. Belief in astrology and omens, psychic mysteries and prophecies, the mystic and occult arts was commonplace. In this social milieu, magicians and jugglers, oracles and seers, prophets and healers had a field day. They promised that they could "put a spell on your enemies," "predict your future," "discover buried treasures," or "restore you to health." Demonology was generally accepted. Prophets or exorcists insisted that they could drive out the possessed, cure the sick, heal the lame, and even bring back to life those who had died.

Lucian bitterly castigated the frauds who duped people by means of trickery. These charlatans appealed to superstitious fears and irrational

12. Morton Smith, *The Secret Gospel* (New York: Harper and Row, 1973), p. 17.
13. See especially Morton Smith, *Jesus the Magician* (New York: Harper and Row, 1978), for an extended discussion and documentation of the view that Jesus was a magician.

hopes and preyed on gullibility and vanity. Since medical science was undeveloped, people desperately sought out anyone who could cure them of their afflictions, whether leprosy, paralysis, blindness, infection, cancer, or various forms of psychiatric disorder. Where else could they turn? So they welcomed faith healers and readily sought out oracles and astrologers, prophets and psychics.

In his essay "The Pathological Liar," Lucian rails against the behavior of such quacks. He pokes fun at "people who exorcise ghosts and cure victims of demonic possession."[14] He graphically describes a Syrian in Palestine who specializes in such cases.

> His patients are the sort who throw fits at the new moon, rolling their eyes and foaming at the mouth. Yet he always manages to cure them, and sends them home perfectly sane, charging a large fee for his services. When he finds them lying on the ground, the first question he asks is: 'What are you doing there?' The patient makes no reply but the devil explains, either in Greek or some foreign language, who it is, where it comes from, and how it got into the man. Then the Syrian starts swearing at the devil and if necessary, threatening it until it goes away.[15]

In his essay "Alexander or the Bogus Oracle," Lucian gives a full-scale account of the famous fraud and swindler who nevertheless made people believe that he had special paranormal gifts, that he was a psychic healer and had other miraculous powers. Alexander of Abonoteichos began as a youthful prostitute who slept with anyone who would pay him. One of his clients was a magician, who taught him the magic arts. He observed that human life is ruled by two tyrants, hope and fear, and that the one thing people want most when they are oppressed is information about the future. With this in mind, Alexander specialized in prophecy. Lucian tells how he concocted a plot to bury in the mud a small snake in an empty eggshell, which he carefully glued together and then contrived to discover before the populace. Alexander uttered some unintelligible sounds as he broke the egg, and the snake slithered onto his palm. This so amazed his audience that they began shrieking and proclaiming it a miracle; they knelt down and prayed. From then on Alexander's career took a dramatic new turn. He installed himself in a home with a large snake (the small one had apparently grown rapidly in a few days!), and began giving readings as an oracle.

Lucian describes the methods by which Alexander fleeced the people who came to consult him. He told each client to write down questions on a piece of paper, fold it up, and seal it with wax. Alexander contrived

14. Lucian, *Satirical Sketches* (London: Penguin Books, 1961), p. 205.
15. Ibid.

ingenious ways to dislodge the paper, read the question, and reseal the wax so that it was undetected. He answered the questions of foolish and naive people, largely relying on his wits and common sense, combined with "imaginative guesswork." His responses, says Lucian, were sometimes obscure, ambiguous, even unfathomable, yet this impressed the people greatly. His reputation spread widely. Not only was he considered good at predicting the future, locating runaway slaves, detecting burglars and buried treasures, but even in some cases "resurrecting the dead." As his fame grew, he sent missionaries out to all corners of the Roman Empire, warning of impending disasters and offering to protect people from these.

Lucian, though otherwise a critic of philosophers, relates how the Epicurean philosophers were critical of Alexander and sought to dissuade people from accepting superstitions and how they attempted to free them from irrational hopes and emotions. The Epicureans sought to get them to face the facts and purify their minds by means of "reason, truth and plain speaking." Alexander hated these critics and condemned their efforts; his fame and influence continued to spread in spite of them. He was clearly a charlatan and magician, yet people believed fervently that he had special paranormal powers. Lucian's account is insightful, for it accurately portrays how psychics, seers, astrologers, and gurus are able to fleece the public even today.

Was Jesus a charlatan? Did he know and practice magic? According to his Jewish and pagan critics, he did indeed. The Talmud maintained that Jesus was a sorcerer and magician who tricked people into believing that he was able to perform miracles. And Celsus asserted that Jesus had lived for a time in Egypt, where he acquired the arts of sorcery. In Origen's *Contra Celsus,* we find the following statement attributed to Celsus:

> (Jesus) hired himself out as a servant in Egypt on account of his poverty, and having there acquired some miraculous powers, in which the Egyptians greatly pride themselves, returned to his own country, highly elated on account of them, and by means of these proclaimed himself a God.[16]

Celsus maintained that the miraculous works performed by Jesus, such as the cures, his resurrection, and the feeding of the multitude, could be compared to the "tricks of jugglers," who profess to do more wonderful things, and to the feats of magic performed by those who have been taught by the Egyptians, and who "in the middle of the market-place, in return for a few opals will impart the knowledge of their most venerated arts," and will "expel demons" from men, "dispel disease," "exhibit expensive banquets and tables and dishes having no real existence and who

16. *The Writings of Origen,* p. 427.

can put into motion what are not real living animals, but which only have the appearance."

Celsus then asks:

> Since, then, these persons can perform such feats, shall we of necessity conclude that they are 'sons of God,' or must we admit that they are the proceedings of wicked men under the influence of an evil spirit?[17]

Was Jesus much like Houdini, Blackstone, and other stage magicians, or like Uri Geller, Eusapia Palladino, and other fraudulent psychics? Is this the proper interpretation of his work? The New Testament abounds with reports of miraculous feats that Jesus accomplished. Indeed, it was not simply Jesus' moral message or theological views that his followers accepted, but it was also his miraculous deeds that overwhelmed them. The message of Christ was considered to be potent because it was Jesus, a charismatic figure, who enunciated it, and it was held to be authoritative because of the incredible things that Jesus could perform.

Miracles attributed to Jesus

According to Jesus' defenders, Jesus was able to perform remarkable miracles that lacked any natural explanation. He could turn water into wine and was able to feed multitudes. He could still a storm, wither a fig tree, walk on water, and even drive a legion of demons into a herd of 2,000 Gadarene swine and send them rushing over a hillside into a lake to be drowned. Jesus also was a healer able to effect cures. This included healing those who were paralyzed, lame, hemorrhaging, dumb, blind, or leprous. He was an exorcist who was able to drive out "unclean spirits" and "demons" from people and restore sanity to those who were possessed. He was able to raise the dead. He also had psychic powers, being precognitive, able to foretell the future; clairvoyant, knowing what was happening elsewhere; and telepathic, able to read other people's thoughts.

Matthew summarizes the kinds of healing power that so astounded his followers:

> . . . curing whatever illness or infirmity there was among the people . . . sufferers from every kind of illness, racked with pain, possessed by devils, epileptic or paralyzed, were all brought to him, and he cured them. (4:23-25—NEB)

They brought him a man who was possessed; he was blind and dumb; and

17. Ibid., p. 475.

Jesus cured him, restoring both speech and sight. The bystanders were all amazed. (12:22-23—NEB)

And now some men brought him a paralyzed man lying on a bed. . . . Jesus said to the man: Stand up, take your bed, and go home. Thereupon the man got up, and went off home. The people were filled with awe at the sight, and praised God for granting such authority to men. (9:2-8—NEB)

Whether Jesus actually cured the people as related in the Gospels is uncertain. We do not have expert medical testimony diagnosing their illnesses. The conditions of the afflicted are never precisely described. Nor do we have any clear-cut evidence that they were cured and that these cures were permanent. The Gospel accounts are somewhat contradictory. For example, Matthew reports that a president of a synagogue approached Jesus and said, "My daughter has just died," imploring Jesus to lay his hands on her so that she will live (9:18—NEB). When Jesus arrived at the president's house, he said, "The girl is not dead; she is asleep." Everyone laughed at him. Jesus had everyone turned out of the room. He went into the room and then "took the girl by the hand, and she got up." This, Matthew notes, was the talk of the country (9:23-26—NEB).

In Mark, we read a somewhat different version, with the president of the synagogue saying, "My little daughter is at death's door." (5:23—NEB). He is not claiming that she is dead. In Mark, someone else sends a message to the president that his daughter is dead, which Jesus overhears. When Jesus goes to the president's house, he says, "The child is not dead; she is asleep" (5:39-40—NEB). Mark also reports that everyone laughed at him. In his version, however, Jesus took the child's mother and father and his own company into where the child was lying, took hold of her hand, and commanded her to get up, which she did (5:35-43—NEB). Luke has the president of the congregation only saying that his daughter "was dying." (8:42—NEB). Thus we do not have a reliable indication of what ailed the girl, only a fragmentary and uncertain report that she was dead or near death, with the strong implication that she rose from the dead. What her illness was or what she was cured of, we do not know. It is clear that Jesus' ability to heal or give the impression of healing strongly impressed those about him. It was taken as a powerful sign of his supernatural powers. According to Mark, those who heard Jesus began to ask one another: "What is this? A new kind of teaching! He speaks with authority. When he gives orders, even the unclean spirits submit" (1:27—NEB). In John, we read the words of Nicodemus: "Rabbi," he said, "we know that you are a teacher sent by God; no one could perform these signs of yours unless God were with him" (3:2-3—NEB).

The Jesus of the Gospels reinforced this veneration, claiming that his powers and words came from God, his Father. Thus he says, "It is the word of the Father who sent me." (John 13:24-25—NEB). And again, "Anyone who has seen me has seen the Father." (John 13:9—NEB). And still again: "I am not myself the source of the words I speak to you: it is the Father who dwells in me doing his own work." (John 13:10-11—NEB). Did he really believe this? Was he deluded? Or did he only half-believe it? Did he also practice magic? Does this explain how and why he succeeded? It is clear that Jesus' miraculous faith-healing abilities assume a central role in his ministry. Without them it is doubtful that he would have attracted any following at all. The Gospels report more than two hundred events about Jesus that involves something unexpected or miraculous. It was these strange happenings that convinced his followers that he had a divine mission, and that he was a healing God of mythic proportions.

Objections to miracles

What are we to make of such occurrences today? Were they miraculous? Did they really happen? Should reports of them be taken as veridical? Are the eyewitnesses trustworthy? Are there other causal explanations that we can assume for them?

David Hume, in his famous essay "Of Miracles,"[18] raises serious epistemological objections to miracles. Our knowledge of nature or history, says Hume, must be based upon the evidences of the senses, including reports of reliable eyewitnesses. Our experience of matters of fact discovers regularities within experience. On the basis of the constant conjunctions of sense impressions, we affirm that there are laws of nature, and these are held on probabilistic grounds. Our explanation, based upon habit and custom, is that uniformities observed in the past will be observed in the future. A miracle by definition "is a violation of the laws or regularities of nature." But Hume says, "No testimony should be sufficient to establish a miracle, unless the testimony be of such a kind that its falsehood would be more miraculous than the fact which it endeavors to establish."

Hume poses the question: Is the evidence for historical miracles strong enough so that we may reject the past and present testimony of the senses that there are uniformities in nature, in favor of these exceptions? He answers in the negative. It is a miracle that a dead man should come to life, he observes. This has never been observed in any age or country in the past. If anyone relates to us that he saw a dead man restored to life,

18. From *An Enquiry Concerning Human Understanding*. In *Hume's Enquiries,* ed. L. A. Selby-Bigge; rev. P. Nidditch, 3rd ed. (Oxford: Clarendon Press, 1975).

we need immediately to consider whether it is more possible that this person is deceived. Our uniform experience about the nature of death is equivalent to a proof that dead people do not come back, and this should stand as a refutation of the miracle.

To accept a miracle, says Hume, the testimony upon which the miracle is founded must be strong enough to overturn our belief in the uniformities of nature that are based upon past experience. However, this has not been the case, he argues, thus far. And he gives a number of reasons why. First, there have never been found in all of history any miracles supported by a sufficient number of reliable persons of unquestioned good sense, education, and learning, so as to guarantee that there is no delusion. Second, he says, human nature is prone to "the passion of surprise and wonder" so that there is a tendency to find satisfaction in such beliefs. There is a strong propensity to accept the extraordinary and marvelous in spite of the fact that history is full of instances of forged miracles and prophecies. Third, such miracles have been chiefly observed "to abound among ignorant and barbarous nations," or to have been transmitted from ignorant and barbarous ancestors to the present. Belief in miracles often has its roots in remote countries, where they have been refuted and exploded, but are held in countries far removed, for their grounds are not examined. Fourth, there is no testimony for any miracle that is not opposed by an infinite number of witnesses. In other words, the miracles of one religion contradict and refute those of all others. The miracles of a particular religious system, such as Islam, differ from those of Christianity or Buddhism. Hume continues his argument by observing that "no testimony for any kind of miracle has ever amounted to a probability, much less to a proof." Belief in religion, he concludes, is thus based upon faith, not reason. And he adds with tongue-in-cheek, "Whosoever is moved by faith to assent to it is conscious of a continued miracle in his own person, which subverts all the principles of his understanding and gives him a determination to believe what is most contrary to custom and experience."

This powerful argument has been used to reject the miracles of the Old and New Testaments *in toto*. Extraordinary claims require extraordinary evidence in order to overthrow the uniformities and regularities uncovered in experience. One has to be cautious that one does not reject on a priori grounds a report of an event that is heralded as miraculous. Anomalies do occur in nature. There are bizarre and unexpected events, unique or inexplicable occurrences—"Fortean facts," so named after Charles Fort, who catalogued a great number of strange events. One cannot simply reject all of them out of hand.

In evaluating anomalous claims there are at least two questions to raise. First, we may ask, were the alleged facts accurately reported or were they based upon mere hearsay, unsubstantiated by expert observation? Second, if they were correctly reported, what is the explanation of them? How may we interpret what happened? Often an event is held to be "miraculous" because people are unaware or ignorant of the natural causes, which they attribute to divine or demonic intervention.

We really have no way of knowing whether all the deeds recorded in the Gospels really occurred. The Gospels, as I have shown, were more likely written by convinced believers attempting to persuade others to accept their religious faith than by objective historians or impartial scientific observers. We may generally doubt that many or perhaps most of the strange events that were reported actually did occur. Many people in the crowds were gullible, all too easily dumbfounded, prone to accept beliefs without proper examination, uncritical in their powers of observation and judgment. They could easily be hoodwinked by a shrewd magician.

Another important point should be made here. Eyewitness testimony, even where it is available, is often unreliable. There have been extensive studies pointing to the inaccuracy of so-called eyewitness reports. Scientific investigations into alleged eyewitness identification of criminals, the reports of crimes by a multitude of witnesses, and the identification of suspects in police line-ups have graphically demonstrated that the memory of witnesses is often prone to error and exaggeration. Judge Nathan Sobel has stated that incorrect eyewitness misidentifications have led to more miscarriages of justice than all other causes combined.[19]

Many researchers have pointed out the pitfalls here. Elizabeth F. Loftus has compiled a great deal of data from recent research to demonstrate the problem.[20] In an experiment staged by students at the University of Washington, two young women entered a Seattle bus terminal, placed their belongings on a bench, and went to the ladies' room. As soon as they were gone, a man rummaged through their bags, put something under his coat, and escaped outside. Upon returning, one of the women noticed that their bags had been opened and screamed, "My tape deck was stolen!" Several bystanders who had witnessed the incident later agreed to give testimony about what was stolen and said they would recognize the man. They went into considerable detail describing the color, shape, size, and even the antennae of the stolen tape deck. As a matter of fact, no tape deck had been stolen; the witnesses had been influenced by misinformation supplied by the experimenters.[21]

19. N. R. Sobel, *Eye-Witness Identification: Legal and Practical Problems* (New York: Clark Boardman, 1972).

20. E. F. Loftus, "Eyewitnesses: Essential but Unreliable," *Psychology Today* (Feb. 1984).

21. Joann E. Rodgers, "The Malleable Memory of Eyewitnesses," *Science 82* (June 1982).

Interestingly, misidentifications may be made not only by unsophisticated people but also by police and detectives at the scene of the crime. A. H. Tinker and E. Christopher Poulton showed a film depicting a street scene to 24 police officers and 156 civilians. The subjects were asked to report any instances of crime and to identify those involved. The results were that the officers reported more alleged thefts than the civilians, but the latter did just as well when it came to detecting actual crimes. There has been case after case of the wrong people being arrested and convicted on the basis of mistaken identification by witnesses.

The results of such studies conclude first, that the memory is often inaccurate; second, that if a situation involves a good deal of stress or excitement, the perceptions are distorted; and third, if people are led to believe that something is the case, this tends to color their observation of the event.

How much more these factors influencing psychological perception apply to the foundations of religion—and in this case to Christianity! This is especially so when the evidence for miracles is passed along by word of mouth, second and thirdhand, in an oral tradition. What may have been a strange or remarkable event to some original observer is further elaborated in time so that it may end up hardly recognizable in its final form.

Some alternative naturalistic explanations

The term *miracle* is invoked often by the incredulous when something strange occurs for which there is no known natural cause. But what they may be confessing is that they are ignorant of the real causes at work. The feats of a magician may be attributed to "magic," for they can be easily explained by sleight-of-hand and other physical explanations; the "healing" of a faith healer need not be attributed to God or occult forces but may also have a perfectly natural explanation.

Thus, rather than claiming that Jesus was of supernatural origin and was able to set aside the laws of nature, we should seek for normal causal explanations. Hume warns us to trust nothing which tradition has transmitted. But perhaps he has gone too far. I prefer to examine how magicians perform their sleight-of-hand or how paranormal readings and healings are done today. These may explain in part what happened in the time of Jesus.

It is not inconceivable that Jesus, whether believing or half-believing that he had a divine mission, also practiced the arts of deception. The hypothesis has at least some credence, not simply because it was the view of many of his critics in antiquity, but because it can be further corroborated by analogy with how similar methods are used today. Thus, what people thought to be miraculous, occult, or paranormal, may have been only natural effects induced by conjuration and the psychological processes of acceptance of the events as supernatural because the causes were unknown.

As we have already pointed out, the ancient world was rife with magicians, some of whom were taken as godmen. Some interesting accounts of their work have survived. Simon the Magician was known to early Christians. We read in Acts (8:9-11—NEB) that a man named Simon in Samaria had swept the Samaritans off their feet with his magical acts. Everyone listened to him. "This man," they said, "is that power of God which is called 'The Great Power.'" He had persuaded many to believe that he was an incarnate divine power. Josephus also reports that the magicians of this period, who were numerous, did things similar to what Jesus had done. "Getting a spirit" (as in Jesus' baptism) was the first step for many magicians. And they practiced "secret rites of initiation mysteries" to enable them to gain powers. They were able to cast spells on those about them who were enchanted by their powers. The story of Jesus being driven by a spirit into the wilderness and overcoming a demon, according to Morton Smith, is a common variant of shamanistic initiation. The ability to work psychosomatic cures follows more easily when there is a strong conviction that the charismatic individual is endowed with supernatural powers.

If Jesus had spent time in Egypt, he could have learned the arts and practices of the priests and magicians who were there in abundance. A case in point is Jesus' use of spitting. In John we read that Jesus encountered a blind man. "He spat on the ground," the Gospel reports, "and made a paste with the spittle; he spread it on the man's eyes, and said to him, 'Go and wash in the pool of Siloam.'" (9:6-7—NEB). Afterwards, the man allegedly could see. Ancient Egyptian papyri contain many allusions to spitting, which was considered to be a religious and magical practice and a curative act. Sometimes the saliva was mixed with sand or water and applied to the nostrils or eyes or other afflicted parts. Jesus here was practicing a form of healing that was not unfamiliar to the ancient world.[22]

Hippolytus gives an extended and interesting account of Simon Magus.[23] In a treatise written about 200 C.E., he described Simon as being skilled in the magic arts; he preyed upon others and even sought to deify himself. Hippolytus also provides insightful accounts of how other ancient

22. See E. A. Wallis Budge, *Osiris: The Egyptian Religion of Resurrection,* vol. 2, chap. 23, "Spitting as a Religious Act" (New York: University Books, 1961), pp. 203 ff.

23. Hippolytus, *Philosophumena,* trans. F. Legge (New York: Macmillan, 1921), pp. 2-3, 40-45.

magicians claiming to have divine powers worked. Marcus, for example, was an "expert in magic" who used trickery and demons to lead many astray and claimed "that there is in him the greatest power from invisible and unnameable places." "Among the things he does by trickery are 'consecrating a cup' and then causing it to appear purple in color and sometimes red 'so that the dupes will think that a certain grace has come down, and has given a blood-like power to the draught.'" He spoke incantations over the cup and thus "distracts the dupe and the bystanders, so that he is considered a miracle worker, he fills the larger cup from the smaller so that it overflows."

The Christian communion service has striking magical overtones. Moreover, the turning of water into wine by Jesus could have been done by magic, by switching one for the other. Even the feeding of the multitude—if it ever occurred—may have been a product of trickery, much as a magician can appear to saw a woman in half. Even walking on water is not unlike wading on a sandbar. One can imagine many possible scenarios of how it could have been done. A curious fact is that when one *believes* that another person possesses superhuman powers, one is more likely to accept the occurrence of any bizarre event and insist that it is inexplicable by natural causes. When such a belief is present, even the slightest deviation from normalcy is taken as evidence of an overwhelming miracle at work.

This phenomenon can be witnessed today. How easy it is for someone to pretend they have psychic powers and to take people in. James Randi, the conjurer and skeptic, in cooperation with two young magicians, Steve Shaw and Michael Edwards, demonstrates how easy it is to deceive people. This is known as project Alpha.[24] Randi was convinced that parapsychologists are too easily taken in by people claiming to be psychics— Uri Geller is a notable example—when the "physics" were merely using standard magic tricks. He planted two young boys in a parapsychology lab in St. Louis. The boys pretended to display psychic powers, using the crudest kind of deception, and the experimenters were easily duped. But, more disturbing, when the press reported that Shaw and Edwards were psychics, people everywhere clamored for their services. It was a frightening phenomenon, they report with amazement, illustrating how easily people can be fooled and how they allow their expectations to color their perceptions of what is happening.

The powerful effect that belief in magical or occult phenomena can have upon people was demonstrated by two psychologists, Barry Singer

24. *Skeptical Inquirer*, Summer 1983 and Fall 1983.

and Victor Benassi.[25] They introduced college students in various psychology classes to a person, named Craig, dressed in a long purple robe who performed so-called "psychic" feats. He bent a metal rod, seemingly by psychokinesis. Blindfolded, he read numbers on a concealed note pad. He was able to transfer ashes from the back of a person's hand to his palm. Although these acts seemed to contradict ordinary experience and our notions of causality, they were simple magic tricks that any good magician can perform. In some of the classes, the professors did not tell the students anything about the performer other than that he claimed to have psychic powers, adding that they personally were not convinced. In other classes, they told the students that the performer was a magician and that he would present a magic act. They were surprised to find that in both the "psychic" and "magic" classes about two-thirds of the students believed that the performer was a psychic.

Relatively few students accepted the instructors' description of Craig as a magician in the two classes where he was introduced as such. Psychic belief was still prevalent; it was strong and loaded with emotion. Some students even covered their papers with exorcism terms and exhortations against the Devil. Many students showed fright and emotional disturbance. Most expressed awe and amazement. By the time Craig was halfway through the "bending" chant, the class was in an excited state. Students sat rigidly in their chairs, eyes glazed and mouths open, chanting together. When the rod bent, they gasped and murmured. After the class was over, some continued to sit still in their chairs, staring vacantly or shaking their heads; others excitedly rushed up to Craig, asking him how they could develop such powers. Singer and Benassi believe they were observing an extraordinarily powerful behavioral effect. If Craig had asked the students to tear off their clothes, throw him money, and start a new cult, they believe some would have responded enthusiastically. Something was happening that they didn't understand. Moreover, many students were experiencing serious emotional disturbance.

Singer and Benassi were so intrigued that they continued the experiments by taking Craig to other classes, changing their introduction to make it clear that they were presenting a magician and that he was doing tricks. This did succeed in reducing psychic beliefs slightly, but never below 50 percent. The most salient result of their test, they said, was their inability to reduce psychic beliefs to any extent even though there were clear explanations that trickery and magic were taking place. Singer and Benassi have concluded that some people will stubbornly maintain a

25. Barry J. Singer and Victor Benassi, "Fooling Some of the People All of the Time," *Skeptical Inquirer,* Winter 1980-81: 17-24.

belief about someone's psychic powers, no matter what evidence is presented to them. Interestingly, at no time did Craig say that he was a psychic or make any psychic claim. Singer and Benassi's tests have since been replicated by others, with similar results. They have concluded that their results, as bizarre as they may be, are "of wide generality" and that the psychological processes they have tentatively identified as being involved in supporting psychic beliefs "are present and active in the general population."

All of this suggests that there is a deep-seated tendency toward magical thinking. Surely the transcendental temptation was as strong in the days of Jesus as it is today. The willingness to believe in occult magic is greatly enhanced when the individual who performs the deed, such as Jesus, is a powerful personality. Such charismatic individuals are able to cast spells of enchantment. Incantation and other rites only contribute to the suggestion that such an individual has divine powers. A strong belief in the efficacy of magic especially applies to the phenomena of faith healing and helps in large part to explain it. People desperately want to be cured. Since many illnesses are psychosomatic in origin, the psychological disposition of the person can influence his behavior. This may provide the explanation of the powerful effect that faith can have on some illnesses. Healing is not God intervening in nature; it occurs because a psychological state influences and creates a change, as Freud noted in hysterical patients, who were affected physically by their neurosis. As we have seen, we have no way of knowing what ailments the people alluded to in the New Testament suffered, what the clinical symptoms were, or indeed whether their illnesses disappeared permanently or only temporarily. The claimed cures are based on fragmentary evidence offered by the Gospel authors. We do not know whether all of the cures Jesus attempted worked. In no case did Jesus cause a limb to grow or do the completely miraculous. Most of the so-called cures seemed to concern psychosomatic illnesses. It is not improbable that many of Jesus' cures may have been effected by methods similar to those attributed to Alexander by Lucian. For example, Jesus apparently utilized seventy "advance men," who were sent ahead in pairs to every town that he intended to visit (Luke 10:1-2). It is possible that these individuals reported back to Jesus the conditions of some of those to be "cured" or of whom he was to make "prophecies"; hence, some of the paranormal events attributed to Jesus could have occurred because of his prior knowledge of the in dividuals involved.

There is an interesting story in John about Jesus' psychic powers. Jesus allegedly was able to tell a woman a great deal about herself. The disciples had left for a town in order to buy food, leaving Jesus alone near a well. Meanwhile, a Samaritan woman came to draw water. Jesus asked her for a drink and engaged her in conversation, uttering a parable about the water of eternal life. She replied that she wanted such water. Jesus said to her, "Go home, call your husband and come back." She answered, "I have no husband." "You are right," said Jesus, "in saying that you have no husband, for, although you have had five husbands, the man with whom you are now living is not your husband; you told me the truth there." The woman is stunned about his knowledge of her and exclaims in the town, "Come and see a man who has told me everything I ever did. Could this be the Messiah?" (John 4:8-30—NEB). Was this a "cold reading" (based upon cues Jesus received from her on the spot) or was it a "hot reading" (based upon facts gathered and reported to him)? Jesus could easily have discovered facts about the woman from advance information supplied by his confidants. If so, then Jesus used familiar techniques used today by so-called psychic readers.[26]

In Luke, we read that there were many who were suspicious of Jesus' powers. After he cured a dumb man by driving out the devil, Luke reports that "the people were astonished, but some of them said, 'It is by Beelzebub prince of devils that he drives the devils out'" (11:14-15—NEB). Jesus was not believed by the people in his own town to be a miracle-worker, but was considered either mad or a sorcerer. Often when they asked for a sign from God, he gave none. Thus we see that Jesus was not always able to perform on demand, which raises suspicion that he needed some advance preparation. Matthew confirms that Jesus did not work many miracles in his own town (13:55 ff.—NEB), so "they fell foul of him." This led Jesus to say, "A prophet will always be held in honor, except in his home town, and in his own family." "And he did not work many miracles there," reports Matthew, adding "such was their want of faith," as if to reinforce the view that miracles depend upon a state of belief in the audience. Where faith is absent or minimal, it is difficult to succeed. Mark also reports that Jesus healed "only a few people" in his own town. And John says that at that time "even his brothers had no faith in him" (7:5—NEB). Luke likewise reports that "others, by way of a test, demanded of him a sign from heaven." To which Jesus replied equivocally, "This is a wicked generation. It demands a sign, and the only sign that will be given is the sign of Jonah. For just as Jonah was a sign to the Ninevites, so will the Son of Man be to this generation." (Luke 11:29-30—NEB). In Mark we also read that the Pharisees demanded a sign from heaven, to which Jesus replied, "Why does this generation ask

26. For a discussion of how psychics do many readings, see Ray Hyman, "Cold Reading: How to Convince Strangers That You Know All About Them," *Skeptical Inquirer* 1, no. 2 (Spring/Summer 1977).

for a sign? I tell you this: no sign shall be given to this generation. With that he left them." (8:11-13—NEB). Thus Jesus could not perform anywhere and everywhere, or on demand, so he may have needed adequate preparation before he could display his paranormal powers.

One famous miracle, which is open to suspicion, concerns his friend Lazarus, whom Jesus allegedly raised from the dead. The entire story may have been borrowed from Egyptian mythology and embellished by the Gospel writers. We can only speculate about what actually happened; the circumstances are curious, to say the least. According to John (11:1-44—NEB), Jesus knew Lazarus and his sisters, Mary and Martha, and "loved all three." The sisters sent a message to Jesus to the effect that "your friend is ill." But instead of proceeding directly to Lazarus' side, Jesus waited two days. Finally, when he arrived at the village of Bethany, he is told that Lazarus has been four days in the tomb. Jesus goes to the tomb, a cave, and asks that the stone be removed from the entrance. He raised his voice in a loud cry, "Lazarus, come forth," and Lazarus complies "with hands and feet and face still swathed in linen bands."

The incident has all of the hallmarks of a subterfuge. Why did Jesus wait two days to visit his friend? This is highly suspicious. Moreover, how do we know that Lazarus was dead when Jesus arrived? It would have been a marvelous feat to perform for the doubters—for Lazarus to feign death so that Jesus could demonstrate that he had miraculous powers. Lazarus, at least (and possibly others), would have had to have been in league with Jesus in deceiving people. Lazarus was a close friend of Jesus, since we hear of him again when Jesus has a supper, at which Martha served and Lazarus sat among the guests of Jesus. The idea that Jesus raised Lazarus from the dead certainly would have been a great boost to his reputation.

The existing Bible is a product of many different authors and of selection and censorship by church councils. Much material has been excised and manipulated. We do not know if what has survived is authentic. We earlier alluded to the Secret Gospel of Mark, according to which Jesus had a special relationship with a youth (Lazarus?) whom he raised from the dead. The youth "loved Jesus," and at one point spent the night with him. Jesus taught him "the mystery of the kingdom of God," which suggests a very close relationship. If Jesus was a magician, Lazarus might have been his shill.

For many, Jesus' raising of Lazarus from the dead was one of the most remarkable feats he accomplished. However, far more impressive was Jesus' own disappearance from the cave tomb in which he was buried and his rising from death. Indeed, the resurrection of Jesus is, for posterity, the single most important event of his entire career.

The crucifixion and death of Jesus

The cornerstone of Jesus' claim to fame rests upon his crucifixion, death, and alleged resurrection. I will not examine the trial of Jesus, the basis of the charges leveled against him, or his conviction. It is what happened after the crucifixion that is essential to Christianity. According to Paul, without the resurrection all of the Christian faith is in vain. Christianity promises eternal salvation for those who believe in Jesus; but if he was not raised from the dead, then the main article of Christian faith is without basis.

> Now if this is what we proclaim, that Christ was raised from the dead, how can some of you say there is no resurrection of the dead? If there be no resurrection then Christ was not raised; and if Christ was not raised, then our gospel is null and void, and so is your faith; and we turn out to be lying witnesses for God, because we bore witness that he raised Christ to life. (1 Cor. 15:12-15—NEB).

What is the evidence that this miraculous event occurred? Jesus was mocked by the bystanders as he suffered on the cross:

> "He saved others," they said, "but he cannot save himself. King of Israel, indeed! Let him come down now from the cross, and then we will believe him. Did he trust in God? Let God rescue him, if he wants him—for he said that he was God's Son." (Matt. 27:40-43 NEB).

Their logic was unassailable. If Jesus was a deity, then surely he should have been able to turn events in a more positive direction. If he was the son of God, why did God not intervene? The great paradox is that Jesus had virtually collapsed in his own defense at his trial—as if he knew he was guilty. On the cross at the final moment of impending death, he cries out: "My God, my God, why hast thou forsaken me?" (Mark 15:34-35—NEB)—as if he had half-believed his own powers and yet had discovered, at the end, that they were for nought. But ironically the ultimate defeat of Jesus is used by his disciples to make it appear that he was victorious. They were able to do so by building up and inflating his resurrection.

What are the facts surrounding Jesus' death and return? Since it is this event—even more than the fictions surrounding his birth—that is crucial for the Christian religion, one would think that it would be well corroborated by unimpeachable eyewitness testimony. Unfortunately, the only "evidence" we have is contained in the New Testament. No independent accounts have survived, if they ever existed, to give credence to this world-shattering event. Even so, the biblical accounts are fragmentary and contradictory. If we read the four Gospels side by side, we are struck by the inconsistencies. This should tend to make us doubt the received story. Was Jesus dead when he was taken from the cross? Or had he simply swooned and was thought to be dead, only to be revived and spirited out of the tomb by his disciples? Had he appeared for a brief period after his crucifixion, partially in disguise, and then escaped in haste, and either disappeared or eventually succumbed to his wounds? There are many possible scenarios that we can speculate about. What is clear is that the case for Jesus' resurrection is extremely weak.

What is especially perplexing is that if Jesus appeared, he appeared only to those who believed in him, not to neutral observers or to those who had opposed him. Moreover, he did not punish those who had persecuted him or had complicity in his trial and crucifixion. Given the fact that God is wrathful toward sinners—helpful only to those who believe in him and are obedient to his will—it is a wonder that he did not come back and inflict divine retribution. What a display of divine power that would have been! Instead, we have him returning secretly. If Christ was really sent by God to proclaim his message, then surely a major public event, staged with great fanfare, would have been far more effective. It could have been staged simultaneously in Jerusalem before Pilate and the Sanhedrin and in Rome before the emperor and the Senate. At the very least, God could have made the sun stand still at midday (as he was reputed to have done according to the book of Joshua) so that every historian in the world would note and record the event.

What is the main outline of the resurrection story according to the four Gospels? Let us first examine Mark (15:25-39—NEB). We read that Jesus was crucified between two robbers at nine in the morning. On his cross was the inscription: "The king of the Jews." At midday, darkness fell over the whole land, lasting until three in the afternoon. Jesus cried out, "Eloi, Eloi, lama sabach-thani?" "Hark, he is calling Elijah," some bystanders said. Using a cane, a man held a sponge soaked in sour wine to his lips. "Let us see if Elijah will come to take him down." "Then Jesus gave a loud cry and died," and the curtain of the Temple was torn in two. The centurion who was standing opposite him saw how he died and said, "Truly this man was a son of God." A number of women were standing at a distance watching; among them was Mary of Magdala, Mary the mother of James the younger and of Joseph, Salome, and several others. He died after six hours, a relatively short duration. In the evening Joseph of Arimathaea bravely went to Pilate and asked for the body of Jesus. "Pilate was surprised to hear that he was already dead." (Mark 15:44-

45—NEB). He sent for the centurion to learn if it was true, and when he heard their report he gave Joseph permission to take the body.

In Luke (23:26 ff.—NEB) other circumstances are introduced. For example, one of the criminals who hung with him taunted him, "Are you not the Messiah? Save yourself and us." But the other rebuked the first saying, "This man has done nothing wrong," adding, "Jesus, remember me when you come to your throne." To which Jesus replied: "I tell you this: today you shall be with me in Paradise." The inscription, according to Luke, read: "This is the king of the Jews." Luke reports that earlier Jesus had said, "Father, forgive them; they do not know what they are doing." Luke has Jesus dying at three in the afternoon and proclaiming "Father, into thy hands I commend my spirit." But Luke has the centurion who saw it also utter the statement, "Beyond all doubt, this man was innocent." Luke reports that Jesus' friends had all stood at a distance, including the women who had accompanied him from Galilee.

In Matthew, we find that the inscription on Jesus' cross read: "This is Jesus the king of the Jews." (27:37—NEB). The inscription is not exactly the same in the synoptics, which makes one wonder how accurate the eyewitnesses were. Matthew says: "Even the bandits who were crucified with him taunted him in the same way." (27:44—NEB). Not only was the curtain of the Temple torn but there was also an earthquake, and rocks split, graves opened, and God's saints were raised from sleep. Matthew maintains that after Jesus' resurrection they entered the holy city, where many saw them. This event is not even mentioned by Mark or Luke. However, Matthew has a centurion and his men watching, filled with awe, proclaiming, "Truly this man was a son of God." (27:54—NEB). Among those standing at a distance was Mary of Magdala, Mary the mother of James and Joseph, and the mother of the sons of Zebedee.

John's account of the crucifixion differs in many important details. For example, Matthew (27:28-29—NEB) gives an account of Jesus being mocked by Pilate's soldiers after the trial and prior to his crucifixion: "They stripped him and dressed him in a scarlet mantle," and placed on his head a crown of thorns. The Roman soldiers then removed these and dressed Jesus again in his own clothes. According to John, however, they "robed him in a purple cloak," not a scarlet one. (19:2-3—NEB). According to John, the crucifixion took place the day before Passover. According to Mark, Luke, and Matthew it occurred the day after. Other facts differ. John says the inscription read: "Jesus of Nazareth King of the Jews" (19:19-20—NEB). Thus we have four slightly different inscriptions. There are other discrepancies in John's account. He claims that there were additional followers of Jesus present and that they were standing nearby: "Near the cross where Jesus hung stood his mother, with her sister, Mary, wife of Clopas, and Mary of Magdala . . . (and) the disciple whom he loved." (19:25-26—NEB). In fact, Jesus converses with his mother. John has someone give Jesus a sponge soaked in sour wine fixed on a javelin (a slender spear, not a cane). One would think that Jesus' last words would be of crucial importance, yet even here we have contradictory accounts. Differing from Luke, John reports: "Having received the wine, he said, 'It is accomplished!' He bowed his head and gave up the spirit." (19:29-30—NEB).

John provides other interesting details. Because it was Passover eve, the Jews were anxious that the bodies not remain on the cross for the coming Sabbath. So they requested Pilate to have the legs broken and the bodies removed. The soldiers broke the legs of the two criminals, but they did not break Jesus' legs, since they found he was already dead. Crucifragium can imitate shock and cause rapid death, but it is not clear whether this was the case with Jesus. An interesting question that might be raised is whether Jesus had drunk a drugged wine that caused him to faint. All sorts of hypothetical scenarios are possible about the collusion of his confederates to provide him with drugs so that he would swoon and then possibly, if taken down from the cross in time, escape capital punishment. If there was a Passover plot, alternative medical explanations of what might have transpired have been made.

According to John, one of the soldiers stabbed Jesus' side with a lance, "and at once there was a flow of blood and water." John maintains that this is "vouched for by an eyewitness, whose evidence is to be trusted." (19:31-36—NEB). Dr. Janet Caldwell, a pathologist, provides a possible post hoc diagnosis of Jesus' condition, maintaining that the piercing of a victim between his ribs was a standard test of death.[27] The emission of water indicates, she says, that there was fluid in the chest. This strongly suggests that Jesus had "tuberculosis pleurisy with effusion." She maintains that this condition is also suggested by an account in Luke, who maintained that when Jesus spent the night in the garden "his sweat was like clots of blood falling to the ground" (22:44—NEB). Moreover, he was unable to carry his own cross to the crucifixion. He had been flogged, which no doubt weakened him; but the account also suggests that he may have had an illness. If Jesus suffered from this condition, Dr. Caldwell maintains that modern medicine can provide an explanation of his pseudodeath, in the form of unconsciousness or coma. To sum up the possible scenario: Jesus suffered pleurisy with effusion in the right chest cavity, which during crucifixion resulted in a state of shock with loss of consciousness and the appearance of death; tapping the fluid from the

27. Janet Caldwell, *Jesus: A Psychobiography and Medical Evaluation* (New York: Carlton Press, 1976).

chest cavity made his recovery possible, aided by his removal from the cross and being put in a horizontal position in a cool chamber, which according to the Gospels is what happened when he was placed in the tomb.

Dr. Pierre Barbet maintains that the agony of crucifixion is caused by the raised position of the arms, which causes the crucified person to have the feeling of progressive suffocation.[28] Such a position causes extremely disagreeable dyspnea, or painful respiration. Oxygenation is not properly produced in the lungs, which are not working efficiently, and the additional burden causes a kind of tetanic condition of the entire body. Thus asphyxiation develops. If Jesus had pleurisy, this condition would have rapidly intensified by alternating asphyxia-respiration during crucifixion, and a state of shock would have been induced fairly rapidly. Thus Jesus might not have been dead when he was taken down from the cross.

The next stage of the Gospel story also has contradictory elements. Joseph of Arimathaea took the body of Jesus (according to Matthew); he wrapped it in a clean linen sheet and laid it in his own unused tomb, cut out of the rock. He then rolled a large stone against the entrance. Mark and Luke agree with this account. John, however, says that he was joined by Nicodemus and that they wrapped the body "in strips of linen cloth" according to Jewish burial customs (19:40-41—NEB).

This raises questions concerning the so-called Shroud of Turin, which many people today believe is the burial shroud of Jesus. John later reports that there were "linen wrappings" and a "napkin, which had been over his head." (20:6-8—NEB). If John is correct, it was not a shroud but linen wrappings. But if Matthew, Mark, and Luke are correct, they contradict John's account, which implies that the body was perhaps wrapped like a mummy. Some versions of Luke have Peter peering into the tomb and observing wrappings (24:12—NEB), which would reinforce John's account of multiple cloths. The existence of a separate face napkin would contradict the legend that there was a shroud.

The persistence of the fame of the Shroud of Turin is a classic illustration of willful belief. It is dismaying to learn that large sectors of the population today believe it to be the authentic burial shroud of Jesus. If this were true, it would be the only corroboration of his resurrection. What are the facts?

There have been numerous reports of relics and other religious artifacts that have been venerated and even accounts of other shrouds. The Shroud of Turin surfaced in Lirey, France, in about 1353. It is a 14-foot-long linen cloth allegedly showing a print of the body of Jesus. The figure is wearing a crown of thorns, with bloodstains on its hands, feet, and right

---

28. Pierre Barbet, *A Doctor at Calvary* (New York: Doubleday, 1963).

side. In 1357 the shroud was placed on exhibition and great crowds of pilgrims flocked to view it. Bishop Pierre d'Arcis undertook an investigation of its authenticity in 1389, and a lengthy report was sent to the Pope; the bishop maintained that the shroud was a forgery. He also said that the motive for exhibiting the shroud was one of avarice and financial gain; moreover, he wrote that it was done by clever sleight-of-hand. Indeed, the artist who painted the cloth came forth and confessed. Meanwhile, the cloth was used to attract the multitudes, and claims were made that it could work miracles, including cures. The bishop complained that certain men were paid to represent themselves as healed at the time of the exhibition. The shroud eventually ended up in Turin as the property of the royal house of Italy.[29]

One would think that if the shroud was genuine—a remarkable fact indeed—some mention of the fact that the image of Christ was impressed upon it would have been made in the Gospels. The authors were surely interested in heralding the divinity of Jesus. If such a shroud had been found, it would have been a powerful support of the miracle of the resurrection. But they are silent about it.

The Shroud of Turin has had a lengthy history, and its authenticity is hotly contested today. There is a team of scientists (or pseudoscientists) who believe in its authenticity. This team is called STURP (Shroud of Turin Research Project). Members claim that no method of transposing the image of the body of a man onto linen clothing was known in ancient or medieval times. The image of Jesus appears like a photographic negative, and this, they maintain, could only have been caused miraculously by a supernatural light, as it scorched the linen at the moment of resurrection.

Exhaustive analysis by Joe Nickell and a team of skeptical scientists show that it is possible to duplicate the shroud by a rubbing technique. Moreover, if Jesus had died, he would have been shaved and washed before the burial spices were applied, following ancient Jewish burial methods. But the shroud depicts him both bearded and bloodied. Microscopic analysis of the red "blood" on the shroud, by Walter McCrone, a Chicago scientist, has shown the presence of ochre, a dye paint pigment. Interestingly also, the individual on the shroud is over six feet tall, whereas first-century Jews were generally short. One question that is raised is how a twenty-centuries-old piece of cloth can be so relatively well-preserved. Although the Catholic church has permitted some testing, it has not given permission thus far for carbon-14 dating,* which might determine

---

29. Joe Nickell, *Inquest on the Shroud of Turin* (Buffalo: Prometheus Books, 1983), pp. 12-13, 23.

*The church has since permitted carbon-14 dating. On October 13, 1988, the results were announced, confirming that the shroud was of medieval origin. This supports the interpretation of forgery.—PK, September 20, 1990.

its age. But even if the shroud were as old as its supporters claim, this would not necessarily prove that it is the shroud of Jesus. Thus the Shroud of Turin cannot be cited as physical evidence for the reality of the resurrection.

The resurrection: What is the evidence?

We must now examine the claims for the resurrection, since this event is the bedrock of Christian faith. Did it occur? What is the evidence for it? Again, we have absolutely no independent supporting data in the extant literature of the period. Nor is there any circumstantial evidence. Thus we must rely solely on New Testament sources. Yet the stories told in the Gospels are so inconsistent that the resurrection story collapses under careful scrutiny. The conflicting eyewitness testimony is so unreliable that it would not stand up to critical cross-examination in any court of law. Yet an entire religious faith is based upon the legend.

According to Matthew, after Jesus was removed from the cross and placed in the tomb, Mary Magdalene and the other Mary sat opposite the tomb. The next day the chief priests and the Pharisees visited Pilate, reminding him that the imposter Jesus said that he would be raised in three days. They asked him to secure the tomb until the third day. "Otherwise his disciples may come, steal the body, and then tell the people that he had been raised from the dead; and the final deception will be worse than the first." (27:62-65—NEB). So Pilate authorized a guard to seal the tomb. Still, at daybreak on Sunday, when the two Marys came to look at the grave, there was a violent earthquake, and Matthew says that an angel descended and rolled the stone away and sat on it. His face shone like lightning and his garments were white. At the sight of him, the guards shook and lay like dead. The angel addressed the women, saying that Jesus was not in the tomb and that he had been raised from the dead. "Come and see the place where he was laid," the angel said, "and go tell the other disciples" (Matt. 28:6-7—NEB). The two Marys hurried from the tomb in awe and great joy and ran to tell the disciples. Suddenly, Jesus appeared and they fell prostrate before him. "Do not be afraid," said Jesus. "Tell my brothers that they are to leave for Galilee, where they will see me."

Mark's account (16:1-8—NEB) diverges significantly from Matthew's. In addition to Mary of Magdala and Mary the mother of James, we have Salome going to the tomb very early on Sunday morning. When they got there they saw that the huge stone at the entrance of the tomb had already been rolled back. They went inside the tomb (not outside), where they saw a youth (not an angel) sitting on the right side wearing a white robe. Jesus' body was missing. They were told to go and tell the disciples and Peter that Jesus had risen and that they would see him in Galilee. But here we are told that they ran away from the tomb "beside themselves with terror." "They said nothing to anybody, for they were afraid," which is the direct opposite of Matthew's account. Moreover, there is no mention of their meeting Jesus. Whose description should we trust?

Reading Luke (23:55-56; 24:1-11—NEB) only exacerbates the confusion. We are told that the women who accompanied Jesus from Galilee took note of where the tomb was. In this version they also went back early on Sunday, and found that the stone had been rolled away. They went inside the tomb and were at a loss that Jesus' body was not there. Suddenly, two men (not one) in dazzling garments were at their side. Remember, they are told, that the Son of Man would be crucified and rise on the third day. Returning from the tomb, they reported all this to the eleven disciples and all the others. There is no hint that they met Jesus, as Matthew claimed. Luke finally enumerates the women present: Mary of Magdala, Joanna, Mary the mother of James, and "other women." But we are informed that the story "appeared to be nonsense," and the apostles did not believe it.

And nonsense it appears to be, for it involves contradictory accounts. To sum up, the synoptic Gospels report:

1. An angel outside of the tomb. A youth inside. Two men inside.

2. The rock is in front of the tomb. It is moved from the cave entrance in front of the women and the guard. The rock has already been removed when they arrive.

3. The women present are the two Marys. Salome in addition. Joanna in addition, and other women.

4. The women are in awe and great joy, hurrying to tell the disciples; they meet Jesus on the way. The women are beside themselves with terror, saying nothing to anybody. The women report the story to the apostles, who don't believe them.

If only the witnesses—whoever they were—could be put on the witness stand. These stories have all the trappings of hearsay, embellished by many retellings. What the original version was is difficult to ascertain. Matthew maintains the women were on their way to Galilee when the guards reported the incidents to the chief priests. The priests offered the soldiers substantial bribes and told them to say that the disciples came by night and stole the body while the guards were asleep. According to Matthew, this story of what had happened became widely known and was current in Jewish circles of his day.

Can John help us solve the mystery of the empty tomb? We have seen that this Gospel often deviates from the synoptics. Was there another cave trick (like the Lazarus cave trick)?

John has *only* Mary of Magdala (and not the other Mary, Salome, Joanna or any other women) come to the tomb early on Sunday morning, while it was still dark. Mary saw that the stone had been moved from the entrance and ran to tell Simon Peter and the other disciple (the one whom Jesus loved) that they had taken the Lord out of the tomb. No men in white are present, nor does she enter the tomb. Peter and the other disciples run to the tomb. They enter and see the linen wrappings and the head napkin lying about. Then the disciples went home, but Mary stood outside the tomb weeping. Finally she peered in and saw two angels in white sitting where the body of Jesus had been. They asked why she was weeping. As she starts to reply, she turns around and sees Jesus standing there; but she did not recognize him. She thought at first he was the gardener. Had Jesus revived from a pseudo-death and was he in disguise? Jesus tells Mary to go tell the disciples that he is ascending to his Father. Mary goes to the disciples with the news that she has seen the Lord. And so here we have a fourth story different in important details from the other three. Which is correct, if any? Should the first-hand report by Mary be accepted, if no one was present to verify it? Did Jesus survive death? In the synoptic Gospels, Jesus tells Mary and the other women to meet him in Galilee. Was he escaping to his own province, a safer place? After having been crucified and just missing a narrow brush with death, in fear and anguish, did he decide to flee, telling the others to rendezvous with him there? Did he subsequently succumb to his wounds? If he had pleurisy, his condition might have so weakened him that he died shortly thereafter. It is, of course, difficult to say.

What happens next? In Matthew (28:16-20—NEB) we read that the eleven disciples made their way to Galilee, to the mountain where Jesus had told them to meet him. They fell prostrate before him, though some were still doubtful. Jesus told them to go forth and spread the gospel and to baptize men everywhere in his name.

In Mark, however, different events are reported (16:9-26—NEB). Not all of the earlier versions of the Gospels have this account, which raises a question as to whether it was later interpolated. In any case, Mark says that on Sunday morning Jesus appears first to Mary of Magdala. But they did not believe her, again raising for us the question of her credibility as a sole eyewitness. Later, Mark says, Jesus appeared in a different guise to two of the men as they were walking. But again no one believed them. This again suggests that Jesus was incognito, or was it someone they mistook for Jesus, some stranger they decided was Jesus because Mary had told them she had seen him? Later, while the eleven were at the table he appeared to them and admonished them for their incredulity. Mark does not report whether this appearance occurred in Galilee. Again Jesus commissions them to carry the word to the world.

In Luke (24:13-53—NEB), we have the report that two of the men were on their way to the village of Emmaus, which was seven miles from Jerusalem, and that on the way Jesus came up and walked alongside them. Luke reports, "but something kept them from seeing who it was." But by the time they had reached the village, as he was about to continue his journey, they pressed him to stay with them as evening was falling. So the man stayed and broke bread with them. Only then does Luke report that "their eyes were opened, and they recognized him." At that point we are told that "he vanished from their sight." So the same question arises: Who was the stranger the two men met? Was it Jesus, in disguise fleeing in fear, or was it someone else they did not recognize but thought resembled him? Or was it their fertile imagination?

The men then returned to Jerusalem (not Galilee), where they found the eleven and the rest of the company. They interpret that Jesus had arisen and appeared to Simon (no one else corroborates this). As they are talking, Jesus then appears and this startles everyone. They think they are seeing a ghost. Jesus asks them for something to eat, and they give him a piece of fish, which he eats. This would be hardly necessary for a dead man but no doubt necessary for someone escaping his executioners. He asks them to carry his message and he leads them out as far as Bethany, when he departs. (Some versions of the Gospels say that he was carried up to heaven.) They return to Jerusalem with great joy.

Thus far in the Gospels, we have Jesus appearing to: (1) Mary of Magdala alone; (2) two men on the road, who do not recognize him and barely make him out; he breaks bread with them but then disappears or departs; (3) the eleven plus others following him from Jerusalem.

This is hardly strong evidence for the resurrection. Jesus does not appear to any who are not his followers, to other Jews or Romans, but only to a silent and demoralized band of believers. The reports of the witnesses are questionable—Mary by herself or two men who are not even certain whom they saw. If Jesus appeared to his eleven disciples, was it a last farewell, as he takes his departure? Or was this a kind of mass self-deception in a small harassed community of superstitious folk, who had followed him and wished to believe in him, now grasping at any shred of evidence that they could. All of this is within the realm of human imagination. What is unlikely is the veracity of the claim that Jesus died and then came back from the dead.

Does the Gospel of John add anything to support the case of a dead and resurrected Jesus? There are significant new additions to the story. One wonders whether the oral tradition had already embroidered the Christ legend so as to strengthen what appeared to be a weak case.

Again, it is late on Sunday evening when the disciples are huddled together behind closed doors. We are now told that Jesus appears to them (20:19 ff.—NEB). The disciples are at first doubtful, but in great joy. John reports that one of the twelve, Thomas, was not with the rest. If Judas was no longer with them or dead, then there should have been only eleven disciples, not twelve. Has John forgotten that important fact? He says that a week later Jesus appeared to the disciples and managed to come through locked doors. Thomas, who doubted the reports of his earlier appearance, is told to touch the marks of the nails in Jesus' hands and the wound in his side.

Some time later, John maintains, Jesus showed himself to several of his disciples by the Sea of Tiberias in Galilee. Again at first they did not recognize him. There Jesus told them where to fish; he ate breakfast with them, including bread and fish, which was cooked on a charcoal fire. This was the third time, said John, that Jesus appeared. Had he made his way to Galilee and was he again in disguise, apprehensive that the Romans or the Sanhedrin might recapture him? And did he appear only fleetingly to his former disciples? After breakfast, Peter saw Jesus leave, and the disciple whom Jesus loved followed him. Did Jesus make his final escape with the help of a companion? Was this disciple his accomplice throughout Jesus' brief career as a magician and conjurer? And was it he who helped him out of the cave after it was discovered that he did not die on the cross and had revived? Or was it his beloved friend Lazarus, as some have suggested? Was it the author of the Gospel of John, who still remains unidentified?

All of this is no doubt pure conjecture. We do not know. But one thing that seems clear is that the evidence in the four Gospels for Jesus' alleged historical resurrection is flimsy. Indeed, it is totally untrustworthy and inconsistent. It may all have been the product of literary imaginations. We don't know whether Jesus did or did not die on the cross and escape from the tomb, or whether his dead body was simply spirited out of the tomb by his followers, as the pagan and Jewish critics of Jesus claimed, or whether the entire tale was concocted by them to save face. What is clear is this: the evidence presented is neither remarkable nor convincing. This momentous event in history—the supposed resurrection of the son of God—is thus unproven. It is not even likely or probable. It remains, in the last analysis, only an article of faith.

The alleged resurrection of Jesus is not unique. The scriptures provide many other accounts of people rising from the dead. As we have seen, Lazarus allegedly rose from the dead. Elijah raised a child from the dead (1 Kings 17:7, 21-22) and Elisha raised the dead son of a Shunammite (2 Kings 4:32, 34-35). After Jesus, Peter raised Tabitha, and Paul raised Eutychus from the dead. Why then, is the resurrection of Jesus so vital to Christianity? Because only Jesus was the son of God. The reality of the resurrection is crucial to the entire Christian theology. It was allegedly based on scriptural prophecy, and we are reminded over and over again that Jesus is fulfilling the prophecies of scripture. Thus, his disciples believed he would rise in three days, more precisely three days and three nights. In Matthew we read: "Jonah was in the sea-monster's belly for three days and three nights, and in the same way the Son of Man will be three days and three nights in the bowels of the earth." (12:40—NEB). Actually, however, if Jesus was hung on the cross at nine on Friday and died at three in the afternoon, after about six hours, and was taken into the tomb on Friday night, it was not after three days and nights that he was raised, but about thirty-six to forty hours until early Sunday morning.

If the four Gospels fail to provide adequate evidence for his resurrection, are there other passages in the New Testament that support the claim? There are, but these do not advance the claim very much. In the Acts of the Apostles (1:1-11), most likely written by the author of Luke, we read that Jesus appeared to the apostles for a period of forty days, a new fact that is not mentioned in the Gospels and is uncorroborated elsewhere. Here we also read that he was lifted up and a cloud removed him from sight. To the primitive mind, God was in the sky and upon entering Heaven, the person departed upward. There were two men in white who appeared at the same time. How reliable is this account? Immediately following it, the author tells us that Judas, who had betrayed Jesus, after buying a plot of land with the payment for his villainy "fell forward on the ground, and burst open, so that his entrails poured out" (Acts 1:18-19—NEB). This again contradicts Matthew, who tells us that after Judas betrayed Jesus, he was seized with remorse and returned the thirty pieces of silver, hurling it into the Temple and hanging himself (27:3-6—NEB). How reliable is the author of Acts, who behaves more like a propagandist for a new faith than a historian seeking to research the factual truth? Luke reports the rapid growth of adherents to the early church, as they were aroused by the mission of Jesus. The resurrection, however ill-founded, was apparently vital to attracting followers to the new creed. Interestingly, people of the early church practiced rites of magic and exorcism, similar to those of Jesus. Had some learned from Jesus the craft of deception, or were they all true believers, deceived as to his divine mission but also practicing the art themselves?

The other major source for belief in the resurrection of Jesus was Paul. As I have pointed out, it is most likely that his letters were written before the Gospels. It was Paul who transformed Christianity into a religion beyond the small band of early Jewish-born believers. For Paul it is *faith* in Christ that is the central tenet of Christianity. Paul never knew Jesus; he was not a direct eyewitness of his ministry or of the resurrection. Yet the entire fabric of Christianity depends upon an article of faith that Jesus is the messiah, the son of God, who came to save mankind, was crucified, and resurrected. Paul knew of the spreading Jewish Christian cult, which he had at first persecuted. He came to accept at face value its belief in the resurrection. He adds a significant new piece of data, however, that does not appear in any other Gospel and is no doubt an exaggeration. Yet it was perhaps a necessary reinforcement, if Paul were to convert the Gentiles to his belief in Christ.

In 1 Corinthians (15:3-7—NEB), Paul says that certain facts had been imparted to him: "that Christ died for our sins, in accordance with scriptures; that he was buried; that he was raised to life on the third day; according to scriptures; and that he appeared to Cephas" (this is the first that we hear of this) "and afterwards to the Twelve" (since the Gospels said Judas was dead, how was this possible?). Paul then adds a remarkable fact, which none of the other Gospel writers cite: "Then he appeared to over five hundred of our brothers at once, most of whom are still alive, though some have died. Then he appeared to James" (this does not appear in the Gospel stories) "and afterwards to all the apostles."

Most of the preceding varies from the accounts already cited where Jesus appears to Mary and two men before the apostles. Then Paul adds the *coup de maître* to support his case: "In the end, he appeared even to me." (1 Cor. 15:8—NEB). This is the famous vision that Paul had on the road to Damascus. In Acts 9:3-9 we have an account of this conversion. Saul, a persecutor of the Jewish Christians, while nearing Damascus, reports that a light suddenly flashed from the sky all around him. He fell to the ground, hearing a voice saying, "Saul, Saul, why do you persecute me?" Saul asks, "Who are you?" The voice answers, "I am Jesus, whom you are persecuting. But get up and go into the city and you will be told what you have to do." According to Acts, the men who were traveling with Saul were speechless. We are informed that they heard the voice, but they "could see no one." Saul got up, but "when he opened his eyes he could not see." Saul "was blind for three days" and took no food or drink. The real question is whether this was a genuine vision, a psychotic reaction, an epileptic seizure, or something else. How did Paul know that those accompanying him were speechless or heard Jesus' voice, if he could not himself see them? There is no independent verification; not even his companions could confirm Jesus' appearance. This revelation from on high is a subjective experience, yet it stands out as possibly the strongest single and most important event in Christendom after Jesus' death. For it is Paul, perhaps more than anyone else, who converts Christianity into a creedal belief for the next generation of people who did not know Jesus, but he based it on his inner soliloquy, a nonevidential excursion into private pathology.

Conclusion

From our study of the historical Jesus we can draw the following conclusions. Although Christianity claims to be a historical religion, built upon the life, ministry, and death of Jesus Christ—the central claim being that God became incarnate in human form—as we examine our historical knowledge of Jesus we find that we cannot say with certainty that he ever existed. But assuming that he did, there are still several alternative, rationalistic interpretations of his life and ministry.

First, there is the traditional Christian view that Jesus was God incarnate sent to save suffering humanity. There is inadequate evidence for this claim and no independent corroboration of his ability to perform paranormal healings or miracles.

Second, even if we could attribute directly and solely to Jesus an ethical code, there is limited merit to the ethic of the Gospel (and it is not exclusive to Christianity), and the Christian code hardly serves by itself as a reliable guide for informed ethical conduct.

Third, if Jesus existed, then he can only be understood in the context of first- and second-century Jewish society, a man influenced by the social, cultural, political, and religious traditions of his day. Neither his appearance nor his message is unique.

Fourth, if Jesus existed and performed any of the feats that the Gospels claimed he did, then it is not implausible that he was a disturbed though charismatic personality who practiced magic, using familiar conjuring techniques. The fact that Christianity has succeeded in vanquishing so many other religions in the world and in challenging the scientific method and rationality for so long suggests that Jesus was extremely successful at his craft, perhaps the most successful magician who has ever existed. The mythology that surrounded his career was hungrily devoured by countless millions of believers, who found in it a message that satisfied their need for salvation and redemption. Others have also succeeded in their craft, but none have been deified with quite the same degree of drama, fervor, and conviction. With the possible exception of Islam, no other religious institution or church has been quite as powerful as the one dedicated to Jesus Christ.

VIII: Moses and the chosen people

Judaism as a religion is also based upon divine revelation as it is recorded in the Old Testament. The Jews, a tenacious and heroic people, have been inspired by this revelation ever since the Pentateuch was written. They have clung to a religious heritage that is harsh and demanding, yet which provided them with a powerful motivation that has enabled them to survive for over three thousand years. The Old Testament is presumptuous, for it arrogantly proclaims the Israelites to be the "chosen people," special favorites of God. This has aroused intense animosity among Gentiles.

The Hebrew God was a possessive deity, demanding unquestioning obedience. He was viewed as favoring the Hebrews in some special way, having delivered them from Egypt and bequeathed them the Promised Land. He nevertheless allowed them to be dispersed and defeated whenever they disobeyed him, and this culminated in the great diaspora after the destruction of the Temple in Jerusalem in C.E. 70. Throughout their history, Jewish prophets, priests, and rabbis taught that their supreme obligation was to obey the law of God as handed down by Moses. Someday, they believed, a messiah would emerge who would deliver them to Israel. This belief has been nourished in open or in secret by the remnants of Israel throughout the ages, as they suffered persecution.

Anti-Semitism is as old as the Jews. In Egypt, before the Mosaic code was enunciated, they were persecuted. Their homeland was later invaded and they were dispersed; some were carried away at different periods, by the Assyrians and the Babylonians. Persecution continued intermittently at the hands of the Romans, who slaughtered great numbers, as did the followers of Mohammed later. They were cruelly persecuted during the Spanish Inquisition, the Russian pogroms, and the Nazi-produced Holocaust. In the twentieth century, many Jews reenacted the exodus from Egypt by returning to Palestine and establishing a new state against the bitter opposition of the displaced Palestinians and the rest of the Arab world.

In Roman days, the Jews were spread throughout the cities of the Mediterranean world. Strabo, the Greek geographer in the first century B.C.E. reported that "it is hard to find a place in the habitable earth that has not admitted this tribe of men, and is not possessed by it."[1] Often

1. Quoted by Flavius Josephus, *Antiquities,* Book XIV, Chapter 7.

intensely religious, practicing their customary worship in isolation from the societies in which they sojourned, a hard core has clung to the Mosaic law and traditions. Since many intermarried with the peoples among whom they lived—in violation of a prohibition against it—it is doubtful today that the Jews who have survived are of the same stock as their forebears. Hitler was mistaken, for modern Jews have mixed blood flowing in their veins. Wherever they settled they mixed with those about them, and they became blondes, brunettes, and redheads as well as black haired, assimilating Egyptian and Canaanite blood at first, but later Greek and Roman, Spanish and Moor, European and modern-day American and Latin American. In medieval times the Khazars of the Russian steppes converted to Judaism, adding yet another strain to the Jews.

Dissident Jews abandoned Judaism and founded the two great competing religions of our day, Christianity and Marxism, both of which, although initially established by the Jews, ended up by persecuting them. Major contributions to world civilization have been made by Jews who transformed or rejected their faith: Paul, Spinoza, Marx, Freud, and Einstein. The Jews numbered an estimated seven million in the Roman Empire of the first century (7 percent of the population). Their numbers have only managed to double, to fourteen million, two thousand years later, given the continued attrition by conversion and the loss of six million Jews in the Holocaust.

The Jews of modern times again live in virtually every country of the world. In the West they provide creative talent and energy in the professions, business, commerce, and the arts and sciences. In every generation, scores leave the faith of their fathers or are assimilated. In some periods they have had enormous families (as among Russian Jews); in others, the birth rate has declined drastically (as in 20th-century United States). Yet an impressive number have loyally clung to the ancient doctrines.

The Jewish saga has taken on special meaning today, for the Jews, in returning to Palestine, have attempted to provide a continuity with their ancient biblical hopes and aspirations. Had it not been for Theodor Herzl and Zionism, which began as a secular movement, and the extermination policies of the Nazis, it is unlikely that the state of Israel would have been established. Most likely, the Jews in large sectors of the world might have further assimilated into the mainstream of other nations. Today there has been a resurgence of nationalistic sentiments among the Jews, reminscent of the Old Testament, and a fervent belief that Israel was promised to them by God and that the Jews should possess it as a divine right. Most Orthodox Jews look upon Israel as God's special creation. Even fundamentalist Christians see the return of the Jews to Israel as fulfilling biblical prophecy. Those who take the Bible literally wait for Armageddon to be ushered in—after the rebuilding of the Temple in Jerusalem—as foretold in Daniel and the Book of Revelation.

Explanations for the continuing persecution of the Jews are many. They were a visible minority wherever they went, following strange religious customs and practices. The Jews have had strong ethnic allegiance, cemented by the memory of their past, strict laws against intermarriage and conversion, and a close family structure, which was rigorously guarded and which enabled them to provide continuity as they survived. Their intellectuality and hard work have given them positions of prominence and wealth in many societies. Interested in defending freedom against the establishment, they often adopted liberal and radical causes and founded or joined protest movements. And so they became easy targets for anti-Semites. They have been the eternal scapegoats for demagogues seeking political power or economic gain. The victims of invective and innuendo, rumor and gossip, they were cruelly persecuted—unfairly and unjustly— to the moral shame of mankind.

Christianity and Islam also present unique systems of belief, and both have their original inspiration in the Old Testament and the Hebrew prophets. However, they are universal missionary religions seeking to recruit new believers. Anyone who embraces the faith can join, and conversions are welcome. The Hebraic message, on the contrary, is specialized to one people and is highly chauvinistic: God intervened in history and endowed the Jews with a special role. Jehovah is the God of the Jews. Thus Judaism is an inward-looking religion only applicable to those born to a Jewish mother. Only grudgingly do the Jews allow those who marry non-Jews to convert. Even then they are not considered to be bona-fide Jews. Moreover, to qualify, males have to be circumcised, a primitive initiation rite that one can impose on infant boys but which is difficult to require of adults.

In one sense, the survival of the Jews over the millennia is based on a neurosis, which has been supported and maintained by the authority of tradition. The entire history of a people has remained out of touch with reality. Their interpretation of their existence is based upon a delusional belief-system that has no basis in fact. This neurosis has enabled the Jews tenaciously to survive, but it meant that they would be condemned to stand huddled together in the midst of hostile groups, nourished by their faith in the Mosaic legends, hopeful of a future messiah. They imposed upon themselves a religious system that strictly regulated their life and a moral code in which repression in the name of God was justified. Out of this developed an addiction to hard work, family fidelity, and an appreciation of intellectual and spiritual values. As a minority, they lived in a state of alienation from the social worlds of their neighbors.

The nations in which Jews settled often appreciated their hard-working abilities, and sometimes gave them the freedom to excel and prosper. But often they would be despised because of their separateness and envied for their achievements. Because they carefully tutored their young in the family and the synagogue that they were God's chosen, a superiority complex developed. If those about them hated or rejected them, they considered themselves special. Being able to read the ancient language of Hebrew, which no one else understood, had profound mystical significance for them as they pondered their sacred books. And it provided them with the support to maintain their separate heritage. They were able to resist the mythologies or beliefs of the alien cultures in which they lived, and they could withstand the conversion entreaties of pagan religions, Christianity, and Islam. Many Jews did convert and were lost to the tribe, for they wished to be accepted; yet a hard core remained and persisted in every age, comforted and sustained by a delusional system of beliefs and aspirations. Although Jews had to learn the language and culture of the lands in which they lived and had to pledge allegiance to their new nationalities, their old ethnic heritage and religion persisted side by side with their new citizenships.

Was Moses an Egyptian?

Basic to Judaism is the Old Testament, especially the first five books, known as the Pentateuch, or the Torah, and central to this is the figure of Moses, which every young Jew, nurtured in the faith of his fathers, learns about. Moses is the greatest figure in Jewish antiquity: the inspiration of Judaism. We know nothing about Moses beyond what has been related in the Old Testament and the written and interpretive traditions based upon it. The Torah was at first the product of an oral tradition, and of the legends that grew up surrounding it. People have a curious veneration for remote antiquity. The great religions have their roots in legend, myth, and some historical facts, so intertwined that it is difficult to separate reality from fantasy; and apologetics are composed post hoc to dramatize and maintain the authority of the tradition. How many of the biblical stories are fiction and how many actually happened is difficult to say. No independent, corroborative evidence of Moses' existence has ever been found. Indeed, there are scholars who have questioned whether Moses even existed.[2] Our chief source is the biblical narrative, and this narrative was no doubt revised, altered, and enlarged upon many times before it was codified in the version that we now have.

2. See especially Eduard Meyer, *Die Israeliten und Ihre Nachbarstämme* (1906).

The books of the Old Testament were written over a period of a thousand years as part of the religious, political, and ideological tradition of the Jews. Biblical scholars have been able to disentangle different accounts in the Bible. The first is called Yahwist, simply because of the relatively consistent use of the name *Yahweh* for the Hebrew God. This is usually called the J form (after the German *Jahweh);* it was thought to have been written down about the tenth century B.C.E. Beside the Mosaic accounts, there are two later sources, designated as the Elohist (E) because it uses the term *Elohim* for God, and much later the Priestly code (P). The orthodox religious tradition, of course, attributes the authorship of the first five books of the Bible to Moses—even the description of his own death at the end of Deuteronomy. But there is no evidence to support this claim of authorship. The Pentateuch in the form we have today was written about 400 B.C.E. after the mission of Ezra, who presented a canonized and revised Torah to the Jewish community. This was accepted as sacred and complete and became the foundation of Judaism as a national religion.

If there was a real Moses, he probably lived in the thirteenth or fourteenth century B.C.E. His main claim to fame was that he was the liberator of the Jews, leading them from bondage in Egypt to conquer and settle in Canaan after they wandered many years in the desert land east of Egypt. The story of Moses' birth and upbringing is similar to that of other great founders of religions: it involves a noble background. Indeed, Moses' origin bears similarities to other hero-child legends. It is similar to that later told about Jesus, but it also bears a likeness to the birth stories of Sargon of Akkad, Gilgamesh, Oedipus, Cyrus, Romulus and Remus. According to Akkadian lore, Sargon's mother cast him adrift in the Euphrates in a pitch-covered basket. A farmer rescued the infant and raised him. Sargon later rose to displace the Sumerian king Ur-Zababa and became king of the Mesopotamian empire. This legend predates the Mosaic story by many centuries.

A common theme in such stories is that a child is born who is viewed as a threat to the throne. A king gives orders to kill the child, and only by God's intervention is he able to grow up and fulfill his divine mission. (In the New Testament Jesus was spared Herod's order to kill all male children and grew up to be called king of the Jews.) According to the story, the Pharaoh was fearful of the high birth rate of the Hebrews and ordered their first-born to be slain. To save him, Moses' mother put him in a rush basket made watertight with clay and tar and left him to float in the Nile, hoping he would be discovered by the Egyptian princess. Indeed, she found him and had him nursed by a Hebrew woman (who turned out to be his mother). Moses was adopted by the royal family and reared in their household. Later he went on to challenge the Pharaoh and thus fulfill his destiny.

The princess called him Moses because, as she said, "I drew him out of the water" (Exod. 2:10).* In the biblical account, Moses was a Hebrew, a descendant of the Levites, though he was raised as an Egyptian. Some biblical scholars have raised the question whether Moses (if he indeed existed) was an Egyptian. The name *Moses* is Egyptian, meaning "child," as in the royal names Thutmoses (Thotmes) and Ramoses (Ramses). The Hebrew philological interpretation thus has been held to be untenable. Sigmund Freud makes this point central in his novel interpretation of the rise of monotheism.[3] There is even some biblical evidence for the possibility that Moses was an Egyptian. Later, when he is living in Midian, he is described by the daughters of Reuel as an Egyptian who rescued them from the shepherds (Exod. 2:19). No doubt Moses had the bearing and dress of an Egyptian and thus would have been so identified. Yet whether or not he was a Hebrew or an Egyptian is still unclear. It would have been a great assault to Hebrew identity to have their national liberator and the founder of their religion to be a foreigner. The Moses that legend invented is reminiscent of other legends; it is made to fulfill an ideological end in order to give a tradition and a purpose to a people.

Freud's radical theoretical reconstruction goes still further. He pointed out that Moses did not discover or enunciate a new religion but rather imposed a version of an Egyptian sect, the Aton religion, which had already been introduced by Ikhnaton in Egypt and propagated in opposition to the dominant polytheism. This new religion did not worship the fire god or animals; nor did it focus on a doctrine of the immortality of souls. According to Freud, Moses was a dissident Egyptian priest who followed the new religion. It is not clear whether Moses believed in the form of strict monotheism later developed by the Jews; certainly one can read the early Mosaic books as defining and defending one god—the god of the Jews—who competes with other gods. While the narrators proclaim that the children of Israel are to have "no other gods" before them, they do not clearly state that the other gods worshipped by other peoples do not exist. Indeed, it is clear that the Pentateuch is as much a nationalistic statement as it is a religious statement; for it defines a national existence written after the fact; and God is first and foremost the god of the Jews.

*All references in this chapter are from the New English Bible.

3. Sigmund Freud, *Moses and Monotheism* (New York: Vintage Books, 1939).

Nor is it even clear at this point that Moses led the Hebrews out of Egypt or that they had a distinct ethnic or racial character before this. Indeed, if one reads ancient sources, one finds a widely held view that Moses led a dissident class of Egyptians, perhaps diseased lepers, from Egypt and that the Hebrews did not even exist at that time. It is generally accepted among scholars that the term *Hebrew* was an appellation meaning "foreigner" long before it was used to identify an ethnic group. It is also interesting to note that circumcision was practiced among the Egyptians and other ancient Bedouin tribes; so that was not unique to Moses. It was subsequently adopted by the Hebrews because of divine sanction, and they gave it some kind of magical signficance.

What do pagan sources say?

Pagan anti-Jewish sources in the first centuries B.C.E. and C.E. held that Moses was of Egyptian descent. In the writings of Pseudo-Manetho, Chaeremon, Apion, Hecataeus, Strabo, Pompeius Trogus, and Tacitus, Moses was held to be an Egyptian priest who was expelled from Egypt along with a group of inhabitants who may have been contaminated.

According to the biblical account, the Hebrews, who became slaves in Egypt, traced their lineage to Abraham, Isaac, Jacob, and Joseph— originally shepherds from Canaan. In the familiar biblical narrative, Joseph was sold into slavery by his brothers. While in Egypt he prospered as an advisor to the Pharaoh and eventually invited his entire family to live in Egypt. As time went on, they proliferated. The Bible says they were in Egypt for 430 years, but it also mentions only four generations (about 80 years), so there is a contradiction on this point. By the time of Moses they were doing all the menial labor and were downtrodden. They were so numerous that they may have outnumbered the Egyptians. The Bible says that some 600,000 male warriors left Egypt, which, if taken literally, would suggest a total population of two and half million. What an enormous exodus that would have been! It is probable that the biblical text is exaggerating, perhaps to bolster national pride. According to biblical sources the Hebrews had a national identity while in Egypt. There are some indications that this may have been invented post hoc by the priests and scribes, who wished to keep alive Hebrew national awareness and to sanctify their own hegemony and control of the populace.

After the exodus from Egypt, the Hebrews defeated other tribes in battle and often married their women. In one case, we are told that the Hebrew warriors took 32,000 virgins as their wives, killing everyone else. This again seems like an inflated figure. Still there is the constant infusion of new tribal and ethnic stocks. The later-day Hebrews were a blend of those who lived in Egypt, fused with Midianite, Canaanite, and other blood. Indeed, Moses' first wife was a Midianite, as were his children. Wherever there is a large-scale immigration and invasion, there is bound to be intermarriage and miscegenation. No doubt the same thing happened to the Hebrews as they formed a national group. Accordingly, it is possible that the original band of followers Moses led were not Hebrews per se. They had no religion to bind them—this was given to them by Moses—or strong traditions. They may even have been a class of Egyptians, including foreigners, who were living in Egypt. Only after the exodus did Hebrewhood begin; it was based on the laws and rituals promulgated by Moses and later elaborated and developed.

As I have said, the pagan literature—Egyptian, Greek, and Roman—held this to be the case. By the second century B.C.E. the Jews had developed a clear national identity; and later, after numerous wars and deportations, they lived throughout the Mediterranean world as well as in Palestine. In particular, there was a large Jewish colony living in Alexandria. The Jews of this period did practice the religion of Judaism, and there was a literature of sacred books, the Temple system, and a rabbinical class, which tried to maintain the observances and keep the Jews together. In this regard, the Mosaic legend became vital, for it not only gave them a past heritage (in which they and they alone were the people chosen by God) but it was also a basis for present cohesion and a future destiny. The promised land, Eretz Israel, was a gift of Yahweh to them. By then, many of the traits that have defined modern Jews had emerged, particularly their exclusiveness and their pride rooted in their religious conviction that they had a special place in the scheme of things. Judaism, in the last analysis, rests on an inflated national pride and theological pretensions that have no basis in fact. These are built upon the claims that Moses delivered the Hebrews from bondage, that he was a prophet of God, that he received revelations from on high, including the Ten Commandments, and that it is the duty of the Jews to fulfill these commandments and maintain a cohesive identity. I find all of these claims false, indeed presumptuous.

Returning to the pagan writers, we may ask: What did these non-Jews say? Many of them no doubt were anti-Jewish, and in the first centuries C.E. also anti-Christian. Accordingly, they would look with considerable disdain and disbelief at the Old Testament, which both the Jews and the Christians accepted. The Christians had adopted the traditional Hebrew litany of the Old Testament. The New Testament was supposed to be its fullfillment. And they accepted the books of the Pentateuch, Genesis, Exodus, Leviticus, Numbers, and Deuteronomy, as the word of

God. They took over in toto the received Hebrew doctrine about Moses. The pagans on the other hand, writing a thousand years or more after Moses, took him to be a renegade Egyptian priest, who abandoned his former religion and counseled his followers, a group of exiles, to despise anything foreign and to destroy foreign temples and religions wherever they went. He gave them a bizarre new religion, which was strange to pagan ears but is less strange to us, for it has captured large sections of the world in its Christianized form.

Hecataeus of Abdera, a Greek writing in the fourth century B.C.E., gave his own account of the Jews and their customs, which differered from the account in the Bible:

> When a serious pestilence arose in ancient Egypt, the populace attributed the cause of the difficulties to the divinity. Inasmuch as many different groups of foreign aliens were living there and followed foreign practices with respect to the temples and the sacrifices, the result was that the traditional worship of the Gods had been neglected. Thus the natives supposed that unless they removed the foreigners there would be no end of their difficulties.[4]

Hecataeus then relates that "the foreigners had been banished." Many of them, he says, went to Greece and other regions; the largest group, however, went into what is now called Judea. This account has the Egyptians expelling all foreigners from Egypt, not just the Jews.

Another account is by the Egyptian author Manetho (about 300 B.C.E.). His story of Moses runs as follows:

> It is said that the priest who laid down for them their constitution and the laws was a Heliopolitan named Osarsiph after the God Osiris in Heliopolis; when he went over to this nation (the Jews) he changed his name and was called Moses.[5]

Josephus, the Jewish historian, writing in the first century, quotes Manetho in a book defending the Jews against their detractors. He writes to refute the claims that the Hebrews were expelled from Egypt to Syria along with 80,000 lepers who allied themselves with them under the priest Osarsiph in order to battle the Pharaoh. Josephus attributed this to "mere legend," but it is interesting that it was a view then current. Josephus' own account of Jewish antiquity is based largely on the biblical account, which he accepts uncritically, miracles and all. He denounced all the pagan stories as fabrications. He insists: "Our fathers were not origin-

---

4. Quoted in John G. Gager, *Moses in Graeco-Roman Paganism* (Nashville, Tenn.: Abingdon Press, 1972), p. 26.

5. *Aegyptiaca* 1.250. Quoted in Gager, *Moses*, p. 114.

ally Egyptian, nor were they expelled, either on account of bodily diseases or any other calamities of that sort" (Flavius Josephus, *Against Apion,* Book II).

In the first century C.E. other stories of Jewish history appeared in Strabo's *Geography*. The Greek text has been attributed to Posidonius-Strabo.

> Although they (the inhabitants of Judea) are thus of mixed origins, the predominant report of those currently held concerning the temple in Jerusalem affirms that the Egyptians are the ancestors of those people who are now called Jews.
>
> For a certain Moses, who was one of the Egyptian priests, held a section of what is called the (lower) region (*chōra*). But he became dissatisfied with the way of life and departed thence, to Jerusalem, in the company of many who worshipped the deity.[6]

Here Strabo was repeating the belief widely held by his contemporaries about the Egyptian origin of the Jews. This was supported by the Roman historian Tacitus (first century C.E.), who lists five different theories about Jewish origins but nevertheless maintains that the most prevalent theory was that Jews were descended from the Egyptians.

Another Roman historian, Pompeius Trojus, wrote in *Historiae Philippicni* about the expulsion of Moses and his followers from Egypt. He claimed that the Egyptians were exposed to "scabs" and "a skin infection," and expelled Moses together with the sick people beyond the confines of Egypt "lest the disease should spread to a greater number of people."[7]

Interestingly, Pompeius offers an explanation for Jewish separateness. It was not because they were the chosen people, but rather:

> . . . because they remembered that they had been expelled from Egypt due to fear of contagion, they took care not to live with outsiders lest they become hateful to the natives for the same reason (i.e., fear of contagious infection). This regulation, which arose from a specific cause, he (Moses) transformed gradually into a fixed custom and religion.[8]

Whether this explanation is correct is hard to say. It was prevalent during that period. Tacitus supports the view when he maintains that the Jews "abstain from pork in recollection of the plague, because the scab which attacks this animal once infected them."[9] The expulsion legend is

6. Quoted in Gager, *Moses,* p. 38.
7. Ibid., p. 49.
8. Ibid.
9. Ibid., p. 83.

repeated by Chaeremon and his contemporary, Apion, two Alexandrian writers who claimed that the Egyptian Pharaoh "collected 250,000 infected people and expelled them."[10]

But we may ask, who were the people who left Egypt? Again a vast number of people, according to the non-Jewish version; they apparently had no common religion or national identity. Why should we accept the biblical narrative as objective history, which already had been transformed into an article of religious faith? Why not admit an alternative account of Jewish history? Perhaps both are biased. Perhaps the truth lies somewhere in between.

The revelations of Moses

Let us now resume our analysis of the Old Testament account of Moses and the exodus. The central thesis of both Judaism and Christianity is that Moses was a prophet appointed by God to lead the children of Israel out of Egypt. This divine mission was entrusted to Moses by means of special revelations. The basic question we must raise is: "Who instructed Moses with this task and what is the evidence for it? How do we know that it was God?" The major testimony is from Moses himself, as recorded in the Bible. There is circular reasoning here. Moses is the prophet of God because he said that he was. How do we know that it was God who said he was? Because Moses (the author of the Pentateuch) tells us he did. At least that is the structure of the argument of the Pentateuch.

The Moses legend of the Bible runs as follows: As a grown man, Moses saw an Egyptian strike a fellow Hebrew. Seeing that there was no one about, he struck the Egyptian and hid his body in the sand. When Pharaoh heard about it, he tried to put Moses to death; but Moses escaped and settled in Midian and there married Zipporah, who bore him a son. We are told in one place that his father-in-law was Reuel and in another Jethro, priest of Midian. There are contradictions in the account, for his father-in-law has still other names: Jether, Cain, Hobab, or Hobabben Reuel.

Of great importance is Moses' first encounter with Yahweh. While minding the flock of Jethro, he went to Horeb, the mountain of God. There we are told "the angel of the Lord appeared to him in the flame of the burning bush" (Exod. 3:2 ff.). Moses noticed that though the bush was on fire, it did not burn. He was on his way to see the wonderful sight when Yahweh called out of the bush: "Moses, Moses." Moses replied,

10. *Aigyptiakē Historia* 288–92. Quoted in Gager, *Moses,* p. 121.

"Yes I am here." And God told him to take off his sandals for he was standing on holy ground. Then God said, "I am the God of your forefathers, the God of Abraham, the God of Isaac, the God of Jacob." Moses covered his face for he was afraid to gaze on God. God next told him that he had witnessed the suffering of the Jews in Egypt under their slavemasters and that he will send Moses to Pharaoh to lead the Israelites, his people, out of Egypt.

Moses demurred, "Who am I, that I should go unto Pharaoh and that I should bring forth the children of Israel out of Egypt?" And God answered, "I am with you. This shall be the proof that it is I who have sent you," and he told Moses that he and the people should worship God on the mountain (Exod. 3:11-12). Then Moses asked what he should say if the people asked the name of God. God replied, "I AM; that is who I am. Tell them that I AM has sent you to them." This quest for the name is based on the age-old belief that knowledge of the right name confers on man some magical power, including power over the deity. But the deity is evasive and will not give his name or power over himself. This belief persists among many Jews even today; they are still unwilling to spell his name. Is God a transcendent being? How is he able to enter into human affairs, influence events, and communicate with special persons?

Now taking the text on its own terms at face value for the moment and leaving aside all historical interpretations, it should be clear that we have the so-called testimony of one man, who says that he encountered God. There were no impartial witnesses on the scene, which provides excellent ground for the skeptic to withhold any kind of assent to the uncorroborated claims of Moses' divine authority. Interestingly, passages of the Old Testament, upon which so much depends, are rooted in uncorroborated appearances or dreams.

The text itself is troubled on this point, for it seeks to reinforce the divine mission. Moses is to tell the elders that he is to lead them out of Egypt to a promised land. Moses himself says, "But they will never believe me or listen to me; they will say, 'The Lord did not appear to you.'" (Exod. 4:1-2). And at this point, the Lord is to provide Moses with the power to work miracles as a sign of his power. God commands him to throw his staff on the ground. He does so, and it turns into a snake. He is told to seize the snake by the tail, and when he does, it turns back into a staff. Moses puts his hand inside the fold of his cloak and when he pulls it out it appears diseased and as white as snow (as if leprous); when he puts it back it returns to health. He also pours water from the Nile onto dry ground; it turns into blood. Thus we have the biblical text supporting the truth of the revelation by reference to signs of miraculous powers that Moses allegedly had. Moses also says that he is bashful and slow of speech, and God tells him that his brother Aaron will be his mouthpiece to plead with the elders and the Pharaoh.

There are two key issues here: First, the use of uncorroborated testimony by the prophets to lend credence to the belief that they had divine revelations, and second, the use of magic to stimulate belief in a credulous audience that they were indeed God's prophets. I will return to the question of whether Moses was a magician, but now I wish to focus on the claim to divine knowledge by means of revelations.

If we look at earlier passages in Genesis, we see that God often appeared to the forefathers in private encounters. At one point, Yahweh speaks to Abram alone and shows him the land of Canaan in the distance, saying that he will give all of the land to him and his descendents forever (Gen. 13:14–16). Later we read, "The word of the Lord came to Abram in a vision" (Gen. 15:1). When Abram was ninety-nine years old, the Lord again appeared to him, making a covenant with him and changing his name to Abraham (Gen. 17:1–6).

Among the most famous stories was the effort of Yahweh to put Abraham's faith and obedience to the test. The Lord commanded Abraham to take Isaac, his only son by Sarah, to one of the hills and offer him as a sacrifice. The biblical narrative has Abraham saddling his ass and taking with him two of his men and Isaac. When they reached the place he said to his men, "Stay here with the ass while I and the boy go over there; and when we have worshipped we will come back to you" (Gen. 22:5–6). Abraham goes alone with Isaac and is prepared to do God's bidding. He tells Isaac to come, and he lays him on the altar on top of the wood. As he takes his knife to kill his son, the angel of the Lord calls from heaven and tells him not to sacrifice his son, since he has proved that he was God-fearing and obedient. Instead, Abraham sacrifices a ram that is nearby. This is an important moral lesson that the Old Testament presents against the barbaric practice of human sacrifice. But it is described as a private revelation to Abraham. We do not even have a word from Isaac to confirm his father's encounter with God.

Later, when Isaac is a grown man, he has a revelation from God while in Beersheba. We read: "That same night the Lord appeared to him there and said, 'I am the God of your father, Abraham. Fear nothing, for I am with you. I will bless you and give you many descendants for the sake of Abraham my servant' " (Gen. 26:23–24). Later, we read in Genesis of Jacob's dream about a ladder, which is resting on the ground with its top reaching to the heavens. The Lord tells him again that the land on which he is lying will be given to his descendants. Jacob wakes up from the dream and says, "Truly, the Lord is in this place, and I did not know it" (Gen. 28:16–17).

Dreams have always intrigued men and women, and they no doubt held a special fascination for the uneducated mind. Were dreams prophetic? Should they be taken as messages from God? Joseph, while in Egypt, is said to have been able to interpret the hidden meaning of dreams. Dreams were cited by the authors of the Bible to justify the existence of God. But subjective experiences are notoriously unreliable, and the stuff that dreams are made of is human desire, imagination, and wish-fulfillment. That God should convey such important knowledge to selected individuals while alone or in a dream state is highly suspect. A careful reading of biblical scripture should make one wary of claims made in God's behalf, whether by Moses or anyone else. Interestingly, the number of occasions when God appears to the multitude are fairly rare in the Pentatauch, in comparison with his personal visitations.

There is an incident involving Moses that is rather strange; it suggests the primordial mythological mysteries from which ancient Judaism emerged. God spoke to Moses in Midian and told him to go back to Egypt, for all those who wished to kill him were dead. Moses took his wife and children, mounted them on an ass, and began his journey back to Egypt. "During the journey, while they were encamped for the night, the Lord met Moses, meaning to kill him, but Zipporah picked up a sharp flint, cut off his son's foreskin, and touched him with it, saying 'You are my blood-bridegroom.' So the Lord let Moses alone. Then she said, 'Blood-bridegroom by circumcision.' " (Exod. 4:24–26). Here not only Moses but also his wife encounters the Lord. Why did the Lord wish to kill him by night after having met him by day in the burning bush? Is it because Moses had invaded sacred ground where Yahweh dwelt? Did Yahweh encounter Moses and seek to kill him, much the same as a demon might seek to preserve his domain from any intruder? And what is the meaning of Zipporah's offering the deity the foreskin of her son in this coarse and obscure story? Does it point to the symbolic role that circumcision is to take in placating the deity from harming his people? The symbols of sacrifice and blood are full of sexual possibilities. Freud considered this to imply the intermixing of a castration complex and the fear of God. Magic and mystery survived this fictitious event.

Perhaps the most important incident in the life of Moses was his encounter with Yahweh on Mt. Sinai. Indeed, Moses' reputation rests on the conviction that he was the lawgiver of the Jews because he had received the Ten Commandments and other laws directly from God. According to the Exodus story, the Jews were wandering in the wilderness. In the third month after they had left Egypt, they reached the Sinai and pitched their tents opposite the great mountain. Moses went up the mountain and received a message from Yahweh, telling him to speak to the

sons of Israel and tell them: "If only you will now listen to me and keep
my convenant, then out of all peoples you shall become my special possession; for the whole earth is mine. You shall be my kingdom of priests,
my holy nation" (Exod. 19:1-6).

Here it is Moses alone who visits with God on the mountain (though
on occasion God also communicates with Aaron and even Moses' younger
accomplice, Joshua of Nun, who would become his successor). Moses has
God strictly forbidding *anyone else* to go up the mountain to communicate
with Yahweh upon pain of death. He says: "You must be barriers around
the mountain," and he warns others, "Take care not to go up the mountain
or even to touch the edge of it. Any man who touches the mountain must
be put to death" (Exod. 19:12). It seems apparent that Moses seeks to
strike terror in the hearts of his band of followers. He *alone* is permitted
to communicate with God and bring his message back, and anyone else
who attempts to see or talk to him will die. Moses reiterates: "The Lord
said to Moses, 'Go down; warn the people solemnly that they must not
force their way through to the Lord to see him, or many of them will
perish" (Exod. 19:21-22). This prohibition also applies to the priests. "Go
down; then come up and bring Aaron with you, but let neither priests nor
people force their way up to the Lord, for fear that he may break out
against them" (Exod. 19:24). "Moses shall approach the Lord by himself,"
we are again told, "but not the others. The people may not go up with
him at all" (Exod. 24:2).

The tale thus has Moses going up and down the mountain. On the
most dramatic occasion, there are heavy clouds and thunder, even the
peals of a trumpet. No one is permitted to go up to the Lord; thus only
Moses can deliver God's commandments and prohibitions. These laws, if
carefully read, create a new priestly class, shower them with sacrificial
gifts, and place Moses and the tribe of Levi in charge of the lives of the
people. Moses has established a theocracy and a new privileged elite—all
in the name of the invisible God, who speaks to *him* alone.

Did Moses really experience the incident of the burning bush? Did he
genuinely hear voices or did he imagine it? Was he emotionally disturbed?
Was his conviction so strong that he had heard the voice of God that he
was able to convince his fellow exiles to heed him, for fear of the wrath of
the Lord. (When Pharaoh allowed the Jews to depart from Egypt, Moses
claimed this was due to great plagues sent by God.) Or did Moses use
deceit to achieve his ends, and did he trick and betray his followers into
believing that he had revelations from on high, simply so that he could
manipulate them? Or was the story a way of giving concrete expression to
a people's ethical growth and political needs?

When we come to the receiving of the Ten Commandments, the

question can be raised: "In what sense did Moses use this crucial incident to frighten his followers into obeying him?" It surely would solidify his rule if the people believed that he was divinely appointed to be the leader-priest of the new theocracy he was establishing. Since the other monarchs of his day, including the Pharaohs, were commonly believed to have been divinely invested, it should not be surprising for Moses also to make this claim. But *his* powers were self-proclaimed.

We have at least four possible hypotheses to consider in interpreting the events of the day:

1. The traditional one: that Moses did speak to God and was divinely appointed to receive the revelation. There is absolutely no evidence for this extraordinary claim, even though the faithful have accepted it ever since.

2. An opposite interpretation has been provided by skepticism, and seems plausible. The Humean-type of argument runs: The stories of Moses were inventions of the superstitious mind. They were transmitted in the oral tradition and embellished by later writers. These fictitious stories are the products of the primitive imagination, especially in regard to the fanciful miracles. Whether any part of these events actually happened, we have no way of knowing, but if they did they can be given naturalistic explanations.

3. Perhaps part of the traditional tale is true. One explanation is that Moses *believed,* at least in part, that he had heard voices; and he thus attributed events in his natural environment to divine forces. Perhaps he genuinely thought that he had a divine mission. This would mean that he was overpowered by religious experiences and that he believed that God was speaking to him and that God had entrusted him with a special task. This has been true of other disturbed individuals historically. Moses undoubtedly was influenced by a culture that accepted the reality of occult forces, and he thus may have been psychologically disposed to believe that some divine force had entered his life. His faith-state, however mistaken in fact, may have been real to him; and it may have moved him to embark with zeal and passion upon his mission.

4. A fourth possible explanation for what occurred was that Moses was a magician-charlatan, with megalomanic tendencies. Interested in power and control, he deceived a group of ignorant peasants to follow him as the prophet of God. Let us examine this hypothesis.

Was Moses a magician?

The close union of magic with religion in ancient times has been well-documented. The priest, like the magician, was capable of wondrous things that seemingly defied natural explanation: in ceremonies and celebrations, he was able to invoke magical powers in which deities and demons seemingly made their presence known. These mystical rites called up forces which, it was believed, enabled men to fulfill their aspirations and allay their troubling fears. Sir James Frazier in *The Golden Bough* maintained that "among primitive peoples, the king is frequently a magician as well as a priest; indeed he appears to have often attained power by virtue of such supposed proficiency in the black and white art."[12] Frazier also points out: "Nowhere, perhaps, were the magic arts more carefully cultivated, nowhere did they enjoy greater esteem, exercise a deeper influence in the national life than in the land of the Pharaohs."[13]

It is likely that Moses would have been acquainted with the arts of magic practiced in the Egyptian religion of his day. Many of the ceremonies performed by the priests of his day were similar to those that Moses was to perform later. The Egyptians feared the wrath of the gods, and they believed that their anger could be deflected by offerings and sacrifices. The gods were often jealous gods who would tolerate no disrespect, but who could be placated by obedient servants. Holy men in particular could communicate with the spirits of the gods, and they used their occult power to impress their devotees.

Was Moses also a magician? Did he and his confederates resort to the arts of conjuration in order to deceive their unsuspecting followers and elicit their obedience? The Bible itself testifies that the Pharaoh thought that Moses and Aaron were using magical spells, and he sought to have his own magicians and sorcerers compete with them.

We read that Moses and Aaron performed various feats in front of Pharaoh as a sign of Yahweh's power. "At this, Pharaoh summoned the wise men and the sorcerers, and the Egyptian magicians too did the same thing by their spells" (Exod. 7:11–12). Moses had Aaron throw down a staff, which immediately turned into a serpent. This the Egyptian magicians did also, but their serpents were swallowed by Aaron's staff. Similarly, Moses and Aaron were able to turn the waters of the Nile into blood; the fish died and the river stank. "But," we are told, "the Egyptian magicians did the same thing by their spells" (Exod. 7:22). Then there was a great plague of frogs. Again the magicians were able to do the same. How much truth and how much fantasy lie in these reports is difficult to say. According to Josephus, who was no doubt aware of the charge that Moses was a magician, Moses approached Pharaoh and told him of the "signs" that were done at Mt. Sinai. The king accused Moses of being an "ill man," who came to him with "deceitful tricks, and wonders, and

12. James George Frazier, *The Golden Bough,* abridged ed. (New York: Mentor Books, 1964), p. 34.
13. Ibid., p. 37.

magical arts to astonish him." The Egyptians were skillful in this kind of learning, said the Pharaoh, and noted that Moses had "pretended then to be divine," and such "wonderful sights could only be believed by the unlearned." Moses replied that "what I do is not done by craft or counterfeiting."[14]

According to the Pentateuch, the Egyptians were again and again visited with terrible calamities: maggots, locusts, swarms of flies, pestilences that infect cattle, boils on man and beast, a violent hailstorm, and famine. The Pharaoh's magicians were unable to replicate these events—all natural occurrences (if they occurred) that Moses attributed to divine agency. The last plague, which struck all the first-born sons of the Egyptians and spared the Hebrew slaves, was the final blow; with that the Pharaoh consented to their departure. Overloaded with treasure given to them by the Egyptians, the Hebrews fled. The Pharaoh had second thoughts, however, and sent an army after them to bring them back. When the army was overtaken by a heavy storm, the Hebrews escaped through the Sea of Reeds, the Egyptians were inundated, and gave up the pursuit. This was the "miracle" of the Red Sea.

The pagan literature, as we have seen, offered another explanation for the exodus. Plagues had struck the population. (Such diseases were apt to occur, given the poor level of sanitation, and it probably decimated large sectors of the population, as the Bubonic plague did in the Middle Ages.) The Egyptians, in response, expelled slaves and foreigners, believing that they had contaminated the population. They sought to placate their gods, whom they thought might have caused such terrible misfortunes. Unable to comprehend the cause and prevention of disease, they were wont to find scapegoats in men or gods, and out of fear they took desperate measures. The interaction of coincidences may have led them and the Hebrews to believe that these events were caused by divine forces. It is only too easy for the animistic mind to read into nature occult powers, which cause human diseases and mete out punishment and reward. The departing escapees inferred that Moses' God was responsible for inflicting these awful things on their Egyptian masters and that they were being saved by Yahweh's hand.

The testimony of the Bible strongly suggests that Moses did indeed use sorcery to impress Pharaoh and his court. The ancient, non-Jewish literature reinforces this view. This appears not only in the literature hostile to the Jews, which was abundant, but also in the ancient literature on magic and alchemy, fragments of which have survived. These parchments point to Moses as a figure of great authority in the field of magic.

14. Josephus, *Antiquities,* Book XIII.

One body of magic papyri was collected and edited by K. Preisendanz.[15] There are also ancient alchemical texts in which Moses is mentioned. They point to Moses as one who had a special relationship to God but who also authored secret magical texts.

Other pagan sources reinforce this view of Moses. Pliny the Elder, in his *Natural History* (first century C.E.), sketches a brief history of magic and writes about a sect of Jews "deriving from Moses" and others who practiced magic. Similar citations of Moses as a magician appear in Apuleius' *Apology* and Numenius' *On the Good.* Celsus also considered Moses to have been a magician and to have made his way by clumsy deceits: "Those herdsmen and shepherds who followed Moses as their leader had their minds deluded by vulgar deceits and so supposed that there was one God."[16] Celsus defended the polytheism of Rome against the monotheism of the Jews and Christians. He said disparagingly of the Jews: "They worship angels and are addicted to sorcery, in which Moses was their instructor."[17] He also rejected the attempt to establish Jewish antiquity by tracing the Jews back to the patriarchs of the Bible, whom he calls the "first offspring of jugglers and deceivers."[18] Harsh words, but is there any evidence in fact?

Origen reports in *Contra Celsus* on a conversation with some Jews about Egypt: How do they meet the charge of the Egyptians, he asks, who rejected Moses "as a sorcerer who appeared to have performed his great deeds by jugglery (magic)."[19] He reports on additional Egyptian criticisms of Moses by saying that "they do not deny entirely the powerful miracles done by Moses, but claim that they were done by sorcery and not by divine power."[20] For Celsus, belief in such miracles is limited to "uneducated people" and those of "depraved moral character."

I will not extensively detail the various miraculous events that allegedly happened to the Hebrews during the wandering in the desert. Whether these events happened as rendered in the Bible is difficult to say. Events that could be given natural explanations were often attributed to divine agency by the ancient mind. Elias Auerbach has attempted to put what happened during the Hebrews' wanderings in the wilderness in a more realistic perspective.[21] He suggests that the number of persons involved were no more than 12,000 to 13,000, not the huge numbers given

15. *Papyri Magicae Graecae die griechischen Zauberpapyri,* ed. and trans. into German by K. Preisendanz, vol. 1 (1928), vol. 2 (1931).
16. Origen, *Contra Celsus,* I:23.
17. Ibid. I:26.
18. Ibid. IV:33.
19. Ibid. I:43.
20. Ibid. III:5.
21. Elias Auerbach, *Moses* (Amsterdam: G. J. A. Ruys, 1953). Eng. trans. (Detroit: Wayne State University Press, 1975).

in Exodus and that they spent most of their time at Kadesh, a large, relatively lush oasis in the desert. Moreover, the refugees probably did not wander continuously for forty years. The wandering tribes are thirsty, and they find water; they are hungry for meat, and they find quail; they need bread, and they receive manna—all thought to have been brought about by divine intervention. Auerbach speculates that the spring which Moses is supposed to have brought forth by striking his staff upon a rock was a preexisting spring at the oasis Kadesh. There is, similarly, a natural explanation for the so-called miracle of quails. It is not rare for migrating birds tired of flying across a hot desert to land suddenly in large numbers at an oasis to quench their thirst. The Bible describes the manna that appeared one morning "like coriander seed, white; and the taste of it was like wafers made with honey." Seeds that fit that description may still be found today in oases; they are the sweet, resinlike product of tamarisks, which are found to fall from trees. To consider manna food for forty years was no doubt an exaggeration of the actual event.

Is it not plausible that Moses consciously deceived his people into believing that he possessed powers from God that worked through him? If Moses was to forge a new nation and a new consciousness out of the ragged tribes that followed him, then invoking the power of the deity, with all of the drama of magic, incantation, and sorcery, was a method that he and his accomplices might use with success. Moses undoubtedly was a charismatic individual who was able to awaken fervent support, but this depended in large part in his being a miracle worker. Here is a mortal man attempting to demonstrate his importance and solidify his power over his flock, constantly needing to convince them of his superhuman status.

We have already discussed the curious sequence of events in which Moses ascends the mountain alone to speak to God and then descends with the Covenant. On one occasion he finds the camp in disarray, the Hebrews dancing, singing, and worshiping a graven image. It is obvious that his people easily revert to their old ways. He needs to again frighten them by invoking a power higher than himself. This he succeeds in doing. But his authority is often contested. There is no constitution and no system of law that governs their behavior. Thus he needs to produce one, which he based on earlier legal codes, such as the Code of Hammurabi, which he could have learned as an educated Egyptian. But he attributes the authority for such a code to a new god, Yahweh, whose name he often elicits.

It is both pathetic and amusing to speculate about the possible devices that Moses used to hoodwink people to engender fear and respect for his sovereignty.

When we view the Pentateuch in this way, Moses emerges as a master of guile, with some unadmirable traits; he was imperious, vain, cunning, hot-tempered, vindictive, all too ready to murder those who opposed him. Unlike Jesus, he shows little compassion for those who differed with him. This appraisal may appear harsh about an individual who has been revered as a great prophet and lawgiver, but whose place in history is shrouded in mystery. Yet a careful reading of the text in modern terms lends credence to my analysis of his personality. I do not deny some important qualities of Moses the lawgiver, who, if the Bible is roughly accurate, had the great courage to challenge the Pharaoh and a dogged persistence to lead his flock through great adversity. The Egypt of his day *was* corrupt; the life of the ordinary man *was* cheap; human sacrifice, even cannibalism, had been practiced. Moses was limited by his times, and he borrowed moral principles from others, but he made important contributions in the laws he gave the Hebrews.

If one examines the biblical record, it is clear that the circumstances surrounding the appearance of the Lord to Moses and the Hebrews are wrapped in obscurity. As we have seen, no one was permitted to ascend to Mt. Sinai "because the Lord had come down upon it in fire." Moreover, there was smoke, which "went up like the smoke of a kiln." "And the sound of the trumpet grew ever louder" (Exod. 19:18-19). Was Mt. Sinai a volcanic mountain with pillars of smoke that Moses knew about because of his early visitations to Midian? And did he use this conveniently for his own ends? Or did he also employ crude props and paraphernalia to achieve his aims? The Egyptians had the technology to build the pyramids. Had Moses mastered some skill that let him practice deception on a vast scale? We do not know for certain. Whatever the explanation, the people were terrified by the scene.

In Exodus we read of the delivery of the Ten Commandments to Moses. The people heard peals of thunder and saw flashes of lightning. What a dramatic stage for Moses to make his appearance on, just when a thunder and lightning storm is threatening. Perhaps an accomplice blew a trumpet or lit a kiln. Cecil B. DeMille could not have done it better! And the people "trembled and stood at a distance" (Exod. 20:18-19). They were fearful to speak to the Lord, since if they were to do so, they believed they would die. Moses kept reiterating dire forebodings. Only Moses was permitted to approach the dark cloud where God was. It was in this dramatic context that Moses delivered the laws that were to govern the Hebrews.

Later, after Moses' power is challenged, he permits Aaron, Nadab, Abihu, and seventy of the elders to approach Yahweh at a distance and bow down (Exod. 24:1-2). But again, Moses alone approaches the Lord and does not let the others come near. The text is obscure as to precisely what occurred. This party went up to see the God of Israel. "But the Lord did not stretch out his hand towards the leaders of Israel." Nonetheless, "they stayed there before God; they ate and they drank" (Exod. 24:9-11). Again the Lord told Moses to come up to him in the mountain. This time "Moses arose with Joshua his assistant and went up the mountain of God." As he left he told the elders, "Wait for us here until we come back to you" (Exod. 24:13-14). Then Moses went up the mountain and a cloud covered it. He stayed there for forty days and forty nights before he brought down the commandments inscribed on two tablets of stone. Apparently, it took Moses a considerable amount of effort to chisel out the commandments, though the faithful no doubt believe that it was the finger of God that engraved the tablets. Joshua emerges in this scenario as a key figure. Was he responsible for the accouterments of the act on the mountain—the trumpet and the smoke? A good magician usually needs an assistant to help with the props.

The numerous rules and regulations that Moses proclaims are designed to ensure that he and his priestly entourage of the tribe of Levi would receive many offerings. It is commanded that food and treasures be offered to Yahweh to placate him, but these are consumed by those in charge of the priestly functions, a new elite that serves in Yahweh's name. In spite of the bravado of his act, it was difficult for Moses to maintain his tight control constantly; even his brother Aaron and sister Miriam contested his hegemony at times.

Moses later effectively struck fear and awe into the hearts of his people with his use of what the Pentateuch calls the "Tent of the Presence" (translated as the "Tabernacle" in the King James version or the "Tent of Meeting" in the Jewish Torah). We read: "Moses used to take a tent and pitch it at a distance outside the camp. He called it the Tent of the Presence, and everyone who sought the Lord would go out to the Tent of the Presence outside the camp" (Exod. 33:7-8). When Moses went to the tent the people would all stand and follow him with their eyes as he entered it. "When Moses entered it, the pillar of cloud came down, and stayed at the entrance to the tent while the Lord spoke with Moses" (Exod. 33:9-10). According to Exodus "there was fire in the cloud by night" (40:38). The burning substance emitted not only smoke but sparks. As soon as the people saw this, they would prostrate themselves, as the Lord spoke face to face with Moses. We then read that Moses would return to the camp, but revealingly, "his young assistant Joshua, son of Nun, never moved from inside the tent" (Exod. 33:11).

We may speculate on what occurred. Did a divine, anthropomorphic entity truly speak to Moses as "one man speaks to another?" Or, were Moses and Joshua able by means of magic to produce a cloud of some unknown substance, which hovered over the tent? Joshua is a central figure, for he had been with Moses since he was a boy, and it was he who assumed the mantle of power after the death of Moses. This same Joshua must have been privy to Moses' secret sorcery and indeed aided him in the conjurer's tricks that they perpetrated on the Hebrews. Indeed, it is Joshua who was responsible for giving Judaism its form after Moses passed from the scene.

The Tent of the Presence plays a central role in Moses' career because, aside from Moses' visitation to the mountain to see the burning bush and produce the Ten Commandments and the laws, it is in or near the tent, the house of mystery and magic that contains the Sacred Ark of the Covenant, that Moses as priest, monarch, and showman performs his act.

As the saga unfolds, Moses took Aaron and his sons to the entrance of the Tent of Presence, where he anointed and consecrated them as priests. Here, a new hereditary priesthood is established by Moses. The Levites in the second year after the exodus are put in charge of the Tabernacle and all its equipment. They are responsible for pitching the tent and taking it down when they move. So fearful does Moses appear to be of being detected that he places a guard. The text reads: "Any unqualified person who comes near it shall be put to death" (Num. 1:51).

At one point there seems to have been a serious danger of detection. Apparently only Moses, Joshua, and Aaron were permitted into the innermost sanctuary of the Tent of the Presence. On one occasion "Nadab and Abihu, sons of Aaron, took their fire pans [offering vessels], put fire in them, threw incense on the fire and presented before the Lord illicit fire which he had not commanded" (Lev. 10:1-2). We are then told that they are burned to death by Yahweh because of that transgression. "Fire came out from before the Lord and destroyed them; and so they died" (Lev. 10:2-3). Moses sternly warns people to be careful in approaching the holy tent and God, and also not to mourn for the sons. Aaron is dumbfounded, but he is left with his surviving sons, Eleazar and Ithamar. The use of fire to kill those who contest Moses' power appears later—as if a crude kind of flamethrower were being employed from within the tent to sear and singe those outside.

Any explanation of such rituals and events in the Pentateuch is difficult and must be based upon conjecture. We have no way of knowing whether the events related actually transpired. They may at best be only distant echoes of real events. Still if we assume they did, then I think we may substitute an alternative hypothesis for the traditional view of Moses. He may have believed that he talked to God—a form of schizophrenia that occurs in those who hear voices or have experiences of revelation—

and he may have used all of the dramatic means at his disposal to convince his followers that he did.

There are some rather interesting challenges to Moses' authority that call for further analysis; for they suggest the use of cunning on his part to interpret events in his way. Things were at times very difficult for the Israelites as they wandered in the wilderness. For example, "there came a time when the people complained to the Lord of their hardships." When Yahweh heard of this, according to the biblical account, he became angry. As a result, fire broke out and raged at one end of the camp. Moses is said to have interceded with Yahweh, and the fire died down (Num. 11:1-3). Moses, in any case takes credit for soliciting Yahweh's help in extinguishing the blaze. The people complain about the lack of abundant food or meat. Moses feels a great responsibility about their burden. How does he deal with their constant pestering? Moses brings seventy elders to the Tent of the Presence. Here we learn that "for the first and only time" the Lord ascended in a cloud and spoke to them. He conferred the same spirit on them that he had on Moses. And they fell into a "prophetic ecstasy" (Num. 11:24-26).

What happened here is difficult to say because the details are so sketchy. Did they have a "religious experience?" Did they imagine that they were able to contact the divinity? We read further that two men, Eldad and Medad, who were part of the seventy, did not go out to the tent, but the spirit alighted on them and they fell into an ecstasy in the camp. When news of this is brought to Moses, Joshua becomes agitated and cries, "My lord Moses, stop them!" as if to say that only he and Moses can claim such a power. Moses, however, admonishes Joshua not to be so jealous and wishes that all of his people could become prophets. The next day a wind from Yahweh drives a huge flock of quail, which falls to the ground and provides food for the people. But they are again struck with a deadly disease, which is attributed to Yahweh's anger.

On another occasion, Moses is challenged by Aaron and Miriam. They ask, "Is Moses the only one with whom the Lord has spoken? Has he not spoken with us as well?" (Num. 12:1-3). Moses becomes furious at them. He summons all of them to the entrance of the Tent of the Presence. There Aaron and Miriam are chastised by Yahweh. Interestingly, he punishes Miriam, who is inflicted with a skin disease and confined for seven days outside the camp.

The people again complain about their life; they are fearful of reports about the men of Canaan, who appear to be of gigantic size, and they are leery about seizing the promised land. They begin to talk about returning to Egypt and choosing someone to lead them back.

The Bible tells us that Yahweh is grieved and says that none of those above twenty years of age (except Joshua and Caleb) will live to enter the promised land (Num. 14:29-30). Evidently, Moses had difficulty in maintaining his hegemony. While he is able to keep his people in a nomadic state without firm roots, he can control them. No doubt he himself is indecisive and/or fearful of settling down—for fear that he will lose his hegemony. And so the Israelites wander in the wilderness for forty years. It is left to Joshua, after Moses' death, to eventually lead a conquering army to establish a new nation in the land of Canaan.

We read of still another insurrection against Moses' authority, this one more serious. It is led by Korah, a Levite, and some 250 Israelites, all men of good rank. They say to Moses and Aaron: "You take too much upon yourselves. Every member of the community is holy and the Lord is among them all. Why do you set yourselves up above the assembly of the Lord?" (Num. 16:1-4). Moses now takes a new tact. As if cornered, he prostrates himself before Korah and his company, saying that tomorrow morning the Lord shall declare who is holy. They must bring offerings to the Tent of the Presence. Each man is to bring a censer (a lamp or fire pan) with burning incense. Apparently the revolt stems from the Levites, although this is uncertain from the text. Since Moses has given a privileged place in maintaining the Tabernacle to the Levites, he is incensed that they wish more equality. Moses sends to others, Dathan and Abiram, for help, but they refuse to come and remonstrate: "Must you also set yourself up as a prince over us?" (Num. 16:13-14). Whatever happened, they ask, to the land of milk and honey? Will we die in the wilderness? "Do you think you can hoodwink men like us?" (Num. 16:14-15). It appears that there is seething opposition among key members of Moses' entourage.

What transpires is the death of Dathan, Abiram, and their entire families in front of their tents. We are told that an earthquake swallowed them up, but it was more likely that Moses had them killed. Meanwhile, Korah and the 250 Israelites presented offerings at the Tent of the Presence. "Each man took his censer (fire pan) and put fire in it and placed incense on it" (Num. 16:18). And they were there killed by the fire. The Lord never appeared, but the assembled are burned to death. It is a clear feat on the part of Moses. How he did it is not clear, but 250 men with robes and flaming fire pans are easily inflammable targets. Was this done by Yahweh, or by the Machiavellian cunning of Moses, priest-king-prophet-magician?

Still the complaints and sedition continue. A plague again strikes the people. We are told that 14,700 die from it. Again it is attributed to the wrath of the Lord. Moses takes credit for ending it.

Another ruse performed by Moses is appalling. Moses assembles the heads of the twelve tribes before the Tent of the Presence. Each chieftain is to bring a staff, on which his name is written. The name of Aaron is inscribed on the staff of the Levites. Each of the chiefs hands Moses his staff, which Moses takes inside the tent. And then we read that the "next day when he entered the tent, he [Moses] found Aaron's staff, the staff for the tribe of Levi, had sprouted. Indeed, it had sprouted, blossomed, and produced real almonds" (Num. 17:8-9). Lo and behold what a miracle! One can imagine Moses and Joshua and Aaron carefully substituting a new staff, identical with the earlier staff of Aaron, inscribed with his name, but cut from the branch of an almond tree. No one save the trusted confederates is permitted in or near the tent on pain of death. One may ask: Was it Yahweh who made the staff sprout or human contrivance?

Moses again has Yahweh cautioning Aaron that he and his sons, together with the members of the Levite tribe, shall be responsible for the sanctuary and the Tent of the Presence. They shall be in attendance at the tent but they "shall not go near the holy vessels and the altar or they will die" (Num. 18:1-4). They shall be careful to perform their priestly duties in everything pertaining to the altar and what is behind the curtain or within the veil, but any outsiders who intrude in it "shall be put to death" (Num. 18.7). And we may ask: What is it that is so secret within the tent that Moses wishes to guard it at all costs and kill anyone who attempts to peek?

Moses' full vindictive fury is unleashed when he finally has his brother Aaron killed, because, according to the text, he "rebelled against my command" (Num. 20:24-25). Moses takes Aaron and his son Eleazer up Mt. Hor, strips Aaron of his robes, and invests Eleazar with them. "There Aaron died on the mountain top . . . as the Lord had commanded" (Num. 20:27-29). Moses descends with Eleazar and the people mourn for thirty days. Though the Bible condemns Cain for slaying Abel, it does not likewise condemn Moses.

Moses is able to rule his flock only by using the bloody fist. An absolute dictator, he allows no questioning of his (or Yahweh's) authority. Anyone who thwarts his will, even his own brother, who stood beside him for years, is liquidated. It is a wonder that the judgment of history has not condemned Moses. Perhaps it was because he began his career as a liberator, and there is a human tendency to view with favor those who seek to free a people from oppression, however oppressive and corrupt they may themselves become in the process and however much they may betray their own code of morality.

Moses the lawgiver

Moses' reputation is based on the assumption that he not only was a national liberator but was also a great lawgiver, handing down a moral code of great grandeur for future generations. If this is the case, it is time that we turn to an analysis of the moral code and the commandments proclaimed by Moses. No doubt, what is contained in the five books of the Pentateuch was developed over the centuries; and so how much was original with Moses and how much was interpolated or codified by later generations is difficult to say. Many people take Judaism basically as an ethical religion, and they tend to minimize its miraculous or supernatural elements, focusing on its moral insights. But in the twentieth century we are entitled to ask: Is the Mosaic code the model of righteousness we have always been told that it is?

Religious believers usually cite the Ten Commandments as an expression of ethical eloquence, but they rarely submit these tenets to careful evaluation, nor do they seem to notice that many of the other regulations are sadly deficient in morality—at least by the standards of the developed moral conscience. No doubt in the context of the times, they may be said to have represented an advance, but this is not unqualifiedly so. The moral code of the Pentateuch has its precursors and analogues in ancient times and is not original with Moses. For example, King Hammurabi of Babylon codified ancient Sumerian and later Semitic laws and anticipated many ordinances found in the Pentateuch. The Hammurabi Code has been dated by scholars as being framed anywhere from 2100 to 1700 B.C.E. Etched on a block of black dorite and standing some eight feet high, it was erected in the public square of the Babylonian city of Sippar. It was uncovered at Susa by archeologists and is now on exhibit in Paris. Interestingly, carved at the top of the Hammurabi stone is a picture of the Babylonian king receiving the laws from the sun god, Shamash, similar to Moses' reception of the law from Yahweh.

Many of the ordinances bear a striking rersemblance to the Mosaic covenant in the books of Exodus, Leviticus, and Deuteronomy. For example, the notion of justice "eye for eye, and life for life" *(lex talionis)* is found in the Hammurabi Code as well as in the Pentateuch. The laws of the code are framed in a similar conditional form. "If *A* is done, then *B* will be the punishment." In many cases similar penalties are imposed for the same offense, such as the death penalty for incest (which was practiced in Egypt by the ruling class). There are also some humanitarian elements in the code. It is particularly strong in its prohibition against defrauding the helpless. It can only be a tremendous deflation to Judaic-Christian pride to learn that the laws presented by Yahweh to Moses and through him to the chosen people had been presented in a similar way still earlier to others.

Other important archeological findings are equally deflating to the Judaic-Christian cultural bias. The defeats and victories of the Israelites or their enemies are attributed to the changing moods of reward or punishment by Yahweh. The discovery in 1868 of a flat basalt slab, called the Moabite stone, shows that a similar theological interpretation of events prevailed in Moab, Israel's neighbor and traditional enemy. According to the inscription on the stone, the king of Moab, Mesha, blamed his country's invasion by the Israelites on the anger of Chemosh, the national god of the Moabites. Similarly, the restoration and revolt of the Moabites against the Israelites is due to the blessings of Chemosh during the ninth century B.C.E.

Similar considerations apply to the Ugaritic texts discovered between 1929 and 1939 at Ugarit, in ancient Syria, on the Mediterranean coast. They are a store of ancient literature that includes several Canaanite stories about Baal, El, and Danel (who later appears in the Old Testament's Book of Daniel). Most of these texts, which date from 1400 B.C.E., show that the Israelites, who settled in Canaan, borrowed many religious and ethical concepts from their Ugaritic and Canaanite neighbors. The law, morality, and theology of Moses thus were clearly not unique or original with the Hebrews.

Let us turn directly to an analysis of the Ten Commandments, which theologians and religionists refer to as sacred. The first thing to note is that the Ten Commandments appear in two places in the Pentateuch. The first version is in Exodus (20:2-17) and the second is in Deuteronomy (5:6-21). These versions differ significantly in the language in which they are expressed. This, of course, wreaks havoc with the simplistic view that the Bible is the literal, unaltered word of God revealed to the prophets. If the Ten Commandments were delivered to Moses already formulated by God, why are the two renditions different? If there is a variation in the most fundamental of all laws—the Decalogue—we may ask how accurate and reliable are other passages of the Pentateuch. The point is that the Bible is a fallible human document recording the experiences, victories, and failures of a people, expressing their hopes and aspirations, but also their mistakes and limitations.

I will illustrate some of the differences in the text by setting side by side the fourth commandment, showing the difference in linguistic expression in the two versions:

Exodus 20:8:11	Deuteronomy 5:12-15
Remember to keep the Sabbath day holy. You have six days to labour and do all of your work. But the seventh day is a Sabbath of the Lord your God; that day you shall not do any work, you, your son or your daughter, your slave or your slave-girl, your cattle or the alien within your gates; for in six days the Lord made heaven and earth, the sea, and all that is in them, and on the seventh day he rested. Therefore the Lord blessed the Sabbath day and declared it holy.	Keep the Sabbath day holy as the Lord your God commanded you. You have six days to labour and do all of your work. But the seventh day is a Sabbath of the Lord your God; that day you shall not do any work, neither you, your son or your daughter, your slave or your slave-girl, your ox, your ass, or any of your cattle, nor the alien within your gates, so that your slaves and slave-girls may rest as you do. Remember that you were slaves in Egypt and the Lord your God brought you out with a strong hand and an outstretched arm, and for that reason the Lord your God commanded you to keep the Sabbath day.

How can we explain the divergent statements other than to recognize that the Bible is the product of many authors and that it evolved over a period of many centuries. What about the content of the commandments themselves? Are they universally applicable? Are they models of moral inspiration? I very much doubt that an unbiased reading of them today would agree.

The first two commandments concern one's religious obligations. These are not necessarily ethical duties, but religious injunctions toward piety and devotion, and they have little to commend themselves to the freethinker or skeptic. A succinct statement of the first commandment as taken from the version in Exodus reads as follows:

1. I am the Lord your God who brought you out of Egypt, out of the land of slavery. You shall have no other god to set against me (Exod. 20:2-3).

This is directed to and applies first and foremost to the Israelites, and it is from their national god. It does not necessarily apply to all men, for they were not brought out of Egypt. This is a highly partial conception of the diety.

The second commandment states:

2. You shall not make a carved image for yourself nor the likeness of anything in the heavens above, or on the earth below, or in the waters under the earth. You shall not bow down to them or worship them; for I, the Lord your God, am a jealous God (Exod. 20:4-5).

Here the commandment is obviously addressing itself to rival religions of the day, which engraved the rival gods they worshipped in animal or human form, such as a sacred cow or bull. The Israelite god could not be easily named, let alone seen face to face—except by Moses, of course. The meaning of this commandment clearly is limited by the cultural context. How would this commandment apply to the many statues and paintings depicting Jesus? Would they violate it? Or the ceiling of the Sistine Chapel by Michelangelo? Is the Roman Catholic church, which fills its cathedrals with images of the holy family and saints, in violation of this commandment? The God of this commandment is not a god of love, mercy, or forgiveness, but is a jealous god, demanding absolute obedience. He is hardly a model of justice, but appears arrogant. This, no doubt, reflects Moses' own personality, as he demands obedience and permits no opposition or dissent. One would think that a rational God would not want his creatures to act like submissive children, but rather to be inquiring, independent, autonomous persons. Why couldn't Yahweh share the beautiful earth with his creations, rather than being so jealous, demanding and possessive.

The rest of this commandment is often not read, and the short published summaries usually omit the following passage. But it is in the Decalogue and seems to me particularly iniquitous. It reads:

> I punish the children for the sins of the fathers to the third and fourth generations of those who hate me. But I keep faith with thousands, with those who love me and keep my commandments. (Exod. 20:5-6)

Now this part of the second commandment seems immoral, for it assumes the validity of collective guilt. God is seen as visiting punishment on the children of sinful parents, even unto the grandchildren and great-grandchildren. In what sense can this express an exemplary form of righteous conduct? It seems especially barbaric to blame a man's children for the sins and omissions of their parents. No civilized community today would ever justify a law of retribution against the innocent children of dissidents or criminals. How can one say that a just God is willing to affix responsibility and blame on them and to consider this a worthy deed? This is a holdover from the barbaric practices of retaliating against a person's whole family or tribe, when one feels wronged. But it goes further, for it applies to their descendants still unborn. What unmitigated cruelty! This part of the second commandment should be rejected by every morally decent person.

The third commandment states:

3. You shall not make wrong use of the name of the Lord your God; the Lord will not leave unpunished the man who misuses his name. (Exod. 20:7).

Again, we can object to the severity of the crime. Freedom of thought by dissenters and nonbelievers is prohibited and punished and those of other religions are proscribed. Later, one reads that those who blaspheme the name of God shall be put to death. The commandment is an expression of religious intolerance and bigotry; it is hardly a paragon of moral virtue.

The fourth commandment requires that we keep the Sabbath day holy. Presumably in a society in which the rights of others are violated, it was an advance to ask for one day a week off, especially for slaves, servants, cattle, and aliens. Today, however, most societies observe a weekend free of labor; and in advanced societies there are two or more weeks of vacation per year. If we were to rewrite this commandment today, it might read: "You shall work no more than five days and have the weekend free for leisure, rest, and cultural activities and enjoyment." My only caveat is that I happen to like to work seven days a week and, indeed, am writing this chapter on Saturday and Sunday, which is a violation of the commandment. Yet orthodox Jews have taken the Sabbath to be so holy that they were afraid to do anything lest they stir the wrath of God.

The last six commandments are more directly ethical and have much to commend them. They should be taken as general moral principles, not universal in form, for how they are interpreted or applied depends upon the context:

5. Honor your father and your mother. . . .
6. You shall not commit murder.
7. You shall not commit adultery.
8. You shall not steal.
9. You shall not give false evidence against your neighbor.
10. You shall not covet your neighbor's house; you shall not covet your neighbor's wife, his slave, his slave-girl, his ox, his ass, or anything that belongs to him. (Exod. 20:12-17).

Should moral principles be taken as *absolute* rules or are they only *general* guides to conduct. As general guides, Commandments 5-10 express considerable moral insight, for they summarize the wisdom of the race. Unfortunately, they sometimes are taken as inviolable. "Thou shall not murder" is a fundamental principle that we ought to obey and we can give reasons to justify it. But there may be some exceptions, such as in time of war, self-defense, or euthanasia. How the principle is to be interpreted depends upon the context. Moses surely violated the commandment (as did Yahweh), and the countryside was littered with those he had slaughtered.

Similarly, for the absolute commandment forbidding adultery. One may say that this is a wise prohibition, particularly where a marriage is viable and is based upon sincerity and trust. However, it is only a general rule and it involves a hypothetical rather than a categorical imperative. Under some conditions, such as those faced by Lady Chatterley (whose husband was rendered impotent in World War I and incapable of making love to her), it may be violated; at least Lady Chatterley thought so and deemed it advisable to take a lover. Moses violated the principle against adultery with impunity. He took a second wife and also distributed the Midianite women to his soldiers (but only after he had their husbands killed). Moral conduct involves the consideration of many other values and principles besides those enunciated by the Ten Commandments. Even Moses recognized this, for he went on to issue hundreds of other commandments.

Most readings of the Pentateuch, however, stop with the Ten Commandments. But there is no good reason to do so, save perhaps the desire to avoid acute embarrassment. If we read Exodus 21, which follows the Decalogue, we find numerous other laws and judgments, as well as in Leviticus and Deuteronomy. Many of these laws are extremely harsh and legalistic, and they are offensive to a developed moral conscience. They are hardly appropriate models for a civilized society today; they were limited by the social milieu in which they were promulgated.

For example, we read that "when you buy a Hebrew slave, he shall be your slave for six years, but in the seventh year he shall go free and pay nothing." If he is married, his wife can leave with him. But "if his master gives him a wife, and she bears him sons and daughters, the woman and her children shall belong to the master" and cannot be set free. If he protests, his ear will be pierced with an awl and he will be enslaved for life (Exod. 21:2-6). Evidently this morality is relative to a hierarchical caste society in which some human beings are considered to be mere chattel. Nor do equal rights apply to women, who are considered inferior to men. Thus, if a man sells his daughter into slavery, she does not enjoy the same rights as a male slave and will not be freed (Exod. 21:7-8). The master should not mistreat her, however. He is allowed to take another woman, but if he does he must still provide for her. Moses himself practiced polygamy.

This same double standard applies to punishment. If a man intentionally strikes another and kills him, then he shall be put to death. However, if he strikes and kills his slave or slave-girl, so that the slave dies on the spot, he "must be punished," but not put to death. However, if the slave survives for one or two days, then he shall not be punished, "because he is worth money to his master" (Exod. 21:20-21). Neither women nor slaves have the same rights as free adult males.

The law of retribution is a key element in the Mosaic conception of justice, and it is especially stringent: "Wherever hurt is done, you shall give life for life, eye for eye, tooth for tooth, hand for hand, foot for foot, burn for burn, bruise for bruise, wound for wound" (Exod. 21:23-25). The application of this principle in some cases and the severity of retribution seem to far outweigh the original transgression.

The relations between children and parents is particularly authoritarian. The fifth commandment obliges all "to honor your father and your mother." Filial piety serves as the basis of the family structure. Reading further in the Pentateuch, however, many will no doubt be shocked as to how it is interpreted: "Whoever strikes his father or mother shall be put to death" (Exod. 21:15).

The penalty is equally severe for lesser infractions. "Whoever reviles his father or mother shall be put to death" (Exod. 21:17). Incredibly, in Leviticus parents are told that if they cannot control their irascible children, that they can have them stoned to death by the community. What more horrendous form of child abuse can be found in literature than the following passage:

> When a man has a son who is disobedient and out of control, and will not obey his father or mother, or pay attention when they punish him, then his father and mother should take hold of him and bring him to the elders of the town at the town gate. They shall say to the elders of the town: "This son of ours is disobedient and out of control; and he will not obey us, he is a wastrel and a drunkard." Then all of the men of the town shall stone him to death, and you will thereby rid yourself of the wickedness (Deut. 21:18-21).

The harshness of the Mosaic code may also be seen by its frequent resort to capital punishment. The death penalty is exacted for murder. But it also applies to the following transgressions: adultery, homosexuality, incest, blasphemy, rape, loss of virginity before the wedding night, sexual relations with animals, failure to keep the Sabbath, sacrifice to other gods, practice of witchcraft, prostitution by priests' daughters, the attempt to contact ghosts or spirits.

Another form of justice that is extremely vindictive concerns children born out of wedlock. The Bible says: "No descendant of an irregular union [i.e., a bastard], even down to the tenth generation, shall become a member of the assembly of the Lord" (Deut. 23:2). So tainted is a person from an illegitimate birth that his entire line of descendants are banished from the Temple for generations to come. Interestingly, the severity of this punishment did not apply to Jesus, who had questionable parentage and clearly came from a highly irregular union, which only demonstrates the ability of the religious temperament to excuse inconsistencies or rationalize hypocrisy.

Many Old Testament regulations lack any expression of human kindness or sympathy. Moses expells from the camp anyone who has a malignant skin disease or discharge, perhaps to maintain cleanliness, but this is hardly the mark of human charity. Some of the regulations are amusing. "No man whose testicles have been crushed or whose organ has been severed shall become a member of the assembly of the Lord" (Deut. 23:1). This means that eunuchs were barred from the congregation, even though they did not consent to be castrated.

No doubt the laws enunciated in the Mosaic code express the irrational phobias of the times. A man may not have intercourse with his wife during menstruation, nor touch anything of hers during that period. A menstruating woman was considered unclean. A woman is also enjoined from wearing any article of men's clothing. Nor can a man wear any article of women's dress. Circumcision is required of all male Jews. Then there is the well-known injunction against eating pork. As we have seen, some of the pagan authors believed that this was due to the fact that the early Hebrews contracted a disease common to swine and for that reason were expelled from Egypt. If so, this might provide some rational justification for the prohibition. Yet most of the dietary prohibitions seem to lack a rational basis, at least in retrospect. If unsanitary pork is the cause of trichinosis, for example, it is prudent to avoid it. But a total abstinence seems ridiculous. Many of the prohibitions are arbitrary. The statement "You shall not boil a kid in its own milk" has been used ever since by Orthodox Jews as a basis for an injunction against eating dairy and meat products together. There seems no reasonable basis concerning health for this phobia. There are also rules governing the ritual slaughter of cattle. One must not eat meat with blood and thus one must drain the blood. Jews are prohibited from eating lobster, oysters, shrimp, and clams, seafood that, if properly prepared, is nutritious and delicious. No wonder that the Gentile world found Jewish dietary rituals strange.

There are also many rules offering advice about how to treat disease, including skin disease and leprosy, much of it based upon the crude standards of medicine then prevalent.

A great number of the regulations concern a person's duty to God by way of prayer and sacrifice; in particular the Bible lays down the mode and method of offerings and gifts. If nothing else, Moses is ensuring that his priestly class will be supported by their vassals. We read: "Every contribution made by way of holy-gift which the Israelites bring to the priest shall be the priest's . . . whatever is given to him shall be his" (Num. 5:9-10). The priest presents mythological opium concerning God and his powers to the masses. The people in return swear undying allegiance to Yahweh and to his interpreters. This conveniently provides adequate economic sustenance for the priestly class. The Jews were not, of course, unique in this, for it was the common practice of most nations to support their god men and to shower them with gifts and power.

The ethics of the Pentateuch is surely not without its redeeming social and moral virtues. One can find passages that are truly humane and that recommend charitable attitudes toward the weak and helpless, as well as righteousness to one's neighbors. In addition to the commandments against stealing, cheating, and coveting, we read that one should not slander or deceive a fellowman. Nor should one oppress or rob one's neighbor. Nor should one keep back a hired man's wages until the next day. Nor should one treat the deaf with contempt, nor obstruct the way of the blind. A person should not prevent justice either by favoring the poor or being subservient to the great and powerful. One should apply justice impartially. One should not harbor hatred against one's brother, seek revenge, or cherish anger against one's kinfolk. We read a form of the Golden Rule: "You should love your neighbor as a man like yourself" (Lev. 19:11-18). And there are other special rules based upon practical common sense that seem eminently fair. If you leave your well uncovered, and if a neighbor's ox or ass falls into it, the owner of the well shall make up the loss.

Most of these principles of moral decency were not invented by Moses, nor were they unique to the Israelites but have their analogues in neighboring nations. They express the common heritage of most civilized communities. The fact that they were proclaimed by the Israelites as divine commandments no doubt gave them added force and sanction. (Though I might add that a mature moral person is one who practices the moral civilities not out of fear of the Lord's wrath, nor out of blind obedience, but because he or she has developed a moral conscience and sympathies concerning others. It is possible to find a basis for moral development without the postulate of the fatherhood of God. Ethical conduct can be justified on the basis of reason, as philosophers from Plato and Aristotle down to Kant and Mill have pointed out. One can thus live an ethical life without benefit of God or clergy. What one ought to do should not depend on the automatic application of absolute rules but on reflective decision in facing moral dilemmas.) Interestingly, Hebrew scholars themselves discovered that the early moral code of the Pentateuch had to be expanded and interpreted, and the later books of the

Bible and the Talmud do just that. The application of moral rules often depends upon unique circumstances, and so some degree of sensitivity and practical wisdom is required. Although this developed within the tradition of Talmudic scholarship, the original inspiration for rules of morality are always traced back to the Torah, whose ultimate authority rests with Moses and God.

The important question that can be raised about the Mosaic code concerns the range of applicability of its moral principles. They were originally intended to apply to the Hebrews. But what about people other than the chosen? It is clear that the Pentateuch extends the ethics of consideration to aliens and strangers: "You shall have one penalty for alien and native alike" (Lev. 24:22). Moreover, "When an alien settles with you in your land, you shall not oppress him." We also read, "He shall be treated as a native born among you, and you shall love him as a man like yourself." And there is the reminder: "because you were aliens in Egypt" (Lev. 19:33-34). Thus there is a clear recognition that one's moral responsibilities apply equally to strangers and aliens. There is also an awareness that the moral law is universal and should be extended to all men. On the other hand, implicit in the concept of "chosen people" is a clear distinction between the Jews, who are supposedly favored by God, and other peoples. And there are limits to moral toleration and civility concerning one's enemies. This kind of chauvinism is not unique to the Jews; other tribes and nations have continually waged war and have resorted to infamous activities against each other—often in the name of the highest principles of patriotism, nationalism, or religiosity. Yahweh affirms in the Bible, "I have made a clear separation between you and the nations . . . [and] between you and the heathen" (Lev. 20:24-27). The Jews interpreted that to mean that the laws governing their conduct within their own house or tribe did not necessarily apply to others living outside.

Interestingly we read in Deuteronomy a contradictory rule concerning money-lending: "You shall not charge interest on anything you lend to a fellow-countryman." It then goes on to apply another rule to foreigners by declaring: "You may charge interest on a loan to a foreigner" (Deut. 23:19-20). The Jews became moneylenders throughout the world, often because they were not accepted as equals and were barred from other professions or land owning in the countries in which they sojourned. Deuteronomy provides a justification and perhaps an explanation for their success in business and finance, though it also advocates honesty: "You should keep true balances and weights."

Most offensive to modern readers, no doubt, are the standards that are often applied to nations at war. Thus we read that if the Hebrews wage war against their enemies and take captives, they could take attractive women as wives, apparently even against a woman's will. "If you see a comely woman among the captives and take a liking to her, you may marry her," we are told, and "you may have intercourse with her." There is a tempering of one's right of possession, however, and a restraint on what a man can do. If a man no longer finds a woman pleasing, he should let her go free. "You must not sell her, nor treat her harshly" (Deut. 21:10-14).

Moses' own attitude toward other peoples who stood in his way was particularly bloodthirsty. The first act we hear about after he is grown is that he kills an Egyptian for striking a Jew and then hides the body. His lack of compassion for the Moabite nation in particular contradicts the view that he was a noble giver of moral laws for humankind. In Numbers, we read of the approach of the Hebrews to the promised land. They battle with other tribes and nations, capturing villages and slaughtering all inhabitants. The reaction of Moses after the battle with the Midianites is particularly cruel. The Hebrews made war on Midian, killed all the men, burned all of their cities, and took captives, spoil, and plunder. The officers returned from the campaign and encountered Moses, who was angry at them. "Have you spared all the women?" he asks, complaining that the Midianite women had seduced the chosen people into disloyalty. And so he demands vengeance, "Kill every male dependent," he orders, "and kill every woman who has had intercourse with a man." They are allowed to spare only the virgins, who, according to the Bible, numbered 32,000 girls (Num. 31:13-35). The Hebrews then divided the spoils, with Moses taking his share, "as the Lord had commanded him."

Moses' barbaric strategy as they advance is to savagely destroy the inhabitants of the lands they take, showing no mercy. Otherwise, he says they "will become like a barbed hook in your eye and a thorn in your side" (Num. 33:55-56). As they proceed, they tear down the sanctuaries and altars of the conquered. There is no tolerance for the deities of competing religions. "You shall demolish all the sanctuaries where the nations whose place you are taking worship their Gods . . . and shall pull down their altars and break their sacred pillars" (Deut. 12:2-3). A scorched earth and extermination policy is justified by Moses. He guarantees his people that the Lord God will "exterminate" the nations "as you advance" (Deut. 12:29). Moses was no peacemaker. Nor was he a paragon of virtue. The Hebrew invasion of Canaan was not unlike the Nazi onslaught in World War II, the invasions of the Huns or Mongols, or of other advancing armies in history. It does not intimate any appreciation for a higher humanistic morality in which all nations and people are considered as part of a world community.

The promised land

Moses never made it into the promised land, nor did most of the adults who had left Egypt with him. Thus he failed in his promise made on behalf of Yahweh that they would settle in a new land, flowing with milk and honey, which was just over the nearby hills. He shifts responsibility for this failure to *their* lack of absolute fidelity to the commandments of Yahweh. One may speculate about the reasons for Moses' indecision and why he failed to move ahead. Was it fear of the unknown enemy on the other side of the frontier? Was it his belief that he could keep his passive nomadic horde under his control by not going further? Did he, as he grew older, harbor doubts about his own mission, knowing that no god had promised him a new land and that it was not available for easy picking. Was Moses killed by his own people, as Freud speculates, with this act of regicide hidden from future generations? In any case, it was left to Joshua and Eleazar to cross the Jordan River and lead the next stage of nationhood. Thus it is Joshua, not Moses, who may be truly said to be the founder of the new nation, although Moses provided the framework and ideology for Joshua to exploit. Under Moses' tutelage, Joshua learned his craft well, and he succeeded where Moses failed. How paradoxical that, in spite of reneging on the most basic promise Moses made to the escapees, he nevertheless is considered the savior of the Jews.

This is not unlike what befell Jesus, who did not live to see the end of the world he predicted, nor the New Kingdom of Heaven. This suggests that religious believers are all too eager to overcome the defeat of their tragic heroes, who die with their promises still unfulfilled by God by elevating them to sainthood. Is the hunger for commitment so strong and blind that no matter how contradictory the message, how false the promises, how lacking in proof the authenticity of the prophet, or how abundant the evidence that he is a fraud and/or disturbed, faith in him will persist and grow in succeeding generations?

The failure of Moses was redeemed by Joshua, who saw to it that the vision of Moses was made secure and that it would not be forgotten by future generations of his countrymen. I will only briefly review the course of events after the death of Moses. Again, it is impossible to reconstruct the historical record, for there are few if any corroborating facts outside the Bible. It is likely that the population that fled Egypt had grown considerably, especially with marriages to captive women. Perhaps the leaders then felt strong enough to invade Canaan. Joshua now had a religion, fashioned for him by Moses, and this he consecrated: the Ark of the Covenant, the laws and commandments, a priestly class, and a tribal structure. All of the males, according to the Book of Joshua, were circumcised. They were now ready to cross the Jordan to engage in new battles. New miracles pave the way: the waters of the Jordan are brought to a standstill and even the sun (and moon) at one point stand still. Joshua has learned well the art of theocratic statecraft, for it is now he who deceives the people into believing that Yahweh talks to him. As his minions engage in battle, they slaughter their enemies and show no mercy. At long last, they take the land they believe was rightfully theirs, divide the spoils, and settle down. Joshua, an old man at the end of the Book of Joshua, proclaims the message of Moses: the Israelites must observe and perform everything written in the books containing the law of Moses. They must obey their covenant with God.

And uncounted numbers of future generations of the chosen people do obey, looking back to their forefathers' years in bondage in Egypt and celebrating their rescue by God. The tragedy of Jewish history is that the Jews have believed fervently in the man named Moses. But it is probable that he neither spoke to God nor received a revelation from on high, and he clearly used cunning and deceit to make his followers believe that he had a supernatural role to fulfill. Moreover, in the last analysis Moses failed in the solemn promise he made to them. The belief that they were a "chosen people" and that they had a convenant with God and a divine mission to fulfill in history has no foundation in fact, and it remains one of the major myths in human history.

Some will respond that the ethnic traditions of the Jews, who have survived throughout history, have an intrinsic beauty and value in spite of the mythology. I am not denying the great contributions of Jewish culture to human history—in literature, science, morality, art, commerce and industry, indeed in all fields of human endeavor. Living as aliens throughout most of their long history, enduring persecution, estranged and forlorn, the Jews were capable of tremendous bursts of creative insight and discovery. Many Jews will assert that the fact that Jewish identity has endured for over 3,000 years—in the face of untold suffering and the hatred of others—only demonstrates the strength of their system of beliefs and practices. To which I add—perhaps there is some saving grace, but what a price to pay for a myth, when one considers the bitter anti-Semitism the Jews have endured, the countless pogroms, and the Holocaust. A religious Jew can only ponder the question: "Eli, Eli, oh God, why hast thou forsaken us?" The cruel paradox of the historical existence of Jewry is the fact that the Jews mistakenly believed that they were the chosen people: that is the basis of their identity but also the seedbed of their alienation and persecution.

Postmodern postscript

What should be said of Jewish history today in this post-modern period? In my view, anti-Semitism cannot be attributed simply to a long-standing and irrational antipathy of the Gentiles against the Jews or the psychological need for a scapegoat. It has its roots in part in exclusive Jewish practices and convictions. And it can be traced back to the mythological basis of Judaism: the very concept of a transcendental deity, selecting the Hebrews as his chosen people. This egregious myth has withstood the heavy blows of time and oppression. It led to a siege mentality, where survival and achievement—in whatever culture the Jews lived —were the highest virtues to be attained. Today it is important that religious Jews reexamine critically the origins of their biblical faith, for their devotion to Judaism is based upon cultural practice, ethnic and genetic loyalty, not divine grounds. The Jews today have come under the influence of modernism in science, philosophy, and literature, and many have been liberated from the blind allegiance to the synagogue and the religious tradition. A distinguished roster of secularized Jews—Spinoza, Marx, Freud, Einstein, and others—have taken new directions and made significant contributions to world culture.

Regrettably, in some societies the Jews who abandoned the faith of their fathers were still labeled as Jews and hated in spite of it. The Nazis decreed that one was still Jewish if he or she had a Jewish grandparent. The Jew was sometimes defined by his enemies rather than himself, in spite of the fact that a human being should be evaluated by what he or she believes and does, his values, aspirations, convictions and commitments. One can abandon Catholicism or Protestantism, Buddhism or Islam and be assimilated into the culture and bloodstream of other societies and nations. New nationalities are constantly emerging; for example, the American today has the blood of many ethnicities and races flowing in his veins. The wandering Jews have survived as a separate cultural identity almost longer than others, but given the growth of secularism, they, as all others, need to become infused into the mainstream of humanity. The Jews have been cosmopolitan and international, citizens of all nations. The barrier against taking a step beyond to full integration and assimilation is often their religious faith—which, under examination, is seen to be simply the faded memory of a national past, based more upon myth than reality.

Had it not been for the enormous horror of the Holocaust and its aftermath, contemporary Judaism might have continued to decline, particularly in secularized Western societies. Bitter memories of persecution during the long saga of Jewish history have reawakened affection for the roots of Jewish ethnicity and inspired intense and often uncritical loyalty to Israel.

The state of Israel today exists as a political-geographical entity. As such, it has as much right to exist as any other national state. Viewed from a humanistic perspective, however, the goal should be to create a humanistic secular state (as the original Zionists wanted), open to persons of all ethnic and religious persuasions. Anything less than that would be a violation of human rights. One must recognize, of course, that few if any Arab states today are secular or democratic, and that most of them are dogmatically committed to the religion of Islam. It would no doubt be an unfair burden to impose an absolute political demand for open borders and immigration on the Israelis until they can be assured of survival.

Modern-day Israelites, fearful of another holocaust, battled Arab armies against insurmountable odds; many thought they were fulfilling the ancient prophecies that this was their land as promised by God. They either drove the Palestinians out or did not permit those who fled to return, much the same as long-settled Jewish populations in neighboring Arab lands either fled or were expelled. This is not unique and has historical precedents. European settlers of the American and Australian continents seized land from the aborigines and Indians; the Russians conquered various ethnic peoples in the vast lands under their hegemony; and the Arabs invaded and imposed Islam on large sections of North Africa. Geographical claims are based upon historical, political, military and economic factors. They are not divine or eternal rights, only rights conferred by residence and the passage of time. In this sense, modern-day Israelites have as much a claim to Palestine today as the former Palestinians now have to Jordan. The Jews living outside of Israel, especially where they have abandoned Judaism, have no special stake in Israel other than sentiment or remembrance of persecution; no more than the Irish, Italians, Greeks, Spaniards, Portuguese, or English who have taken ships to North or South America, Africa, Asia, or Australia and become residents of virgin lands with new opportunities, have a stake in their original homelands.

The wider moral challenge is for all separatist communities to out-grow archaic ethnic or religious fixations and to develop new and more in-clusive commitments to humanity as a whole. The common history of hu-mankind is rooted in the faded memories of ancient civilizations. Jerusa-lem, Athens, Rome, Damascus, Mecca, and Peking have all contributed to world culture; but they have no independent political status; they have provided the culture bequeathed to civilization in general. They are today the common and nonexclusive heritage of all mankind. The new *chosen* are citizens of the global community; the humanistic mission is to create a world beyond race, nationality, religion or ethnicity, in which a person's right to equal dignity and value is based only upon the fact that he or she is a member of the human race.

Summary: Some humanistic reflections

What conclusions do I draw from the preceding analysis? In the place of the Ten Commandments, I propose the following ten points:

First, Moses, if he existed, may have been an Egyptian.

Second, he was probably a magician.

Third, we have no evidence that he communicated with God or had any revelations from on high, though he may have believed that he did.

Fourth, the Jews are no longer a clearly identifiable ethnic stock, but have the blood of many people flowing in their veins.

Fifth, they are not the "chosen people." This myth was proposed by Moses and perpetuated by latter-day prophets and priests.

Sixth, Jewish identity, neurosis, and alienation is in large part rooted in this myth.

Seventh, the moral code of the Old Testament is limited by the time and place in which it was proposed, and it needs to be radically modified and supplemented, or else rejected.

Eighth, Jews and others who live in Israel and have some sort of political, cultural, or linguistic identity should be permitted to do so.

Ninth, it would be more meaningful if the Jews (and Judaism) were assimilated into the mainstream of world humanity, and similarly for all other ethnic groups.

Tenth, the true prophets of the Jewish people in the modern world today are those who emerged after the Enlightenment—the scientists, artists, writers, and philosophers. Their new ethical mission is the liberation of humanity from blind obedience to divine authority and the cultivation of rationality, creativity, and secular universalism beyond frontiers.

IX: Mohammed: The prophet of Islam

The religion of Islam was founded by Mohammed in the seventh century. It is based upon revelations received by Mohammed during the latter part of his lifetime. Mohammed's revelations, forthrightly proclaimed and recorded in the Koran, so impressed those around him that they followed the prophet with dedication and devotion. Mohammed began attracting large numbers of believers under his banner during his lifetime. Incredibly, in less than a century after the death of Mohammed, the armies of Islam conquered and converted vast areas of the world, stretching from Spain and Morocco in the west, all the way to the Indus River and the Punjab in the east. Today Islam incorporates about a billion Muslims in Africa, Asia, the Middle East, and elsewhere, who are committed to Mohammed's precepts, and it vies with Christianity for the hearts and souls of men and women.

The term *Islam* means "submission to the will of God." A *Muslim* is a believer in the faith established by Mohammed, that is, one who "surrenders to God." *Mohammedan* is a misnomer, for the Muslims do not worship Mohammed nor do they consider him divine. He did not work miracles, as did Jesus, but was simply a prophet of Allah (in Arabic, "the God") who had received revelations similar to those of the prophets of the Old Testament.

Mohammed accepted the prophets of the Old Testament, and he traced the lineage of the Arabs back to Ishmael, eldest son of Abraham. He also accepted Jesus, not as the eldest son of God but as one of the most important prophets. Muslims today insist that if one believes that God revealed himself to the biblical prophets, then one must, to be consistent, also accept the revelations to Mohammed as divinely inspired.

Background

The life of Mohammed and the rapid growth of the religion founded by him is a fascinating tale. Although the exact details of his early life are somewhat obscure, it is commonly believed that he was born in 570 in Mecca, a city in Arabia. His father, Abdullah, died before he was born.

His mother died when he was six, and so he was brought up by his grandfather, and then his uncle, Abu Talib.

Mecca is situated in a gorge of a range of mountains about forty-five miles inland from the Red Sea. It was strategically located at the midpoint of the Arabian peninsula, and became an important trading center which was visited by numerous caravans. Mecca was also some sort of ancient religious center, for it was the home of the Kaaba, a sacred shrine containing the Black Stone. This had magical and superstitious significance, and numerous pilgrims were already visiting it from all over Arabia by Mohammed's day. The Quraysh people who lived in the Mecca area believed in a form of animism; spirits (Jinn) lived in the land and occupied animals or objects. Polytheism prevailed, as different tribes had their own local deities. Slavery was commonplace. Slaves were obtained either from war or the plunder of hostile territories, or they were the offspring of slaves already possessed. Aside from a few oases and trading centers, Arabia was a barren desert, and the life harsh and demanding. The inhabitants pursued either an agricultural occupation, where the soil permitted, or they lived as nomads. There was no unifying central political order. The various tribes were composed of family units, each headed by a tribal chief or sheikh, who would protect all of those who lived under his tutelage. Raiding parties were a common source of income. The only protection came from blood ties. Each tribe guaranteed the safety of its members, and vengeance and the law of retribution governed the loosely knit fabric of society.

Mohammed, as a poor orphan, was taken in by his uncle, who as head of the Hashim clan, was highly esteemed in the area. Mohammed began his adult career as a camel driver, participating in the rich camel trade. At the age of twenty-five, he married a wealthy woman, Khadijah, who was fifteen years older than he. Before the marriage, she tested his abilities by dispatching him as her agent on a caravan to Syria. He was successful in her employment and she proposed marriage to him. A man of some business acumen, Mohammed was able to prosper in the commercial life of Mecca. He sired several children by Khadijah. All of his sons died young, but he was able to marry off his four surviving daughters to well-to-do husbands. Moreover, the marriage eventually freed Mohammed from the pressures of earning his daily keep, and it also provided him with some leisure to follow other interests. No doubt the moment of greatest significance for Islam was the call that Mohammed first received when he was forty years old. This marked the beginning of his profound religious conversion and of the revelation that he was supposed to have had.

Historical documents

One can raise an important question about the origins of the three great monotheistic religions. We have already seen the difficulty in determining the veracity of the Old and New Testament texts, and particularly of the accounts of the prophets portrayed in them. This is much less the case with Mohammed and Islam. Indeed, a vast literature has been developed by Muslim scholars and historians. In recent years, Western scholars have attempted some kind of historical reconstruction. Like Judaic and Christian sources, the Islamic historical record is pregnant with legends and myths. Moreover, there does not seem to have survived—if indeed there ever existed—any extensive heretical or critical literature written by Muslims or religious groups they overran and superseded. Accordingly, there is a compelling need today for the application of scientific tools of investigation to the study of the origins of Islam.

Let us examine the two major sources we now have. First, the Koran itself. According to fundamentalist Muslims, every syllable of the Koran is of divine and eternal origin. Actually, however, the method by which this book was compiled is well-known. The Koran is comprised of *surahs* or chapters, which are divided into *ayat* or verses. It consists entirely of the commands and revelations of Mohammed, which he said were received either from the angel Gabriel or directly from God. Mohammed would be "inspired," and at that moment or shortly thereafter, the passages received were recited by Mohammed to relatives, friends, or followers who were present. These sayings were later committed to writing by someone who could write (Mohammed couldn't): on leather, stone, palm leaves, or whatever crude materials were available. The revelations were numerous, beginning when Mohammed was forty and continuing for twenty-three years until his death. We do not know in what state the original writings were preserved. Many of the revelations had only a temporary purpose, growing out of a specific practical situation. Whether all the revelations were retained is uncertain, but it seems unlikely that when Mohammed died there was a complete collection of all the original transcriptions.

The method of communicating such revelations was by means of oral recitation, an art which Mohammed himself cultivated. Members of his entourage would commit these recitations to memory and repeat them to others. In later years they were read in the mosques and intoned in the religious schools and public places. Since the words of Mohammed had aroused such deep awe and reverence, great efforts were made to see to it that the oral recitations were accurately rendered, and the reciters were proud to be able to memorize and deliver the recitations intact. It is common knowledge that an oral tradition is often elaborated upon, however slightly at first, as it passes from mouth to mouth. So we have no guarantee that the Koran preserves the revelations exactly as they were first delivered. In any case, it was only after the death of Mohammed that his early disciples—close relations and friends—assembled all that they could locate of the written record. This became known as the Koran, which means "recitation."

The first individual to assemble the written documents was Zayd, Mohammed's adopted son and one of the first converts. Zayd attempted to gather whatever he could locate within two or three years after Mohammed died. The *surahs* were not arranged in chronological order; generally the longer *surahs* were put first and the shorter ones last. Since Zayd had often recited the Koran in the presence of Mohammed, it was thought to be a reliable version. Varieties of expression soon crept into Zayd's edition, and different versions appeared. Accordingly, another effort to compile a faithful reproduction was made by Othman, an early disciple, who several years after the death of Mohammed assumed power as caliph. As best as can be determined, it is this version of the Koran, perhaps assembled twenty years after Mohammed's death, that Muslims now possess; and scholars believe that it is more or less an accurate rendition of a good part of Mohammed's revelation.

Unfortunately it is often difficult to date precisely when the revelations occurred or to ascertain the context in which they were uttered. Many passages of early revelations were abrogated or contradicted by later passages. Thus one cannot always interpret the meaning of the revelations or determine the precise intention of the author. It is believed that the shorter verses at the end of the Koran were from the earliest days of the prophet. Nevertheless, the Koran remains as a unique book, able to arouse intense feelings of devotion for countless generations of believers, moved by its eloquent poetry and powerful message.

The second source of early Islamic history which helps interpret the Koran and the life of Mohammed is known as the Hadith ("the tradition"). This source is more difficult to accept as reliable. It consists of recollections of the sayings of Mohammed and tales about him told by his friends and followers, handed down from generation to generation by oral means and eventually collected and recorded in writing. A'isha, the favorite wife of Mohammed, was alone responsible for 1,210 traditions allegedly uttered by him. The first followers of Mohammed were evidently deeply impressed by his forceful and charismatic personality. Known as the Companions of Mohammed, they were largely simple nomads or warriors, mostly illiterate. It is doubtful that their testimonies can be accepted without qualification, particularly since, like the authors of the New Testament, virtually all of Mohammed's companions were propagandists for the new faith; after the death of Mohammed many assumed positions of power in the expanding Islamic empire, and might have been tempted to inflate the Mohammed legend to justify their own positions. It is most likely that the second and third generations of Muslims embellished on these early stories still further, as was surely the case with the Jesus legend.

Intense veneration developed about the person of Mohammed, and any tale about him, even a shred of his hair, was considered to be of inestimable worth. As Islam expanded into Africa and Asia, so did the storytellers, known as Collectors, who wove legends about the prophet. The Hadith thus took on majestic and supernatural dimensions, so that by the second century after the death of Mohammed a large body of fictional material had been gathered.

One of the main problems in scholarship is to separate fact from fiction. An insuperable difficulty faced by the objective scientific inquirer is that Islam imposed strict limits on free inquiry, and the penalty for any dissenting interpretation was sometimes death. We know that Mohammed faced great opposition when he first proclaimed his message; his critics thought he was mad. Reverberations of this appear in the Koran and the Hadith. But we do not have many details of their skepticism nor their arguments as to why they rejected the revelations. Thus there is a dearth of skeptical Muslim accounts of Mohammed's life, though there were numerous Jewish and Christian critics in later centuries.

The earliest biographies of Mohammed were written 120 to 200 years after his death, and these abound with tales of miraculous events, which are difficult to corroborate. Leaving aside these obviously exaggerated extrapolations, it is nonetheless possible to fathom some of the key details of Mohammed's life, and even of his revelatory experiences.[1]

First revelations

Mohammed's revelations began under curious circumstances. He was accustomed to visiting a rocky hillside just outside Mecca, where he spent much of his time alone in mediation. Sometimes he would spend several days and nights in seclusion in a cave at the foot of Mount Hira. At times his faithful wife (or family) would accompany him on his sojourns.

---

1. The Moslem calendar (A.H.) begins with the *Hijra*, the flight of Mohammed and his followers from Mecca to Medina in 622, ten years before his death. Among some of the early Muslim historians whose works have survived and are often cited as reasonably reliable are Mohammed Ibn Ishak (who died about A.H. 151), Ibn Hisham (A.H. 213), Al-Wakidi (A.H. 207), his secretary Ibn Sai'd (A.H. 230), and At-Tabari (A.H. 310).

As far as we can tell, Mohammed was experiencing a time of great inner turmoil, and he apparently suffered intense depression, perhaps a midlife crisis. Tradition has it that he seriously contemplated suicide by hurling himself from a rocky precipice. During this period, his vivid dreams and mysterious visions disturbed him greatly. Mohammed was apparently undergoing a profound religious conversion, questioning the deities and morality of his polytheistic culture, and groping towards the kind of monotheism expressed in ancient biblical scriptures. It is said that he could not read or write, but he was aware of the religious traditions of the Jews and Christians, whom he encountered on his caravan treks, and of their beliefs in divine revelation. If these people had great prophets who received messages from God, why not the Arabs? And so Mohammed began to have his own revelations, expressed in rhapsodic poetic form.

Most historians consider the first and one of the most important revelations to be Surah 96, lines 1-5. In this he tells of his encounter with a divine presence. One day, after wandering among the peaks of Hira, perhaps lost in deep meditation and intense anxiety, a vision suddenly appeared before him. It was a mighty being coming out of the sky, who descended very close to him. Muslim tradition identifies him as the archangel Gabriel, a messenger of God. This mighty being repeated the word *iqraa* (recite, read, proclaim) three times. "What shall I recite?" asked Mohammed in exhaustion. Finally the following words were uttered:

> Recite! In the name of the Lord who created—
> Created man from a mere clot of congealed blood
> Recite! For the Lord is most bountiful.
> It is He who has taught (to write) with the pen—
> Has taught man that which he knew not.[2]

The voice was forceful, commanding him to proclaim a message sent to him from on high. The effect on Mohammed was overwhelming. He fell to his knees and dragged himself along while the upper part of his chest was trembling. He eventually recovered and rushed home to tell Khadijah about his terrifying experience. According to the Hadith, Khadijah comforted him and assured him of his sanity. She also reinforced the conviction that he was receiving prophecies. She consulted her cousin Waraqua, who knew both Jewish and Christian scriptures, and he said that Mohammed was receiving prophetic revelations. Some commentators believe that were it not for Khadijah's moral support, Mohammed would

2. The text that I have used is *The Holy Qur'an,* translation and commentary by A. Yusuf Ali (Washington, D.C.: Islamic Center). I have modernized some of the spellings.

never have presumed to don the mantle of the prophet.[3]

Although the exact personal chronology is uncertain, there was apparently a hiatus of several months after this first encounter, during which Mohammed did not experience another visitation from Gabriel, though he kept having inexplicable experiences. This raised his doubts anew. Undergoing considerable psychological distress, he needed reassurance that he was not mad or possessed by the Jinn (evil spirits). At last, Gabriel again visited him, proclaiming him to be a true prophet of the

---

3. The following account of Mohammed's first religious soliloquy is from Al-Wakidi: "The first beginnings of Mohammed's inspiration were *real visions*. Every vision that he saw was clear as the morning dawn. These again provoked the love of solitude. He would repair to a cave on Mount Hira, and there pass whole days and nights. Then, drawn by the affection of Khadija, he would turn to his home. This went on till the truth burst upon him in the cave. It happened on this wise. Wandering in the hills around, an angel from the sky cried to him, '*O Mohammed, I am Gabriel!*' He was terrified, for as often as he raised his head, there was the apparition of the angel. He hurried home to tell his wife. 'Oh, Khadija,' he said, 'I have never abhorred anything as I do these idols and soothsayers; and now verily I fear lest I should become a soothsayer myself.' 'Never,' replied his faithful wife; 'the Lord will never suffer it thus to be,'—and she went on to speak of his many virtues, upon which she founded the assurance. Then she repaired to her cousin Waraka, and told him all. 'By the Lord,' cried the aged man, 'he speaketh truth! Doubtless it is the beginning of prophecy, and there shall come upon him the *Great Namus,* like as it came upon Moses. Wherefore charge him that he think not aught but hopeful thoughts within his breast. If he be raised up a prophet while I am yet alive, surely I will stand by him.'

"Now the first Sura revealed to Mohammed was the 96th, verses 1-5, *Recite in the name of the Lord,* etc.; and that descended on him in the cave of Hira. After this he waited some time without seeing Gabriel. And he became greatly downcast, so that he went out now to one mountain, and then to another, seeking to cast himself headlong thence. While thus intent on self-destruction, he was suddenly arrested by a voice from heaven. He looked up, and behold it was Gabriel upon a throne between the heavens and the earth, who said: '*O Mohammed! thou art the Prophet of the Lord, in truth, and I am Gabriel.*' Then Mohammed turned to go to his own house; and the Lord comforted him, and strengthened his heart. And thereafter revelations began to follow one upon another with frequency." Quoted in William Muir, *The Life of Mohammed: From Original Sources,* ed. T. H. Weir (Edinburgh: John Grant, 1912), pp. 49–50.

The following outline from Ibn Hisham and At-Tabari of Mohammed's first inspiration is at some variance from Al-Wakidi's, since Mohammed is not alone in the cave: "On the night whereon the Lord was minded to deal graciously with him, Gabriel came to Mohammed as he slept with his family in the cave of Hira. He held in his hand a piece of silk with writing thereon, and he said *Read!* Mohammed replied, *I cannot read.* Whereupon the angel did so tightly grip him that he thought death had come upon him. Then said Gabriel a second time *Read!* And Mohammed, but only to escape the agony, replied, *What shall I read?* Gabriel proceeded:—*Read* [recite] *in the name of thy Lord, etc.,* repeating the 96th Sura to the end of v. 5. When he had ended the angel departed; and 'the words,' said Mohammed, 'were as though they have been graven on my heart.' Suddenly the thought occurred to him that he was possessed of evil spirits, and he meditated suicide; but as he rushed forth with the intention of casting himself down a precipice, he was arrested by the appearance again of Gabriel, and stood for a long time transfixed by the sight. At last, the vision disappearing, he returned to Khadija who, alarmed at his absence, had sent messengers to Mecca in quest of him. In consternation he threw himself into her lap, and told her what had occurred. She reassured him, saying that he would surely be a prophet, and Waraka confirmed her in the belief." Ibid., p. 50.

Lord. Thus we read in Surah 68:1-5:

> . . . By the pen and by the (record) which (men) write—
> You are not, by the grace of thy Lord, mad or possessed.
> Nay, verily for you is a reward unfailing.
> And you (stand) on an exalted standard of character.
> Soon you will see, and they will see
> Which of you is afflicted with madness.

Mohammed now began to believe that the voices he was hearing were real, and he became convinced that they came from outside himself; he was certain that he could distinguish between his own inner thoughts and the revelations which came from God. At first he attempted to keep these events secret, confiding them only to his wife, who kept reassuring him that he was receiving divine messages. Gradually, believing that he was indeed a prophet of God, he became bolder and began to express his convictions openly: he had a divine commission to spread the word of God, to preach and summon people to faith in Allah. They must believe in one God only, the source of all creation and the judge of men.

Eventually, Mohammad began to preach in public. According to the Koran, he was taunted at first by those who heard him. They considered him to be possessed by demons and accused him of sorcery and magic. Tradition has preserved various stories about the fact that he continued to experience revelations. One is that on one occasion, fearful of his sanity, Khadijah tested the character of a spirit confronting Mohammed, by making him sit first on her right knee, then on her left. Mohammed kept experiencing the apparition, no matter what his position. She then took Mohammed in her lap, removing her veil or uncovering her body, at which point the spirit disappeared—which seemed to demonstrate that the being was virtuous or modest. Khadijah said: "Rejoice, my cousin, for by the Lord! it is an angel and no devil." On another occasion, terrified by such an experience, Mohammad asked Khadijah to cover him, which she did, constantly administering to his fright.[4]

A psychophysiological diagnosis

What are we to make of Mohammed's experiences? Were they caused by God, or is there some physiological explanation? It is not easy, given this late date, to offer a comprehensive diagnosis of his condition. Nonetheless, some of the reports of his earlier revelations suggest (1) that he may have

---

4. From Ibn Hisham and At-Tabari, in Muir, *Life of Mohammed,* p. 51.

suffered from hallucinations, which indicate pathological episodes, and (2) that he may also have had a condition similar to epilepsy. In the first case, he thought that the revelations were communicated by Gabriel from God. In the second he suffered a "brainstorm" or some kind of epileptic-ecstatic seizure. Epilepsy was known to ancient civilizations; some even considered it to be "the divine madness."

Most cases of epilepsy begin in early life. There are some intimations that Mohammed may have suffered this condition in early life, even as a child, but it is not certain that this was the case, since there is little reliable testimony about him before the age of forty.[5] Epileptic seizures come on suddenly and without warning. They are characterized by a loss of consciousness and usually involve convulsions, biting of the tongue, muscle stiffness, and arrest of breathing. The individual has no memory of what transpired during the attack and is apparently exhausted by it.

The testimony we have about Mohammed's condition and the symptoms of his affliction according to his defenders do not follow the classic epileptic syndrome exactly. Usually his attacks came on suddenly and without warning. He would often fall into a deep faint and manifest a high fever, since he was reported to have been sweating profusely. Some reports were that he trembled or shuddered violently and that he was exhausted by his attacks. The following account of what occurred during one such seizure late in his life is revealing, for it clearly suggests a physiological disorder. According to Ibn Sai'd:

> At the moment of inspiration, anxiety pressed upon the Prophet, and his countenance was troubled. He fell to the ground like an inebriate, or one overcome by sleep; and in the coldest day his forehead would be bedewed with large drops of perspiration. Even his she-camel, if he chanced to become inspired while mounted on her, would be affected by a wild excitement, sitting down and rising up, now planting her legs rigidly, then throwing them about as if they would be parted from her. To outward appearance inspiration descended unexpectedly, and without any previous warning to the Prophet. When questioned on the subject he replied: "Inspiration cometh in one of two ways; sometimes Gabriel communicateth the Revelation to me, as one man to another, and this is easy; at other times, it is like the ringing of a bell, penetrating my very heart, and rending me; and this it is which afflicteth me the most."[6]

5. There is some suggestion that he may have had an epileptic attack at the age of six, being "visited by angels" even then. Abu Talib is reported to have said, shortly after taking Mohammed under his care: "I fear that the boy may have had an attack. Take him back to his family before his disease declares itself." Ibn Hisham, *Sura, Das Lebens Muhammeds,* ed. F. Wüstenfeld (Göttingen, 1859–60), p. 105. Quoted in Maxime Rodinson, *Muhammed* (New York: Pantheon Books, 1971), p. 56.

6. From Ibn Sai'd, in Muir, *Life of Mohammed,* p. 52.

Describing the character of a seizure and what followed, 'Abd ar-Rahman relates that many years later in returning from Al-Hodeibiya, when Mohammed was fifty-eight, suddenly people began to urge their camels on. What was the reason for the rush? "Inspiration has descended on the prophet," was the reply. Al-Rahman also hurried on his camel and reached Mohammed. When Mohammed noted that a sufficient number of people had gathered around him, he began to recite Surah 40.[7] According to the Hadith, Mohammed was generally unaware beforehand that inspiration was about to overcome him. After the seizure, Mohammed was apparently able to compose himself sufficiently to recite a surah. Later in life, Mohammed is reported to have remarked that the white hairs that began to appear in his beard were hastened by these experiences. He is alleged to have replied to his devoted friend Abu Bekr that the "terrific" surahs were responsible for his gray hairs.[8]

Tradition indicates that when Mohammed awoke and was being questioned about his experience, he reported his revelation. An important question to be raised is whether Mohammed underwent a hallucinatory revelation while he had his seizure (he seemed to have been unconscious), or if he had no conscious memory of his experience, had he engaged in a ploy to spin out a story to deceive his followers after the fact, since they had already accepted a supernatural interpretation of his seizures.

Many commentators who are sympathetic to Islam deride this explanation, and maintain that Mohammed was sincere. They say that he genuinely came to belive that his revelations were from God. How do we test the sincerity of a man whom countless millions have honored and revered as a great and noble prophet? Mohammed must have been a man of enormous personal magnetism, able to bring those about him under his sway. Critics have said that his character showed a calculating cleverness, premeditation, and a predilection for using sly methods of intrigue—a view which is hard to reconcile with the belief that he was of a peerless and honorable character.

Was Mohammed's religious ministry dominated by a lust for power? Perhaps not at first. But it is clear that he wielded great power when he finally was in a position to do so at the end of his life. In time he may have come to believe that his visions and strange seizures were indeed from God; since his reports were accepted by those about him as authentic, they could only reinforce his own delusions. Later on, as we shall see, Mohammed (like Moses earlier) was able to turn some of his revelations off and on at will, to meet urgent personal and political needs, and these revelations were taken by those about him as the final law of God.

7. Ibid., p. 52.
8. Ibid.

Let us examine his experiences more closely by consulting further details supplied by the Hadith. According to Ibn Sai'd, A'isha, Mohammed's favorite wife, relates: "Once I witnessed how the revelation came to Allah's apostle on a very cold day. When it was completed his brow dripped with perspiration." According to Abdullah Ibn'Umar, who asked the prophet, "Do you know when the revelation comes to you?" he replied, "I hear loud noises and then it seems as if I am struck by a blow. I never receive a revelation without the consciousness that my soul is being taken away from me," which strongly suggests a loss of consciousness.[9]

Mohammed seems to have had visions of Gabriel only at the beginning of his career. After that his revelations were generally auditory. Some witnesses tell of his intense suffering and physical pains at the time of inspiration, which again suggests some kind of epileptic seizure. "When the revelation came to the prophet, they pressed hard upon him and his countenance darkened." Falling to the ground as if intoxicated, it is said that he even "groaned like a camel's colt."[10]

These descriptions seem to confirm that Mohammed suffered from a serious affliction. When he first encountered Gabriel, he may have been suffering from hallucinations brought on by fasting and exhaustion. He may also have suffered from epileptic-estatic seizures, which incapacitated him. Unable to explain his condition in medical terms, he interpreted it in miraculous terms. In any case, Mohammed and those about him came to believe that it was a divine calling. The paradox is that one of the great religions of humanity is based upon ignorance of the natural causes of the phenomena that may have afflicted its key prophet, and that a supernatural cause is invoked to explain it. The transcendental temptation is so strong that there is a willingness to read a divine cause into anything in spite of no credible corroborating evidence.

The elementary criteria that believers completely overlook is this: Why accept uncorroborated, firsthand, subjective reports as true? Perhaps Mohammed's claims were taken at face value because of the success of his enterprise. Certainly as he gained adherents, his fame and influence spread, although he had many enemies who considered him insane and opposed him from the beginning. His critics were eventually defeated by the persuasive appeal of the new faith and the power of the sword.

The question the reader may very well raise is whether there was any trace of fraud in Mohammed's career. The Koran itself relates some of the accusations made against him by the Meccans and later by the Jews: that he practiced sorcery and downright fraud and that he got his early

    9. Quoted in Tor Andrae, *Mohammed: The Man and His Faith,* trans. Theophil Menzel (New York: Harper & Row, 1960), pp. 49–50.
    10. Ibid., p. 50.

ideas from some foreigner—some Jew or Christian—with whom he had
conversed daily. It is difficult to ascertain with precision what did occur,
since negative comments generally have not survived, though I did locate
one suggestion of this. The *Musnad,* or the collection of Traditions of Ibn
Hanbal, who died in 855 (A.H. 241), reports that one of the scribes
employed to take down Mohammed's sayings became convinced that
Mohammed was an impostor, and abandoned Islam as a result.[11]

Was Mohammed feigning revelations as a result of his seizures or
could he eventually produce a revelation at will without having had a
seizure? The best evidence that I can uncover suggests that later in his life
Mohammed deviously resorted to "making up" revelations to suit his
political and private purposes.

The first converts

What happened to Mohammed's career following his first visitation from
Gabriel in the cave at Hira is remarkable. Glimmers of light had no doubt
been struggling within the dark recesses of Mohammed's soul as he sought
to make sense out of his experiences. And this he did by attributing them
to a divine source. Gradually he began to believe that God was the sole
creator, judge, and ruler of humanity. The people of his society were
hopelessly mired in idolatry and heathenism. He felt that he had a divine
mission. His great task was to teach them by means of his new religious
awakening. He would impart to them an appreciation of God's majesty
and their utter need to obey him. Only by this means could they achieve
salvation. This was not novel, for it was essentially the Judaic-Christian
monotheistic message. What was new was that he interpreted his own
visions as a sign that he, Mohammed, was appointed by God to be the
last prophet, bringing the final message. This was made directly relevant
to the Arabic cultural context, and it led to needed moral and spiritual
reform. Mohammed's convictions were expressed in impassioned poetry,
always warning his fellow countrymen to find God. His pronouncements
were couched in words of great beauty and force. Surah 95 reads:

> I swear by the fig tree and the olive,
> By Mount Sinai and by this land inviolate.
> Verily we made Man of the choicest creation,
> Then we rendered him the lowest of the low—
> Except such as believe and work righteousness
> Unto them shall be given their reward that fadeth not away.

11. Cairo, 1890, vol. 3, p. 212. Quoted in David S. Margoliuth, *Mohammed and the
Rise of Islam* (Freeport, N.Y.: Books for Libraries Press, 1972), p. 89.

Then, after this, what shall make thee deny the day of reckoning?
What, is not God the justest of all judges?

Mohammed most likely expressed these feelings to those in his inner
circle before he ventured forth into the world to proclaim them. The
friends and relatives of Mohammed were no doubt perplexed as they
listened to his reverential admonitions. Revelations continued to descend
upon him, and he insisted that he was a prophet of the Lord. It was only
after three years, when Mohammed was forty-three, that he ventured
forth to announce openly his message in the streets of Mecca. It is clear
that if a cult is to grow and eventually gain mastery, it must be built on
solid foundations. Who the first converts close to the master are and how
willing they are to sacrifice, even die for the cause, is crucial to its success.
Among his first converts to Islam were his wife Khadijah and two young
members of his household, Zayd, formerly a slave who was adopted as a
son by Mohammed, and Ali, his cousin who was only thirteen or
fourteen.

The neighbors who first heard Mohammed's teachings mocked and
taunted him. He was considered a lunatic. But he soon added other
believers to his little circle of followers. First there was an old friend of
Mohammed's, a man by the name of Abu Bekr, about the same age, a
successful merchant who shared Mohammed's religious concerns. Abu
Bekr's role was very important, for he helped bring in some of the earliest
converts, including Othman, and later he even purchased slaves who
would convert to Islam. Abu Bekr and others were to play powerful roles
after the death of Mohammed, but they were also crucial in the first
stirrings of the new faith. The slaves of Mecca were especially sympathetic
to the preachings of Mohammed, as were dissatisfied young men and
women, generally people without power or influence. After three years of
public preaching, it is estimated that Mohammed had attracted nearly
forty persons to his cause.

A noticeable change now occurred in the relationship of Mohammed
to the people of Mecca. At first they considered his teachings to be rather
harmless, but as his influence grew, their hostility was aroused; believers
in Islam began to be harrassed. The initial cause, no doubt, was that
Mohammed seemed to threaten the attachment of the Quraysh to the
Kaaba. Mohammed abused their pagan idol worship. He preached the
unity of God, the judgment after death of good and evil acts, the resurrection of the dead, prayer, charity, fasting, and obedience to the prophet.
"Allah-o-Akbar!" he would cry aloud—"God is great!" And people would
become angry. Although they were familiar with the idea of a monotheistic deity, such as confessed by the Jews and Christians in their midst,

the ruling council no doubt feared that they would lose control if Mohammed prevailed. The persecution of the Muslims thus intensified.

All during this period, Mohammed was protected by his uncle, Abu Talib, though other members of Mohammed's family were incensed by his behavior. Another uncle, Abu Lhab, disdained his appeals and even insulted him. As the sect grew in numbers, Muslims were further ostracized and castigated. Converted slaves and strangers from the lower classes were especially vulnerable, and some were imprisoned or physically attacked. Yet the more they were persecuted, the stronger became their fidelity to the prophet's message. A psychological factor was operating: the more true believers were oppressed for their belief, the more intense it became. The fact that Mohammed was accused of deceit and fabrication only reinforced his followers' dedication to him. During this period Mohammed lived in virtual isolation in one quarter of the city, protected by his uncle. He occupied a house owned by an early convert, Al-Arkam, yet he was surrounded by his spiteful wrath of his neighbors.

In time, Mohammed was able to convert some men of considerable standing, such as Hamza and Omar, both of whom had previously abused him. His followers then approximated one hundred persons. A turning point in Mohammed's ministry occurred in 619, with the death of Khadijah, and about the same time his uncle died. (Incidentally, Abu Talib had remained an unbeliever until the end.) The death of his wife of twenty-five years was an especially grievous blow to him. Abu Talib was succeeded as head of the Hashem clan by Abu Lhab, who at first agreed to extend his protection to Mohammed, but later angrily withdrew it when Mohammed warned him that Abu Talib was suffering the torments of hell because he was an unbeliever. From then on, the situation of the Muslims rapidly deteriorated. Defections now set in, and there were few new converts. Another base had to be found for Mohammed's group if they were to survive. If Mohammed had not found new soil, then his small cult, like the countless thousands that have existed before and after, would no doubt have declined and disappeared. But fortunately, the possibility for expansion presented itself elsewhere, and Mohammed capitalized on this new opportunity.

Medina: The sword of Islam

The observation that a prophet is without honor in his own town proved to be true for Mohammed. For it was only when his message was taken to Medina, where it began to take root, that a dramatic change occurred. Medina (or Yathrib) was a town—or more accurately a cluster of houses on an oasis—180 miles north of Mecca. It had been embroiled in blood feuds between contending tribes for many years, and it was at the point of erupting again into wholesale slaughter. Mohammed had visited the area as a child. Moreover, his father had been buried there and he had relatives living there as well. The people of Medina needed some peace and tranquility, particularly since they depended upon agriculture and harvests for their livelihood. Some of them had already visited Mecca to partake in the pilgrimage to the Kaaba. Mohammed was able to win some of them to Islam. He even dispatched one of his disciples to Medina to recite verses from the Koran and to convert new believers. At one point seventy-five citizens of Medina came south to Aquaba to meet with Mohammed. He promised them that Allah, through him, would end internal tribal conflict and arbitrate future differences. They agreed to pledge their protection to him if he would move to their city.

The migration of Mohammed and his followers from Mecca to Medina occurred shortly thereafter. This was known as the *Hijra,* a decisive moment in Islamic history, since all calendars are dated from that point forward. Thus the year 622 in the Christian calendar is designated A.H. 1. Once Islam was taken to Medina, there was a rapid growth in the movement Mohammed assumed leadership of the area. He established his own army which became vital to the success of his efforts. It now became clear that it was no longer the revealed word by itself but also the sword that would extend his horizons. A transformation of great significance occurred. Mohammed was no longer a weak prophet, imploring his followers to virtue or pointing the way to salvation by example; he could impose his will by force. It was not pure revelation itself that established the new religion—though obviously this was the dramatic factor that appealed to men's hearts and their longings for Paradise. Rather, it was the union of religious fervor with political and military power that catapulted the Muslim faith forward.

The separation of church and state that ultimately came to be so cherished in Western democracies is foreign to Islam; for it was the union of mosque and state that laid the foundations of a new holy empire that was to emerge. If a religious leader cannot convince believers to accept the faith, he can compel them not to reject it and can dominate their lands. This is not unique to Islam, for the establishment of Christianity as the official religion of the Roman empire by Constantine in the fourth century had already transformed the Bible into an ideological weapon. No doubt without the messianic idea at the heart of their new faith, the sword could not have cut so sharply or deeply. Without the belief that God had sanctified the Prophet to perform his will, the legions of the Prophet might not have dashed forth so rapidly. The victories that

Mohammed achieved in the next decade of his life were breathtaking; and they revealed him to be a complex personality, deeply motivated by religious impulses, capable of generosity and moral virtue, yet also willing to use consummate political strategy and ruthless methods when deemed necessary to achieve his ends.

Mohammed was welcomed to Medina with open arms. The immediate attention of Mohammed and his followers was turned toward building a mosque and homes for themselves. The first major problem to be faced, however, was economic sustenance. The new Muslims in Medina did not have the money to purchase land or flocks to support themselves. Mohammed resorted to a traditional Arab solution. He assembled a raiding party. The target was the rich Quraysh caravans from Mecca, which passed only sixty miles from Medina as they made their way north to Damascus. After some successful raids, the Quraysh became incensed and began to raise an army against the Muslims. The first major battle was at Badr. Mohammed had assembled a party of 300 men, the Meccans fielded 900. A pitched battle ensued, and the Meccans were defeated by the Muslims, who were no doubt spurred on by Mohammed's promise of Paradise. The spoils were divided equally among the Muslims, though Mohammed himself took only one-fifth of the booty (instead of one-fourth as was customary in such affairs). The Muslim victory, with inferior forces, was taken as a sign of God's favor. This belief rallied other converts to the holy cause.

Meanwhile, the Quraysh, disheartened by their defeat and intent on revenge, were determined to put an end to the Muslim threat. They now raised an army of 3,200 men and cavalry. Mohammed could field only 1,000 soldiers, a number quickly reduced by defection to 700. This time Mohammed himself was wounded in battle at Uhud, and the Muslims were defeated, with many dead and wounded. However, the audacious Prophet rallied his forces, and set about in hot pursuit of the Meccan army, lighting many night fires so that they would think that his forces had greatly increased. The Meccan army had blundered by not pursuing the retreating Muslims to finish them off. Nevertheless there was no reengagement, and the position of Mohammed was now challenged by his opponents. If the victory at Badr was a sign of God's grace, did not the defeat at Uhud mean that he had lost God's favor? Mohammed attempted to quiet doubts by proclaiming a new revelation. God had not abandoned the faithful; this was only his way of testing those who truly believed and separating them from those who professed the faith but in the inner recesses of their hearts were not believers (Surah 3:1-19). The Quraysh resolved that they had to destroy the Muslims once and for all. They now marshalled three armies of 10,000 men and 600 cavalry. Mohammed was able to bring together only 3,000 men. His people dug a deep ditch with high ramparts around the perimeter of the oasis at Medina, which they heavily fortified. The Meccans' siege was unable to break the barrier, resulting in a stalemate; the Meccan forces eventually withdrew. This was interpreted as a victory for the Muslims, for it demonstrated that Mohammed could not be defeated by force.

All during this period, Mohammed's relationship with the Jewish tribes in the Medina area grew worse. They rejected his message, and he questioned their loyalty to him. He believed that he had to put an end to their opposition. The relationship between Islam and Judaism during the long history of these two religions is far too complex to be fully summarized here. I will focus only on Mohammed's use of Jewish scriptures— at first, to gain the Jews' allegiance, and when this failed, his later effort to defeat the Jewish tribes in the area. It is clear that a good part of his theological outlook was derived from the Old Testament. Mohammed was familiar with Jewish customs and beliefs, for there were a great number of Jewish settlements in Arabia. The early part of his message drew heavily on Jewish sources: first and foremost, he accepted monotheism. Early in his ministry, he had his followers face Jerusalem to pray. He observed the fast of Yom Kippur (the day of atonement). He instructed the faithful to pray on Fridays. He also set aside a time for prayer in the middle of the day, as did the Jews, and he bade his followers to adopt similar dress, hairstyles, and dietary rules. Mohammed revered Abraham as the founder of both Judaism and Islam, and he even mentioned that the Kaaba had been built by Abraham for the worship of God. He expected the Jews to accept him as the last in a long line of Hebrew prophets, as he had accepted their scriptures. God had sent other prophets to bear witness to his divine glory, including Moses, Noah, Abraham, Lot, Ishmael, Isaac, Jacob, and David. Mohammed taught that his new revelations must supersede all others. But when the Jews rejected his call and remained "unbelievers," content in their practices and scriptures and believing that they were God's chosen people, he turned against them.

Thus a revelation was introduced by Mohammed in 623 to mark a decisive shift in his religious policies. The circumstances were as follows: About sixteen or seventeen months after arriving in Medina, when there seemed no longer any hope of reconciling Judaism and Islam, Mohammed made a change in the method of praying. According to the Hadith, one day as he was praying, turning his face upward to Jerusalem, a divine revelation came upon him unexpectedly:

Turn then thy face in the direction of the sacred Mosque:
Wherever you are, turn your face in that direction (Surah 2:144).

Mohammed had already made two prostrations with his face toward Jerusalem. When he received this revelation, he suddenly turned around, looking to the south toward the Kaaba in Mecca, and he finished the service in that direction. As he did, all of the worshippers in the mosque followed his example and turned south as well, and that has remained the Muslim practice ever since. The circumstances of the revelation did not require that he go into a trance; he merely uttered the revelation in the mosque.

Meanwhile, Mohammed's controversy with the Jews erupted into open warfare. Some of the pagan tribes of Medina who welcomed Mohammed had been formerly allied with the Jews in their internal conflicts. They had hoped that Mohammed would honor their longstanding alliances. This, Mohammed now refused to do. His uncompromising attitude was reinforced when he learned that some of the Jewish tribes had made formal alliance with the Meccans against him. He considered the existence of such tribes a threat to his control. The first conflict arose with the Jewish clan known as Qaynuqa. A trifling incident between a young Arab girl and some Jewish youths was blown out of proportion, and retaliation led to retaliation. When Mohammed's army attacked the Jews, they sequestered themselves inside their fortress, which was blockaded for fifteen days until they surrendered. Mohammed was prepared to condemn all the inhabitants to death, but he relented when their Medinan allies demanded their freedom. Not wishing to break relations with his newly found friends in Medina, he lifted the death penalty but insisted that the Jews evacuate Medina within three days, leaving behind all of their goods to the victors.

A bloody massacre of a Jewish tribe occurred in May of 627, after the battle of the Great Ditch. It was with the Jewish tribe Qurayza. Mohammed's army laid seige to their fortified village and demanded that they surrender, which they did, unconditionally. Mohammed offered to spare them if they embraced Islam, which they refused. A terrible penalty was exacted: All adult males were executed, the women and children sold into slavery, and their property divided. The men were imprisoned in a yard separate from their women and children. They spent their last hours of darkness in prayer, chanting passages from the scriptures. Meanwhile, Mohammed had trenches dug during the night across the marketplace of the town. In the morning, he ordered the male prisoners to be brought forth five or six at a time. Each group was forced to sit down in a row at the edge of the trench. They were then beheaded and their bodies pushed into their grave. The killing that began in the morning continued all day and into the night, lit by torchlights. The blood of an estimated 600 to 800 victims bathed the marketplace. All during these proceedings Mohammed was a spectator. He gave orders to fill in and cover over the trenches, and then distributed the spoils. The land, possessions, cattle, and slaves were all divided, Mohammed keeping one-fifth of each for himself. He presented slave girls and female servants to his friends as gifts. The rest of the women and children were sold into slavery. He kept for himself, however, Reihana, a beautiful Jewess whose husband and male relatives had perished. Although he asked her to be his wife, she declined, but having no choice, lived with him as his concubine or slave.

What happened to the messages of the Lord? Why in so brief a time was Mohammed transformed from a lonely prophet seeking the truth of God into one willing to massacre those who refused to accept his rule? The same kind of brutality that Mohammed resorted to against the Jews he used against his pagan Arab enemies in quelling opposition. Mohammed was not unique in his religious vindictiveness, for Moses and Joshua had also massacred those who opposed them in their quest for power over conquered tribes and territories. It does raise questions about the sincerity and nobility of character of some of those who believe that they talk to God and seek to impose his law on humanity.

Mohammed later subdued other Jewish colonies, one by one, but he was not as harsh. They were allowed to keep their property, and he spared their lives on the condition that they pay taxes. This became a model for future armies of Islam, who after conquering native people often only exacted tribute from them.

Mohammed revised his attitude toward Judaic practices and beliefs still further. The fast of Ramadan, for example, replaced the feast of Yom Kippur, and most Jewish dietary practices were rejected. The Old Testament was much less important than it had been in the past, for this daring new religion now relied primarily on a single prophet, namely Mohammed. All opposition from the Jews in Arabia was now crushed, and they would no longer be an impediment to the expansion of Islam.

The next major effort of Mohammed was to subdue Mecca. The leaders of Mecca, including Abu Sufyan, apparently decided that it was in their own self-interest to reach an agreement with the Muslims. At one point they even permitted the pilgrims from Medina to enter Mecca to visit the Kaaba. They were impressed by their discipline and apparently believed that they could reach an accommodation. They reopened negotiations. Meanwhile, Mohammed was able to raise a large army of more than 10,000 men, including many new recruits, and he again marched south. The Meccans panicked. Abu Sufyan again met with Mohammed, and was converted to Islam. Mohammed presented his terms. The lives and properties of the inhabitants would be respected if they permitted the Muslim army to enter peacefully.

Thus, on January 11, 630, eight years after he had fled Mecca with his small band of Companions, Mohammed and his army reentered the city in victory. He immediately visited the Kaaba, touched the Black Stone, and cried out "Allah-o-Akbar!" He destroyed all of the pagan idols in the shrine, and urged the Quraysh to swear allegiance to him as the true prophet of God. In the succeeding two years before his death in 632, Mohammed engaged in battles with other tribes, particularly the Hawazin, and was able to unify the entire Arabian peninsula under his rule. His personality matured, and he was given to beneficence and kindliness. In the later surahs, he emphasized the virtues of obedience to God, honesty, and charity toward others.

After his death, Abu Bekr, an old friend and confederate, was named his successor and became the first caliph. Bekr died two years later and was succeeded by Omar, another early companion of the prophet. He was succeeded in turn by Othman in 644, and he by Ali, Mohammed's cousin, now married to Fatima, his daughter. Ali ruled until 661. All these rulers were either close relatives or friends of Mohammed. Much of the Hadith was derived from their memories. Whether their reminiscences of Mohammed were influenced by the exigencies of maintaining power is difficult to say, but it is not unlikely that this occurred. There were bitter disputes about succession within the Muslim world, which have not been resolved even today, yet the caliphs succeeded thirty years after the death of the prophet in extending Muslim control beyond the borders of Arabia. They conquered Syria, Palestine, and Egypt, which formerly had been under Christian hegemony, and Iran; and within eighty years of the death of Mohammed, the message of the prophet and the sword of Islam ruled all the way from Spain in the West to the Indies in the East. In successive centuries, four great Muslim empires flourished in turn—the Seljuk and the Ottoman, based in Istanbul, the Safavid in Iran, and the Moghuls in India—all in the name of Allah and his messenger.

Prophet of Allah

Let us return to analyze more carefully the personality of Mohammed. As we saw, he began his life as a poor orphan. Later wedded to Khadijah, he succeeded as a prosperous merchant. At the age of forty, he received his religious calling, probably having suffered from hallucinations and/or epileptic seizures. The latter part of his life was spent as the prophet of Islam but also as its military and political leader.

Shortly after the death of Khadijah, Mohammed, who was then fifty, was persuaded to remarry. He took two new wives: the first, Sauda, was a widow and one of his early converts, and the second, A'isha, the daughter of Abu Bekr, was only six or seven years of age (their marriage was not consummated for three years). Since polygamy was widely practiced among the Arabs, Mohammed was not unusual in this respect.

After eight years, Mohammed divorced Sauda, who was growing older, spent most of his time with A'isha, who had become his favorite. Sauda promised to yield one of her days (Mohammed spent a day with each of his wives in turn), if he would take her back. She wished to rise on Judgment Day as his wife. Mohammed assented, revealing a new surah (4:128), which allowed a husband and wife to seek mutual agreement:

> If a wife fears cruelty or desertion on her husband's part
> There is no blame in them
> If they manage an amicable
> Settlement between themselves;
> And such settlement is best.

How convenient it was for Mohammed to have God deliver a new commandment for mankind, just when Mohammed needed one. The revelations that flowed from the prophet now clearly had concrete, practical ends to fulfill, those that satisfied the private and political needs of the prophet. Still, his decision about Sauda was a charitable one.

In 628 A'isha accompanied Mohammed on an expedition, but was inadvertently left behind. A'isha managed to catch up with Mohammed's party with the aid of a young man who happened along. Suspicions of adultery grew, and Mohammed was urged to renounce his wife. The prophet was reluctant to do so, and he resolved the quandary by a new revelation. The circumstances under which such a revelation came to be pronounced is related in the Hadith, an account, no doubt, based largely on the testimony of A'isha. Apparently the scandal about A'isha's virtue dragged on for several weeks, and was intensified by Ali, a member of Mohammed's inner circle, who thought that Mohammed should punish his wife for adultery. Mohammed was apparently weighed down by sorrow because of the quarrel. According to tradition, he visited the chamber of A'isha, who was sitting with her parents. She was overcome with grief. Mohammed apparently told her that if she were guilty, she should repent. She burst into a flood of tears, declaring her innocence.

Then, as all were silent, the prophet appeared to fall into a trance. Those present covered him and put a pillow under his head. A'isha was confident that he would vindicate her. In a little while he recovered, sat up, wiping the sweat from his forehead, and told his wife to rejoice, for God had found her innocent. Mohammed then went out to the people and recited a new commandment concerning adultery.

The Koran provides that those who are guilty of fornication should each be flogged one hundred strokes. Mohammed adds the qualification, however, that one needs four witnesses to prove the charge of adultery, which makes it a very difficult charge to prove. The Koran then continues:

> And those who launch
> A charge against chaste women
> And produce not four witnesses
> To support their allegation
> Flog them eighty stripes
> And reject their evidence. (Surah 24:4)

Since there were not four witnesses at the time A'isha was suspected of having committed adultery, she was to be considered innocent.

Mohammed, like the Arabs and other peoples of his time, was an extreme sexist in his attitude toward women, an attitude that dominates Muslim customs and law today.

All told, Mohammed took ten wives and had at least two concubines. Some of these marriages had a political motive, but others no doubt satisfied his sexual proclivities. Tradition tells us that Mohammed liked three kinds of pleasures best: women, good food, and scents. His taste for women was well known, and indeed even scandalous. One story in particular illustrates this point, for it concerns the wife of Zayd, Mohammed's adopted son. Zayd's loyalty and devotion to Mohammed throughout his life was total. One day, Mohammed called at Zayd's home and happened to see his wife, Zaynah, who was scantily dressed. Mohammed left the home in confusion, having been overcome by her beauty. Upon learning of this encounter, Zayd immediately offered to divorce his wife and give her to the Prophet if that was what he wished, since apparently it was Mohammed who had earlier arranged Zayd's marriage to Zaynab. Mohammed turned him down, but Zayd divorced Zaynab anyway and Mohammed went on to marry her—but not until he secured permission from God in the form of a revelation (Surah 33:36). Many members of the community were displeased by the marriage. Since Zayd was Mohammed's adopted son, they considered the marriage to be incestuous. Mohammed was able to get his way and abort their criticisms by receiving a specific revelation from on high. In the Koran we read the following:

> Then when Zayd had dissolved (his marriage)
> With her, with the necessary (formality),
> We've joined her in marriage to you:
> In order that (in the future) there may be no difficulty
> To the believers (in the matter of marriage)
> With the wives of their adopted sons

When the latter have dissolved with the necessary (formality)
(Their marriage) with them. (Surah 36:37)

Mohammed concludes the revelation, "And God's command must be fulfilled." To make sure that everyone would understand his marriage to Zaynab had divine sanction, he issued the following revelation:

There can be no difficulty to the prophet in what
God has indicated to him as a duty
It was the practice (approval) of God
Amongst those of old that have passed away
And the command of God is a decree determined. (Surah 33:38)

That Zayd would willingly give his wife to Mohammed is itself a cause for considerable amazement, and that he would accept Mohammed's justification as divine borders on the incredible. This case illustrates the character of many of Mohammed's latter-day revelations. They were often contrived for his own gain and advantage. Interestingly, Mohammed seemed on occasion to be able to induce them at will, which suggests that even if he had genuine epilepticlike seizures, on some occasions the seizures may have been feigned. They were used to resolve his marital problems and amorous desires. That they were taken as eternal dictums in order to govern the affairs of all men and women under Islam is hard to believe. It reinforces the observation that once a forceful individual of determined character is able to convince his flock that his rule is divine, he can get away with almost anything, however transparent his motives may appear to history.

The revelations of Mohammed were also introduced to serve political ends. They were resorted to in order to pronounce general orders in military campaigns and to provide the regulations of a theocratic government concerning the treatment of allies and the disaffected, the formulation of treaties, and the acceptance of terms. They contained elements of a civil and criminal code. Legislation is likewise enunciated for the care of orphans, marriage, divorce, wills, usury, and other practical concerns. In this regard, the Koran is not unlike Leviticus and Deuteronomy, which Moses gave to the children of Israel to govern their behavior and to exalt himself as the prophet of the Lord.

The Koran evidently gave Mohammed a powerful weapon with which to intimidate his followers. Mohammed's word alone was final and constituted Islamic law. All disputes must be referred to him for resolution. Although he was not to be considered a deity, he nonetheless ruled like an absolute sovereign. When he appeared on the scene, everyone in the assembly rose and gave place to him and his entourage. Muslims were required to approach him reverentially, speak quietly in his presence, not to crowd around him. They were to keep their distance. They could not visit him uninvited. His wives were to be shielded from public gaze. As the Favorite of Heaven, he had special privileges not due others. Revelation, it turned out, was the instrument by which he first established and then enforced absolute obedience to his will. God alone had anointed Mohammed to be his messenger. All others must obey or else.

Thus from the first vague, indecisive visions of Mohammed, there developed a body of explicit prescriptions that became the method by which he could have his wishes and political and ethical notions fulfilled. His followers were never allowed to question Mohammed's word, for it was the word of God. Revelation in this sense expresses a form of paranoia. A disturbed individual acting out his fantasies, persuading others to accept his distorted images of the world as God's truth, now becomes an all-powerful being, himself ruling the lives of men and women. Mohammed's quest for power—submit or ye shall die—depended on the other side of the social transaction; those who believed that his message had a transcendental source abandoned all their claims to critical intelligence and agreed to submit to his pronouncements.

Obedience to God—the highest virtue

Are there any redeeming virtues to Islam and the ethical political system that Mohammed bequeathed? Surely an ethical-socio-political system that has endured for almost 1,400 years and still continues to exert a profound influence on the lives of hundreds of millions of Muslims must have some deep functions to perform. First, it established a degree of order and harmony in those areas where it predominated. It substituted for the haphazard custom, anarchy, and chaos that had existed before a system of law and regulation. Second, it provided an ethical code that has governed the lives of countless generations of human beings, given stability to social life, and provided parameters for guiding conduct. Third, Islam gave new existential meaning to men and women, who had lived in harsh and ignoble conditions. It helped to negate the pain and anxiety endured in life, soothed the dread of death, and it thus provided some solace for suffering souls who looked forward to Paradise as a relief from their worldly troubles. Fourth, Islam has inspired beautiful art, poetry, and philosophical and intellectual inquiry.

The Islamic religion represented in its time an advance over both Judaism and Christianity. If one accepts the unity of God, then Mohammed's form of monotheism, unlike Judaism (which is directed only toward the "chosen people"), is more universal in scope, for it is available to all peoples, irrespective of race, ethnicity, or nationality. It thus implies a form of equality: it applies to the rich and the poor, men and women, rulers and their subjects. Unlike Christianity, Islam did not develop a supernatural mythological tale about a risen deity who wrought miracles. Moreover, it has not developed a strong priestly class or ecclesiastical hierarchy, except recently in Iran. The religion has no complicated rites and rituals, nor is it mired in mystical abstractions, incomprehensible to the ordinary person. The Muslim must pray five times a day, by prostrating himself before God; he must fast during the daylight hours during the month of Ramadan; he must, if he can, make a pilgrimage in his lifetime to Mecca; and he must contribute to charity and be righteous. Although there are strict regulations against alcohol, gambling, usury, and adultery, Islam did not rail against the pleasures of the body or of sex, as did Christian ascetics and priests (down to our own time), but it tried to balance physical enjoyment with spiritual devotion. Paradise was not an austere place where one sang hymns throughout eternity (Hell could not have been worse), but a place where one could continue to enjoy the familiar earthly pleasures. This simplified religion had broad appeal, and enabled Islam to grow so rapidly that it supplanted Christianity in many areas of the world.

Unfortunately, Islam has serious negative features. It has enshrined a set of moral values of which not all are any longer relevant to the postmodern world. The primary failure of the system, for the humanist at least, is its first premise: that we ought to submit obediently to God. This is rooted in faith in the Koran and its message. but we have no reasonable evidence that God/Allah exists, that Mohammed received revelations from God, or that there is an afterlife. Why should we entrust our entire destiny to a false doctrine and submit to its authority?

What are we to say about an ethical system that applauds total psychological submission and prostration, and that has insufficient faith in the ability of human beings to use their powers to achieve a good life? Adoring God as the source of all wisdom and law demeans man. Elevating faith deprecates critical intelligence. Insisting that we are dependent undermines independence. Seeking to derive ethical obligations from the fatherhood of God impairs efforts to create autonomous ethical principles and values based upon reason and tested by their concrete consequences for human good. In enshrining what a self-proclaimed prophet said 1,400 years ago, Muslims limit the application of free intelligence allied with critical scientific inquiry to develop ethical values and to reform social institutions appropriate to the changing contexts of life.

The Muslim tradition attempts to regulate all aspects of life; and this leaves little room for independent freedom of choice or conscience. Thus it denegrates the ability of individual persons to discover their own resources of truth, beauty, and virtue, independent of the religious tradition, or to adopt alternative lifestyles. It restricts any moral freedoms not recognized by the Koran or Islamic law. Conforming to religious conventions blots out individuality.

In its most narrow application, Islam prevents women from sharing equally the opportunities for enjoying and realizing their full powers in the world. Sexist discrimination is codified by Islam, because the Prophet so decreed it. However, it is granted that Mohammed helped in his own day to emancipate women from some barbarous customs that had prevailed. For example, fathers who preferred male offspring would sometimes bury their female children alive, and Mohammed forbade this to continue. And although Mohammed practiced polygamy, women were given some rights, such as the right to own or inherit property, and men were enjoined not to mistreat their women. But still, most women lived shielded from the world, and were unable to participate fully in economic, social, and political life. True, Judaism and Christianity likewise expressed the sexist prejudices of the times in which they were being developed, but insofar as one submits to an orthodox reading of the Koran, it is difficult to achieve recognition of the rights of women, to grant them equality in dignity and value as persons. So deep is obedience to Islam that not only do men continue to follow these outmoded customs but in many Muslim countries women also willingly submit to the hegemony of their husbands and fathers—since God decreed it.

The right of free thought and freedom of conscience, as basic human rights, are insufficiently recognized by Islam. Granted that after the death of the Prophet, Islam developed a rich philosophical and intellectual civilization, in which learning and the arts flourished. Indeed, were it not for Islam, much of the great literature of the pagan world, interpreted and elaborated by Islamic scholars, would have been lost to Christian Europe. It was the rediscovery of these works in the monasteries of the twelfth and thirteenth centuries that contributed to the revival of learning and the eventual emergence of the Renaissance.

Since Islam means the comprehensive governance of the entire life of a society, which must be dedicated to God, there is little or no appreciation for the principle of the separation of church and state, which developed in the West and which permitted the secularization of culture and the emergence of religious liberty and the free mind. It is true that Islam did develop a measure of tolerance for the other religions in its midst, and so there is a kind of *modus vivandi* extended to Jews and Christians, who were usually permitted to follow their own religious practices. But this toleration did not extend to those who were born into the Muslim world. Heresy and the right of dissent are not recognized, and the greatest sin for a Muslim is apostasy or unbelief. This has led to a new spate of fanaticism among Shiite Muslims, in Iran and Lebanon.

Until the Islamic world undergoes its own Renaissance and Enlightenment, the respect for human freedom enjoyed elsewhere will have difficulty in developing. And this means an appreciation for religious freedom, including liberty of thought and conscience—for unbelievers as well as believers. Pehaps the secularization and democratization of values and the recognition of the right of dissent will emerge as Islam comes into contact with other areas of the modern world by means of trade and communication. Perhaps continued economic and technological developments will make clear the necessity for also developing scientific research, which can only flourish under conditions of free inquiry.

These important changes will be facilitated only when those within the Islamic world come to recognize the uncorroborated nature of the claim that Mohammed was the prophet of God and realize that it is not one's obligation to obey all of his alleged revelations.

X: Sundry prophets: Greater and lesser

Since the origins of the chief monotheistic religions are dimly clouded in a distant past, it is often difficult, if not impossible, to corroborate the claims made on behalf of their exalted personalities. I have noted nevertheless that there are similar psychological and sociological processes at work in the historic religions. Although Moses, Jesus, and Mohammed all proclaimed some kind of divine credentials, it is possible to consider alternative interpretations of their ministries. Perhaps they were disturbed individuals, half-believing that they had a divine mission yet cunningly using subterfuge and conjuring tricks to convince their credulous followers that they had had revelatory experiences and spoke in the name of God. Critics may say, of course, that my interpretive hypothesis has not been verified and that, although there may be some evidence to support it, it remains a speculative though suggestive hypothesis.

The question that should next be raised is: Do we have any recent or contemporaneous evidence of analogous processes, which would support my thesis? There are many cases of God intentional individuals who have made revelatory claims, but their influence has not survived their day and religions were not founded in their names. Either the absurdity of their claims were exposed or else they never enlisted the disciples who could effectively carry their message into the future. The historical record demonstrates that of the many competing prophets, most lost out in the great struggle for souls. Surprisingly, some have managed to succeed. Often their claims were found preposterous in their own day, yet they survived their critics and their detractors. We are close enough to the historical record of some of these newer religious sects to apply the skills of historical analysis and to test their claims of prophetic revelation.

North America, for a number of reasons, has spawned a wide variety of cults, many of them so bizarre that they flaunt all standards of logic and evidence, and yet they have persisted, and have even thrived. Perhaps it was because the United States was a frontier society to which countless immigrants came; they were rootless and on their own, seeking meaning and guidelines, and hence they fell easy prey to absolutist doctrines. At least four major "prophets" and the sects that developed around them in the nineteenth century come to mind: the most influential were Joseph

Smith and the Mormon Church, Ellen G. White and the Seventh-Day
Adventists, Charles Russell Tate and Jehovah's Witnesses, and Mary
Baker Eddy and Christian Science. All of these churches claimed to be
based on the Bible, yet they went far beyond it by proclaiming the latter-
day revelations of their founders.

It is puzzling for the skeptic to note that the intense religious revivals
on the frontier during the early nineteenth century followed swiftly after
the period of the American Revolution when the winds of rationalism,
deism, and anticlericalism were blowing strong. After the establishment of
the Republic a reaction set in. What does this say about human nature?
Are people so unsatisfied spiritually by skepticism and rationality that
they will embrace absolutistic doctrines that claim to answer the questions
they pose about the meaning of life?

Joseph Smith and the Book of Mormon

Joseph Smith, money-digger

I will devote the major part of this chapter to a case study of Joseph
Smith and the Book of Mormon. The similarities to the career of Mohammed and the growth of Islam are striking. The Mormon religion was
founded by Joseph Smith, a young man living in the Palmyra area, a
small community in upstate New York near Rochester. Born in 1805,
Smith was a charismatic and highly controversial personality. His most
important visions were revelations from Moroni, an angel of the Lord,
who appeared to him and told him where to uncover gold plates; from
these he "translated" the Book of Mormon, which Smith published in
1830. He continued to have revelations throughout his life, and he gathered a dedicated band of true believers around him, who grew in number.
Often compelled to move, he forged an army of apostles ever ready to
work mightily for his word, even to die for it. Smith was eventually shot
to death by an angry mob in Carthage, Illinois, in 1844, but not before he
had aroused the intense devotion or hatred of tens of thousands of people.
Interestingly, Mormonism did not perish with the murder of its prophet
and founder, but it has continued to grow, for a long period of time even
in the face of tremendous opposition and persecution. His beleaguered
band of followers trekked westward to Salt Lake City, Utah, which was
not yet a state, to build a new Zion in the wilderness. Today the Mormon
Church has an estimated six million adherents worldwide and is one of
the fastest growing religions, with missionaries fanning out all over the
globe recruiting new converts. Moreover, it has spawned about fifty dissi-

dent offshoots, the largest being the Reorganized Church of Jesus Christ of Latter-Day Saints (RLDS)

Mormonism is important for the thesis of this book because it provides us with an opportunity to empirically investigate the birth of a relatively new religion and study its growth in subsequent generations. The Mormon Church has been transformed from a radical cult persecuted by the establishment of its day into today's defender of a new orthodoxy. Dedicated Mormons are regarded as industrious and highly moral individuals. They do not drink liquor, coffee, or tea; they do not smoke. They are models of the traditional moral virtues and are highly patriotic. Mormonism is the most nativistic of American religions, for it grew out of and adapted to the frontier conditions, as it attracted new settlers who were arriving from Europe. Moreover, it found a place for Native Americans and the new continent in the divine scheme, things that had been left out of the Bible.

One may ask: Was the founder of Mormonism a divine prophet? Had he received a unique set of revelations from God or was he a clever imposter, paranoid personality, and lecherous rogue, as the critics of his day charged?

The life and works of Joseph Smith have been heavily researched and documented, both in his own day and today. There are the official church histories, including Smith's own account of his ministry.[1] But there are also critical biographies, based upon painstaking research, many of them by Mormon or ex-Mormon scholars. I have in mind especially the works of Fawn Brodie, Ernest Taves, Jerald and Sandra Tanner, and E. D. Howe. I shall draw upon their biographies heavily in the following discussion.[2] Scholars and historians who have published skeptical studies were excommunicated from the Mormon Church, including Fawn Brodie and the Tanners (at their request), and other forms of pressure are exerted on dissenters. Official church histories and translations have often been modified, interpolated, and even fabricated by church leaders in order to preserve the faith. This is somewhat analogous to what happened earlier in the historical development of Christian dogma.

Joseph Smith had fairly humble beginnings. His family changed resi-

---

1. See Joseph Smith, *The Book of Mormon, Doctrine and Covenant,* and *Pearl of Great Price;* Smith, *History of the Church of Jesus Christ of Latter-day Saints,* ed. B. H. Roberts, 7 vols. (Salt Lake City, 1932-51. See also James E. Talmage, *The Articles of Faith,* 16th ed. (Salt Lake City, 1930).

2. See Fawn Brodie, *No Man Knows My History: The Life of Joseph Smith,* 2nd ed. (New York: Knopf, 1971); Ernest Taves, *Trouble Enough: Joseph Smith and the Book of Mormon* (Buffalo: Prometheus Books, 1984); Jerald and Sandra Tanner, *The Changing World of Mormonism* (Chicago: Moody Press, 1980); E. D. Howe, *Mormonism Unveiled* (Painesville, 1834; reprinted by the Tanners' Utah Lighthouse Ministry, Salt Lake City).

dences several times in search of a better living. They moved from Vermont to western New York in 1816, and lived there during the building and opening of the Erie Canal. The atmosphere of the day was heavily charged with constant revivalist and spiritualist ferment. Unfettered religious interests bred bizarre cults.

Ann Lee (1736-1784), a mystic, had claimed that she was the reincarnated Christ. She preached celibacy and cleanliness and helped found Shaker communities in the American colonies. The Shakers had originated in mid-eighteenth century England among the Society of Friends. The group was known as "Shaking Quakers," because of the physical manifestations of spiritual fervor, especially during public worship. The Shakers advocated a community of shared possessions, separation from the world, sexual equality, and they engaged in singing, dancing, and marching during worship services. Ann Lee claimed to have received a revelation that proclaimed that the maternal element of Christ's spirit was resurrected in her. She emigrated to New York state in 1774. After her death many Shaker communities were established. The cult believed in the dual nature of deity; the male principle appeared in Jesus, the female in Mother Ann Lee. Since the group preached celibacy, it eventually disappeared.

Many other gurus appeared after the Revolution. Isaac Bullard, clad only in a bearskin girdle, led a band of "pilgrims" and championed free love and a primitive form of communism; he considered washing a sin and wore the same clothing for years. Jemima Wilkinson believed herself to be the Christ, said she was eternal, and governed her flock by means of revelations from heaven. Abel Sargent likewise received revelations and talked with angels. He toured New York state in 1812 with twelve women apostles, claiming that he could raise the dead and preaching that if one were pious one could exist without eating. There were faith healers, bible thumpers, and evangelists galore, as one awakening after another inundated the state. There was also a renewed interest in occult beliefs and practices.

Joseph Smith began his career as a "necromancer," a "glass looker," and "treasure hunter." He and his father attempted to make a living helping people uncover lost treasure for which they employed a form of dowsing. Joseph's father used a divining rod. Rumors spread great excitement about there being hidden Indian treasure, Spanish gold, and silver mines in the area. Joseph's primary method of treasure hunting was to use a "seer stone," which was something like crystal gazing. He maintained that he could see things by looking into the stone. Interestingly, Joseph Smith was indicted and tried in 1826, at the age of twenty-one, for being "a disorderly person and an imposter." He was convicted on the testimony offered, including his own admission that he practiced the magic arts, spells, and incantations. He was invariably unsuccessful in his diggings; when his clients thought that they had struck something, the treasure would move down still further. He was unsuccessful, that is, until he uncovered a new lodestone, the Book of Mormon.

Joseph Smith maintained that on September 21, 1823, when he was seventeen, an angel by the name of Moroni appeared to him and revealed the whereabouts of gold plates, which were conveniently buried on the Hill Cumorah, not far from his home. These plates were translated by him and became the Book of Mormon. Much later in life, in his auto-biographical writings, Smith maintained that he had had a still earlier revelation at the age of fourteen, this time right across the road from his house in a grove of trees, now called the "sacred grove." In this revelation, Smith said that two personages "whose brightness and glory defied all description" stood above him in the air. He asked which of the numerous contending sects he should join, and the reply was as follows:

> I was answered that I must join none of them . . . . that: "they draw near to me with their lips, but their hearts are far from me, they teach for doctrines the commandments of men, having a form of godliness, but they deny the power thereof."[3]

This revelation proved to be a basis for justifying the foundation of a new religion.

One may today visit Palmyra and Manchester, New York, to see the various sites of Joseph's revelations. The Mormon Church has purchased Joseph Smith's home, the land surrounding the Hill Cumorah (where a statue of an angel has been erected), the "sacred grove," and other land-marks, and has turned them into Mormon shrines. Thousands of Mormons make their pilgrimages to the area. Yet a critic is stunned by the presumptive belief that Joseph's reports of his revelations are veridical. We have no corroborative evidence that Joseph had any visions on the hill or in the grove; or if he did, that they were divinely sent. Like other reports of prophets and mystics, his accounts are purely subjective and without witnesses.

Western New York in the early nineteenth century was replete with Indian burial mounds, and the white inhabitants of the region speculated about their origin. Joseph Smith provided a truly fanciful solution. He maintained that the gold plates he dug up had been in the ground since C.E. 421. Described at a time when things Egyptian were in fashion

---

3. Joseph Smith, *History* 1:11-19.

throughout Europe and North America, these plates were engraved in "reformed Egyptian" characters. They revealed that the American Indians were descended from two remnants of the lost tribes of Israel (Ephraim and Manasseh) and from the tribe of Judah, who left the Middle East and set sail for the new world about 600 B.C.E. There were two main peoples, the Jaredites from the Tower of Babel and later the Nephites and Lamanites from Jerusalem. The Jaredites who came first spread out over the entire American continent and built great towns and cities. After many generations they destroyed each other because of their pride and sin. The Nephites and Lamanites, descendents of the prophet Lehi, fled Israel before the Babylonian invasion. They too spread over the face of the land. The Lamanites "dwindled in unbelief." Their skin became "dark and loathsome" and eventually they killed off the industrious and pious Nephites. In this account, Christ visited America in C.E. 34, and after his resurrection and ascension manifested himself to the Nephites, performing miracles. The Book of Mormon provided for the salvation of the Indians and a new revelation to supplement the Bible. Moroni, the last remaining Nephite and the son of Mormon, who had collected the stories of his people's prophets, sealed up the record and buried the plates. It was the same Moroni, then an angel, who visited Joseph in 1823 and subsequent years. It was not until 1827, however, that Joseph was finally allowed to receive the plates.

The question that has been raised is whether Smith's account is true. Where did Joseph Smith get the Book of Mormon? His story appears to be a far-fetched tale, for it has not been confirmed by archeological or anthropological investigations. There is no independent, verifiable evidence for the existence of the tribes mentioned or their Middle East origin.

In Joseph Smith's day it was commonly believed that the Native American Indian tribes were related to the lost tribes of Israel. Many critics of Joseph Smith have pointed out that a book with a similar thesis had been published in 1825 by Ethan Smith in Poultney, Vermont. It was entitled *A View of the Hebrews,* and likewise argued for the Hebraic origin of the Indians. Commentators have shown many textual analogies between the Book of Mormon and *A View of the Hebrews.* Most likely, Smith was familiar with the commonly held nineteenth-century beliefs referred to in the book.

Those who are devotees of the Church of Jesus Christ of Latter-day Saints (the Mormon Church) nevertheless maintain that, like the Bible, the Book of Mormon is the word of God and that it was given directly to Joseph Smith to be translated from the ancient gold plates. Mormon apostle Orson Pratt wrote in 1851:

The Book of Mormon claims to be a divinely inspired record written by a succession of prophets who inhabited ancient America . . .

This book must be true or false. If true it is one of the most important messages ever sent from God to man . . . . If false, it is one of the most cunning, wicked, bold, deep-laid impositions ever palmed upon the world. . . .

The nature of the message in the *Book of Mormon* is such, that if true, no one can possibly be saved and reject it.[4]

Although the Mormon Church accepts the Book of Mormon in toto as the revealed word of God, it is difficult for noncommitted historical researchers to do likewise today. There are many contradictory and unsubstantiated claims surrounding the origin of this work. According to Joseph's account, he first had a visitation from Moroni on September 21, 1823. The angel appeared to him after he had gone to bed:

A personage appeared at my bedside, standing in the air. . . . He had on a loose robe of most exquisite whiteness. . . . I was afraid; but the fear soon left me. He called me by name, and said unto me that he was a messenger sent from the presence of God to me, and that his name was Moroni; that God had a work for me to do; and that my name should be had for good and evil among all nations, kindreds, and tongues. . . .

He said that there was a book deposited, written upon gold plates, giving an account of the former inhabitants of this continent, and the sources from whence they sprang. He also said that the fullness of the everlasting Gospel was contained in it as delivered by the Savior to the ancient inhabitants; also that there were two stones in silver bows—and these stones, fastened to a breastplate, constituted what is called the Urim and Thummim. . . . The possession and use of these stones were what constituted "Seers" in ancient or former times; and. . . . God had prepared them for the purpose of translating the book.[5]

According to Smith he was warned not to show the plates and the Urim and Thummim to anyone under pain of death. (This is reminiscent of Moses' warning to the children of Israel never to ascend Mount Sinai, the residence of Jehovah, or they would be struck dead.) Moroni returned twice that night. The next day, Smith went to work in the fields, but since he looked pale, his father told him to return home. On the way he climbed a fence and fell unconscious. Did he have an epileptic seizure like Paul or Mohammed may have had? Moroni again appeared and told him to tell his father, Joseph Smith, Sr., what had occurred. His father told Joseph that he must do what the angel had requested. The Smith parents were

---

4. Quoted in George D. Smith, "Joseph Smith and the Book of Mormon," *Free Inquiry* 4, no. 1 (Winter 1983-84); from *Orson Pratt's Works* (Liverpool, 1851), p. 1.

5. Smith, *History* 1:28-50; quoted in Taves, *Trouble Enough*, p. 27.

both believers in supernatural events and even witchcraft, and his father had many paranormal visions and dreams; and so both adults were no doubt receptive to the tales related by their son. Joseph said that he visited the Hill Cumorah each year for four years until he was finally informed by the angel that he could remove the tablets.

According to Ernest Taves in his study of Joseph Smith, other accounts of these early events were available at that time. Joseph's father told a nearby neighbor, Parley Chase, that on Joseph's first visit to the plates, the angel Moroni was not present but that "a toad-like being that transmogrified into a man" was. Parley Chase said that "they scarcely ever told two stories alike."[6] Another critic maintained that the story of the gold scriptures was begun as a "speculation," i.e., for profit. And still another said that they intended to use the profits gained from the sale of the book to continue their money-digging business. Another reported that when Joseph returned to the plates a year later with his wife, a "host of devils" screeched and screamed at him.

After an early follower, Philastus Hurlburt, defected, he gathered a great number of defamatory testimonials about Smith from his neighbors, friends, and ex-Mormons. This material was published in a book in 1834, edited by E. D. Howe and entitled *Mormonism Unveiled.* The book contains a statement by Peter Ingersoll, who maintained that in 1827 he was a close confidant and friend. Joseph allegedly told Ingersoll, in a humorous vein, the following story. His family was eating dinner and asked what he had in the frock. "So I very gravely told them it was the Golden Bible."[7] Much to his surprise the family believed him. Anyone who looked at it, he warned, would be struck dead. "Now," Joseph informed Ingersoll, "I have got the damned fools fixed, and will carrry out the fun."[8] Ingersoll continued his account:

> Notwithstanding, he told me he had no such book, and believed there never was such book, yet, he told me that he actually went to Willard Chase, to get him to make a chest, in which he might deposit his golden Bible. But, as Chase would not do it, he made a box himself of clapboards, and put it into a pillow case, and allowed people only to lift it, and feel of it through the case.[9]

This story is corroborated by Willard Chase. The Mormons reply that out of envy and hostility all kinds of stories were made up about Smith, i.e., people were persecuting him.

6. Taves, *Trouble Enough,* p. 28.
7. Ibid., p. 31.
8. Ibid.
9. Ibid.

How do we know that Joseph Smith received a visitation from Moroni? We only have his word for it. Was he self-deceived or did he make up the entire story? And once it was out, was he so inspired by the gullible reaction of those who believed him that he continued the deception? Fawn Brodie speculates that Joseph Smith began as a "bucolic scryer" using the rude techniques of magic common to his area and time, and only later as he attracted a following, did he develop his skills as a preacher-prophet.

From what we can learn of Joseph Smith, he was an affable man, striking in appearance, charismatic to those he attracted, and also a man of fertile imagination. A resident of Palmyra at that time, Daniel Hendrix, describes him as follows:

> He was a good talker and would have made a fine stump speaker if he had the training. He was known among the young men I associated with as a romancer of the first water. I never knew so ignorant a man as Joe was to have such a fertile imagination. He never could tell a common occurrence in his daily life without embellishing the story with his imagination; yet I remember that he was grieved one day when old Parson Reed told Joe that he was going to hell for his lying habits.[10]

As we saw, Smith had been tried and convicted for his sting operation. Fortunately, we have a court record of the trial, March 20, 1826. Several witnesses reported on how he performed. Joseph would look into a hat in which he had a seer stone, claiming that he could find lost treasure. Two witnesses, Arad Stowell and A. McMaster, said that they went to see the prisoner (Joseph Smith) to be convinced of his skill, but that they "came away disgusted, finding the deception so palpable."[11] According to them, Smith pretended that he could discern objects at a distance by holding a white stone to the sun or a candle, but he declared that at that time looking into his hat hurt his eyes. In this regard Smith is not unlike present-day fortune tellers or clairvoyants, who claim they can see things from afar and that they possess paranormal powers.

We also have the testimony of Joseph Smith's father-in-law, Isaac Hale, of Harmony, Pennsylvania, about Smith's career. Joseph had sought the hand of Emma Hale. Peter Ingersoll went with Smith to help him move Emma's household furniture from Harmony to New York state. According to his account, Emma's father reproached Smith for stealing his daughter. "You spend your time digging for money—pretend to see a stone, and thus try to deceive people."[12] Ingersoll reports that

10. Ibid., p. 16. Hendrix's view of Smith is stated in a letter he wrote to the *St. Louis Globe-Democrat,* Feb. 2, 1897.
11. Brodie, *No Man Knows My History,* p. 428. This record of Smith's trial was first unearthed by Daniel S. Tuttle, Episcopal bishop of Salt Lake City.
12. Ibid., p. 433.

Joseph wept, admitted that he could not see in a stone, and that his former pretensions were all false. And he promised to give up his habits of looking into stones and digging for money.

The question that can be raised about a paranormal, occult, or revelatory story is: Who are we to believe? Are we to believe the sworn statements from almost a hundred neighbors and former friends in and around Palmyra and Harmony, Pennsylvania, that Hurlburt collected? No doubt many were biased against Joseph Smith because of his trial and conviction and the fact that his religious views were unorthodox. We must recognize that a prophet is often unappreciated in his own town. Yet Howe published several statements by people who went to the trouble to attest to the widely held view in the area that the Smiths were unsavory. Their statements all point to the view that Joseph Smith and his family were considered to be of "questionable character," engaging in "dubious" money-making schemes.

All of the evidence seems to support the view that Smith drew deeply on folk magic and that he skillfully merged the occult arts with biblical interpretation. He used divining rods, seer stones, and ritual magic early in his career, but the aura of magic persisted throughout his colorful life. The striking similarity of Joseph Smith with latter-day mediums and psychics is apparent. He claimed not only to have special revelations from on high, but to possess unique paranormal powers and gifts. He so aroused the transcendental temptation in those about him that they were inspired to follow him and, indeed, to build a new religion. The close affinity between magic and religion, the magician and the prophet, is especially evident in the first stirrings and early development of Mormonism. For Smith appears to be using the arts of the magician and the conjurer and to draw upon his elementary knowledge of the methods of deception to enhance his image. His ontological world-view is surely akin to the magical world-view of the wizard and the seer. His earlier efforts at treasure dowsing proved to be abortive and did not earn him much money. A profound shift in his life occurred when he transformed his alleged paranormal talents into a new religion, and this had a far-reaching effect not only on his own life but on those who followed him in this new venture.

An uncharitable writer would conclude that the most profitable phase of Smith's life was his fabrication of the Book of Mormon; for this proved to be a tremendous boon to his career, launching him on the road to prophet-preacher fame and eventually martyrdom.

Was there *any* corroborative evidence to substantiate the divine origin of the Book of Mormon? Following Joseph Smith's personal account of his revelation, we have the testimony of a total of eleven witnesses. These are reported in the Book of Mormon.

> Be It Known unto all nations, kindreds, tongues, and people, unto whom this work shall come: That we, through the grace of God the Father, and our Lord Jesus Christ, have seen the plates which contain this record, which is a record of the people of Nephi, and also of the Lamanites, their brethren, and also of the people of Jared, who came from the tower of which hath been spoken. And we also know that they have been translated by the gift and power of God, for his voice hath declared it unto us; wherefore we know of a surety that the work is true. And we also testify that we have seen the engraving which are upon the plates; and they have been shown unto us by the power of God, and not of man. And we declare with words of soberness, that an angel of God came down from heaven, and he brought and laid before our eyes, that we beheld and saw the plates, and the engravings thereon; and we know that it is by the grace of God the Father, and our Lord Jesus Christ, that we beheld and bear record that these things are true. And it is marvelous in our eyes. Nevertheless, the voice of the Lord commanded us that we should bear record of it; wherefore, to be obedient unto the commandments of God, we bear testimony of these things. And we know that if we are faithful in Christ, we shall rid our garments of the blood of all men, and be found spotless before the judgment-seat of Christ, and shall dwell with him eternally in the heavens. And the honor be to the Father, and to the Son, and to the Holy Ghost, which is one God. Amen.
>
> Oliver Cowdery
David Whitmer
Martin Harris

> Be It Known unto all nations, kindreds, tongues, and people, unto whom this work shall come: That Joseph Smith, Jun., the translator of this work, has shown unto us the plates of which hath been spoken, which have the appearance of gold; and as many of the leaves as the said Smith has translated we did handle with our hands; and we also saw the engravings thereon, all of which has the appearance of ancient work, and of curious workmanship. And this we bear record with words of soberness, that the said Smith has shown unto us, for we have seen and hefted, and know of a surety that the said Smith has got the plates of which we have spoken. And we give our names unto the world, to witness unto the world that which we have seen. And we lie not, God bearing witness of it.

Christian Whitmer	Hiram Page
Jacob Whitmer	Joseph Smith, Sen.
Peter Whitmer, Jun.	Hyrum Smith
John Whitmer	Samuel H. Smith

Do the preceding statements satisfy the need for corroborative testimony? Three men saw the plates in a vision and eight family members supposedly confirmed their existence. Isaac Hale, Smith's father-in-law, maintained "that the whole 'Book of Mormon' (so-called) is a silly fabrication of falsehood and wickedness, got up for speculation, and with a design to dupe the credulous and unwary—and in order that its fabricators may live upon the spoils of those who swallow the deception. . ."[13] Lucy Harris, the wife of Martin Harris, later maintained that Harris first entered the Golden Bible business to make money, though "he believed that Joseph could see in his stone anything he wished."[14]

The question to be raised is what it is in the psychology of believers that enables them to accept as genuine what seems to be a patent fabrication and to abandon all standards of critical intelligence. Was Smith, for example, such a compelling personality that his followers were unable to detect that he was a flim-flam man and a consummate actor? Is there something in the psychology of true believers that tends to break down their defenses and cause them to swallow whatever is fed them? Is there a need to do so? Is there something in the cultural milieu that makes them receptive to mythological doctrines? Or is psychological craving so intense that it blocks any doubts that may arise about incredible tales?

Joseph Smith, the prophet

The next stage of the Joseph Smith story is fascinating. Here we have an evidently known "impostor," with an apparent "personality disorder," attempting to convince a close circle of family (were any of them in on his act?) and friends. It is no doubt difficult to make a clinical diagnosis of an individual who died more than a century ago. Fawn Brodie suggests that Smith may have suffered from "pseudologia fantastica."[15] She quotes the psychiatrist Dr. Phyllis Greenacre to show that celebrated imposters are not mere ordinary liars but people of intense conflicts. Quoting the psychiatrist further, she sees in such people a struggle between two strong tendencies: the strongly assertive and temporarily focused imposturous one and the often crude and poorly-knit one from which the impostor developed. There seems to be an extraordinary and continued pressure for the imposter to live out his fantasy, and this may assume the form of a delusion—though there may be some formal awareness that the claims are untrue. The imposter has some sense of reality, especially a keen sense of guarding himself against detection. His overall sense of reality is nonetheless impaired.

13. Brodie, *No Man Knows My History*, p. 440.
14. Howe, *Mormonism Unveiled*, p. 255.
15. Brodie, *No Man Knows My History*, p. 418 ff.

Great imposters rely on "omnipotent fantasies." They are invariably good showmen, absolutely dependent on an audience. "The impostor," writes Dr. Greenacre, "cannot be sustained unless there is emotional support from someone who especially believes in and nourishes it." There seems to be a transactional effect. This characterization would seem to apply also to some kinds of political leaders, to economic con men, gurus, and various kinds of imposters from many walks of life. The imposter feeds his flock's fantasies by promises; and he defends himself adroitly against detection and misadventure. They in turn grant him money, power, adulation, and even extraordinary reverence, as in the case of a religious prophet. This only feeds the imposter's ego further. There is no going back. Once a ruse is perpetrated, its success may be such that he cannot admit his earlier deception. He may even half-believe or fully believe that he possesses special powers or that he has been appointed to carry out a unique mission. The imposture, if successful, then feeds its further development, having a life of its own: this is a network of deceit and subterfuge. At some point the deception may lead to a distorted picture of oneself, megalomania, and corruption. Imagine a poor un-educated man, secretly harboring doubts about his own ability, discover-ing that if he plays the role of a prophet, many people will be duped and will attribute superhuman powers to him. His ego-involvement may be-come so great that it needs constant reinforcement and reassurance. And so he continues in his fantastic illusion. If we examine Joseph Smith's letters and personal journals, we realize that Smith apparently came to believe in Jesus' second coming, viewing ordinary events in his own life in apocalyptic terms.[16]

But what of those in the immediate circle who accept the illusion, knowing the criticisms of the man and his idiosyncratic imperfections and faults and yet managing to overlook them? Are they so blinded by his magnetism, charisma, and his heady rise to prominence that they manage to suppress their doubts? Are their lives so tied up with his that they too cannot turn back? And what of the next circle of admirers he attracts? How much of the deception are they aware of? How much do they really believe? Can the imposter be on guard all the time, or are there cracks in his armor? Is it because his followers are given power and influence that they are sucked into the masquerade, overlooking its comic features? In time, the message of the imposter may be carried far and wide, beyond his immediate circle of contacts, cronies, and original disciples. Standing at a distance, these new converts may find it difficult to evaluate whether or not his claims are true. There may soon develop around such a

16. Dean C. Jessee, *The Personal Writings of Joseph Smith* (Salt Lake City: Deseret, 1984).

personality a mystique, much of it exaggerated, and yet it compels men
and women to follow any sort of commandment. This phenomenon is
true not simply about religious leaders, but of great generals, such as
Alexander the Great and Napoleon, who led men to their death in battle,
or political figures, such as Hitler and Stalin. Demonic figures though
they be, while the world hated them, they were adored by their legions.

We have quoted eleven early witnesses, who each affirmed the truth
of the revelation of Joseph Smith. How reliable was their testimony? In
no case do they say that they had themselves actually read the gold plates,
which were engraved in a "reformed Egyptian script." (It was about this
time that Champollion, the great French linguist was hard at work attempting to unlock the key to the Rosetta Stone and thus translate the
ancient Egyptian demotic and hieroglyphic scripts.) Nor was any effort
made to determine whether the plates were accurately rendered into
English by Joseph Smith—no doubt a difficult task, since Egyptian script
was still indecipherable. All that the witnesses testified to was that they
saw gold plates. Does this authenticate the Book of Mormon or its
meaning? It would have been easy for Smith to fabricate the alleged
plates. If he kept them hidden and allowed them only to be peeked at,
they may never have been examined carefully and critically.

After Joseph announced his "find," in 1827, considerable interest in it
developed in the Palmyra area. He then began the task of translating the
plates. Like Mohammed, Joseph dictated his major revelations; his wife,
Emma, was his first scribe. Joseph placed his magic stone over his eyes,
covered his head with a hat, and dictated. He also used a breast plate,
which he called the "Urim and Thummim" (mentioned in the Bible) and
allegedly had found with the plates, for the first 116 pages of translation.
Joseph never let Emma look at the plates. She said that he did not
consult the plates as he dictated and that she came to believe they were
authentic. (Emma was a tragic figure, for she stayed by his side until his
death, even after he had taken several dozen other wives. After Joseph's
death, she refused to follow Brigham Young and held fast until her son
could take over for the Reorganized Church.) Even Joseph's father apparently believed that God was speaking through his son.

Smith's second scribe was Martin Harris, a successful farmer who
lived down the road from Joseph. Harris became a convinced devotee. In
the translation process, Smith invariably sat behind a makeshift blanket
curtain, the box containing the plates beside him or wrapped in linen on a
table. He read out the alleged contents of the plates, his scribes writing
down what he dictated. Joseph was uneducated in the art of writing,
though he had a highly developed imagination and knew the King James
translation of the Bible. As in the case of Mohammed, his friends and

relatives asked how one so unlearned could compose such a book, if it were not divinely inspired.

Martin Harris was intrigued by the discovery of the plates and later provided the funds to get the manuscript published. In a recently discovered letter written by Harris to William W. Philips on October 23, 1830, Harris describes his relationship to Smith and the translation process. He said that Joseph "found some giant silver spectacles with the plates, he puts them in an old hat and in darkness reads the plates." Did he ever have any doubts that the Book was authentic? His wife, Lucy Harris, bitterly protested that Harris was being taken in, that he gave Joseph money, and neglected his otherwise prosperous farm. Harris asked Joseph to let him view the plates, but Joseph adamantly refused to do so. Harris was permitted to lift them several times in their clapboard box; he calculated that their weight was about forty to fifty pounds. He was never allowed to open the chest or examine them directly, although he signed the statement attesting that he had seen the plates.

Harris then asked Joseph for a transcript of some of the inscriptions, which Joseph finally consented to copy from the plates and give to him. Harris took this to Professor Charles Anthon, a classicist at Columbia University in New York City. Official church history reports that Anthon attested that the characters on this sheet were "true Egyptian, Chaldaic, Assyriac and Arabic," and that the translations were "correct." This is odd since it is highly unlikely that Anthon could have read Egyptian, Chaldaic, or Assyriac script, much of which had not been translated at that time. Harris said that during his visit to Anthon, the professor first wrote a certificate to testify to the accuracy of the translation. As he was about to depart, however, Anthon inquired about the history of the symbols. When Harris reported that they were revealed by an angel of the Lord, Anthon retrieved his certificate and tore it up. Anthon suggested that the gold plates be brought to him for translation. Harris told him that this was strictly forbidden. Anthon replied, "I cannot read a sealed book." When Harris related the details of his interview with Anthon, Joseph leafed through his Bible. He came to Isaiah 29:11-12:

> And the vision of all is become unto you as the words of a book that is sealed, which men deliver to one that is learned, saying, Read this, I pray thee: and he saith, I cannot; for it is sealed: And the book is delivered to him that is not learned.[17]

Harris was astounded, for he thought that Joseph had truly fulfilled a

17. Brodie, *No Man Knows My History,* p. 52.

prophecy. What convoluted logic! Harris was now willing to risk his farm to finance the Golden Bible.

Upon learning that his name was being used to authenticate the Book of Mormon, Professor Anthon denied it and stated in a letter dated February 17, 1834, that "the whole story about my having pronounced the Mormonite inscription to be 'reformed Egyptian hieroglyphics' is perfectly false." He went on to state that he came to the conclusion that someone was attempting to perpetuate "a hoax upon the learned" or a "scheme to cheat the farmer of his money."[18] Anthon's statements nevertheless had no effect in dissuading Harris from his folly. Later, after the Book of Mormon had been published, Anthon reported that Harris visited him again, bringing a copy which he tried to sell him. When Anthon declined to buy it, Harris asked permission to leave the book anyway. Anthon again told Harris that a "roguery" had been practiced upon him, and he advised him to get a magistrate and have the trunk containing the gold plates examined. Harris responded that the "curse of God" would come upon him if he were to do this, but finally said that he would open the trunk if Anthon would take the curse of God himself. Anthon said that he would do so. Harris then left, but Anthon never saw the plates.[19]

A copy of the sheet given to Anthon on the first visit was discovered in an old family Bible. It is clear that these characters do not resemble Egyptian, Chaldaic, Assyriac, and Arabic script and that they were mere gibberish. Martin Harris was willing to grasp at any possible shred of evidence that the translation by Joseph Smith was authentic, and so he dismissed Anthon's later denials that it was accurate.

Harris asked to take the first part of the translated manuscript (116 pages) home with him, obviously to placate his skeptical wife. The manuscript disappeared and could not be located. Had Mrs. Harris hidden or destroyed it out of rage? Smith was upset by this turn of events and had to start over. If the old translation were found and his new translation was not the same, someone, in comparing the two versions, might accuse him of a fabrication. Smith devised a pretext. He told Harris that he would begin an entirely new translation. The earlier translation, he informed him, was only an abridgment and he was providing a fuller account. Harris accepted this explanation. Now there was no way that anyone could claim that the two versions did not match, should the first be discovered. How naive must Harris have been! He was now firmly convinced that the Book was of divine origin. Joseph Smith would never again let it out of his possession until a second copy could be made.

A third transcriber, Oliver Cowdery, who had heard of the discovery

18. Ibid.
19. Howe, *Mormonism Unveiled,* pp. 270-72.

of the plates, showed up. Cowdery, a third cousin to Smith, was also born in Vermont. He befriended Joseph and volunteered to help with the work. Cowdery was a twenty-two-year-old blacksmith and school teacher. He apparently was impressed by Joseph's demeanor and had confidence in his powers, though later in life, Cowdery admitted that he had had misgivings about how Joseph could translate by means of the seer stone, without the plates in sight. Sometimes Joseph would not even pretend to use the stone but simply close his eyes and dictate. Joseph Smith would render the first version, which would be written down by Cowdery, who then read it back to see if there were any corrections. They worked at a fast pace. Cowdery did not put in periods or commas, but wrote down verbatim what Joseph dictated. It was the typesetter in Palmyra who set the book in proper grammatical form. A friend of Cowdery, David Whitmer, was also brought in and watched the translation process. Like Cowdery, he later claimed that "the Book of Mormon was translated by the gift and power of God, and not by any power of man."[20] Note that Cowdery and Whitmer also claimed to have seen the plates.

Sometime during 1828-1829, the career of Joseph Smith underwent a radical transformation. No longer would he be interested simply in transcribing the golden plates; nor was he intent on publishing them in book form for a profit (the first edition had listed Joseph Smith as "owner and proprietor"), as was his original motive. He now began the process of building a new church, for apparently he had so duped Harris, Cowdery, Whitmer, and others into believing that a supernatural dispensation had been conferred upon him that he took the giant step of founding a new Aaronic priesthood. He must have been amazed at the credulity of his original disciples and the fact that they actually believed his words had a divine sanction. Martin Harris trusted Joseph so completely that he was willing to provide him with all the money needed to finance his efforts. During this period Smith and his family were in financial straits, and this newly found friend provided them with support. Harris agreed to mortgage his farm for $3,000 in order to print 5,000 copies of the Book of Mormon; later he had to sell some of his land to pay for the note that came due. The enthusiasm and adulation of Cowdery and Whitmer were also apparent.

Is there an important lesson to be learned about the founding of a new religious faith? Some poor souls are so easily duped that they willingly give everything to one who claims to be a divine prophet and acts out the role with masterful deceit. Perhaps something latent in their personalities was watered by Joseph, and they were encouraged in their belief and

20. Taves, *Trouble Enough,* p. 43. David Whitmer, *Address to All Believers in Christ* (1887), p. 12.

hope that they were part of his inner circle and would be given a special place in the divine scheme. He baptized and ordained Oliver Cowdery into the priesthood, who in turn did the same for Joseph. Shortly thereafter others were added to the inner circle of priests: David and Peter Whitmer, Hiram and Samuel Smith, and later Martin Harris and others. After Whitmer broke from the church, he reported that Joseph had early expressed a strong desire to be the leader of a new church. He abandoned the peep stone and thenceforth uttered revelations as if they came directly from God. His new vocation was now solidly launched. And with his new role sanctified, he began to attract numerous followers.

What can we make of their claim quoted earlier from Harris, Cowdery, and Whitmer that they saw the golden plates? They were repeatedly denied the opportunity by Joseph. Eventually Joseph realized that he would have a credibility gap, unless he could show others his golden plates. Finally, he said, that he would, if they demonstrated sufficient faith. At last, he led the three men into the woods to pray.

Joseph Smith apparently had a magnetic influence over his associates; Cowdery and Harris were caught in his spell. Later, Cowdery described Joseph as having a "mysterious power," which he could not fathom. He seemed to have an intuitive understanding of those about him, which led them to think that he had psychic powers and could read their minds. Was his personality so strong that he became at times hypnotic, causing his followers to see visions? Both Whitmer and Cowdery reported that as they were engaged in prayer in the woods, they beheld a bright light in the air and that an angel stood before them. In his hands the angel held the plates, turning the leaves one by one, so that they could see the engravings. The voice informed them that the plates were revealed and translated by the power of God. Meanwhile, Harris had left the others to pray alone. Joseph went after him. The same vision, we are told, was beheld by Harris who cried out, " 'Tis enough; 'tis enough; mine eyes have beheld." Interestingly, according to the local newspaper, the three witnesses subsequently related somewhat different versions of their visions. Was it the power of faith or simply imagination at work? Were they hallucinating? Did Joseph somehow stimulate their visions? Why had he not shown the plates to anyone during or after the process of translation? Since the plates are so crucial and so holy, why were they not preserved? Why have they disappeared? We have seen in an earlier chapter the unreliability of eyewitness testimony. Where there is a strong predisposition to believe, people will imagine that they have seen almost anything.

The eight additional witnesses who attested to the existence of the plates were friends of Joseph or members of his immediate family, including his brothers and father. According to later accounts, the eight were set to continual prayer and spiritual exercises. Joseph finally produced a box, which he insisted contained the plates. The lid was opened and they looked within but could see nothing. After two more hours of intensive prayer on their knees, they finally claimed to have seen the plates. Was there a kind of contagious mass hysteria? The eyewitness testimony of witnesses can become clouded if the belief-state is allowed to interfere with the powers of observation. Did the great anticipation and desire to see these venerated objects or hallucinations brought on by exhaustion finally lead them to believe that they had seen the golden plates?

Did Joseph Smith construct some counterfeit plates—easily done—and store them in a box, allowing his followers at some point to peep in? Some of the witnesses talk about the weight, size, and metallic texture of the plates. In any case, whether or not the plates ever existed, they eventually disappeared; Joseph informed everyone that the angel who had revealed them to him carried them back to heaven!

As best we can reconstruct, we have the following scenario: (1) Joseph repeatedly refused direct access to the plates to anyone during the translation process or afterwards, on pain of death, thus instilling fear. (2) He claimed that he eventually allowed eleven witnesses to see the plates or have a vision of them, but at no time did anyone pore over them leaf by leaf or check the plates as they were being transcribed. (3) Joseph did not directly read the plates as they were being transcribed. (4) When their usefulness ended, they conveniently disappeared. Was this a hoax?

Two of the original witnesses, Cowdery and Whitmer, were later excommunicated from the church by Joseph Smith. Martin Harris, the third witness, later left the church in a dispute with Smith, whom he accused of "lying and licentiousness." The Mormon leaders in a journal edited by Smith, in turn charged that Harris and others were guilty of "swearing, cheating, swindling, drinking, with every species of debauchery."[21] Cowdery in the same year accused Joseph of "adultery, lying, and teaching false doctrines." Whitmer maintained that "all of the eight witnesses who were then living (except the three Smiths) came out" of the church. Cowdery and Whitmer were attacked by Smith and other Mormon leaders and accused of having "united with a gang of counterfeiters, thieves, liars, and blacklegs of the deepest dye, to deceive, cheat, and defraud the saints out of their property."[22]

Harris was especially unstable. He was said to have been converted to various sects (Quakers, Restorationists, Baptists, Universalists, Presbyterians, Mormons, and Shakers), changing his religious position frequently. Later in life, he accepted the Shaker holy book (*Sacred Roll and Book,*

21. Quoted in Tanner, *Changing World of Mormonism,* p. 96, from the *Elders Journal,* August 1838, p. 59.
22. Tanner, p. 97, from David Whitmer, *Address to All Believers in Christ,* pp. 27-28.

published in 1843) as divinely inspired. The Shakers believed that Christ had made his second appearance on earth in the form of Ann Lee. Interestingly, like the Book of Mormon, the Shaker holy book has a section in which eight witnesses maintained that they saw a holy angel standing upon a house top holding the *Sacred Roll and Book*. Martin Harris became such a firm believer in Shakerism that he said that his testimony in its favor was greater than it was for the Book of Mormon.[23] Harris' first wife had divorced him a few years after he had mortgaged his farm. His second wife moved out to Salt Lake City, as did Harris at the end of his life. He never took back his earlier testimony about the divine origin of the Book of Mormon however, nor did Cowdery or Whitmer; indeed, Harris and Cowdery were later rebaptized into the church.

The Book of Mormon was most likely the product of a number of sources that influenced the fertile imagination of Joseph Smith. It contains many contradictions and inaccuracies. George D. Smith points out that it incorporates many passages taken directly from the King James translation of the Bible, though they are often used in the wrong historical context. Several Old Testament prophets (for example, Malachi and Isaiah) appeared earlier than they lived. Nephi is supposed to have left Jerusalem about 600 B.C.E. for America, but Old Testament passages of a later era are placed in the wrong time-frame. Moreover, New Testament sources appear in an Old Testament context. Smith was no stickler for details, and his imagination embellished the historical record.[24]

The main thesis of the Book of Mormon—that America was settled by the Hebraic peoples of the Middle East and that Jesus visited the American continent—has not been proved by any scientific evidence, though the Mormons have gone through great efforts to try to do so. There has been inadequate archaeological evidence pointing to the existence of the Nephi, Jared, or Lamanite peoples. If there had been such vast civilizations, as Joseph Smith maintained, surely some of their artifacts would have been discovered. On the other hand archaeologists and anthropologists believe that the most likely explanation of the American Indians is that they came from Asia over the Bering Straits ice bridge some 30,000 years ago and settled in North and South America. Smith has many other errors. He attributed horses and the use of steel to the Indians, long before they were introduced. According to Smith, the Book of Mormon was compiled by numerous prophets. Yet Ernest Taves has demonstrated by stylometric analysis that the various parts of the Book of Mormon (and later the Book of Abraham) did not have multiple authorship but were probably written by the same author.[25]

23. Tanner, pp. 106-107.
24. George D. Smith, "Joseph Smith and the Book of Mormon."
25. Taves, *Trouble Enough*, Part Two.

Many devout Mormons, when confronted with these discrepancies, respond in a way similar to those who believe that the Bible is the inspired word. They appeal to faith, or claim that the Book of Mormon is "inspired allegory," or say that they do not wish to question "the mysteries."

Building a new church

There were so many flaws in Joseph Smith's story about the divine origin of the Book of Mormon that we can only be amazed at the credulity with which those about him accepted his account. But they were not the first to be taken in; and similar psychological patterns most likely have been repeated in other cults by guru-leaders who applied their talents to deceiving their disciples and recruiting new followers. Smith's new Church of Latter-day Saints quickly grew. As it did, he was virtually hounded out of western New York. Suddenly thrust into prominence by the publication of the Golden Bible, he began to attract followers but also adversaries and enemies, who claimed that the Golden Bible was a product of "fraud, blasphemy, credulity and hocus pocus." Within a month Joseph had forty followers, but for every person he baptized, others were reminded that he had been tried and convicted as a money-digger and they accused him of being a false prophet. Fawn Brodie perceptively observed that Smith used the persecution of the Mormons to his advantage. As he became a martyr, his mission took on added significance to those who believed in him, for they were given a special role in the church he was building. He had ingeniously created a new chosen people, who would fight fervently to convince others of the divine origin of the revelations. Smith appealed not to the literati or to the intelligensia but to common folk like himself without sophistication, education, or training in logic. They were dedicated to advancing their station in life.

The main body of believers was forced to move to Kirtland, Ohio, where Sidney Rigdon, leader of another church, and his entire congregation were converted to the new religion. The early critics of Mormonism speculated that Joseph did not compose the Book of Mormon at all, maintaining that it was an artful plagiarism and that Rigdon was in on the scheme from the start. His foes could not believe that this unschooled man had composed it himself. It was rumored that another unpublished manuscript written by Solomon Spaulding two decades earlier presented an outline similar to that contained in the Book of Mormon. At least, some people who had read Spaulding's manuscript claimed that there were striking similarities and implied that Smith had cribbed his book from Spaulding. This theory, though intriguing, has never been confirmed; for the suspected manuscript has never been found, even though another manuscript by Spaulding was later uncovered. According to the official Mormon account, it showed no similarity to the Book of Mormon. Some writers have since claimed to find similarities between Spaulding's *Manuscript Story* and the Book of Mormon. Vernon Holley suggests that Spaulding's work might have been the basis for the Book of Mormon.[26]

In any case, Joseph Smith now concentrated on developing his talents as a preacher-prophet in Kirtland. He was able to arouse an audience to passionate conviction. His voice was so powerful and eloquent that he held his listeners spellbound. People reported that he had a lively sense of humor, yet he manifested dignity and he behaved *as if* he truly was a prophet of God. Reports also began to circulate about his powers of healing and exorcism. Fasting, sleep deprivation, chanting, and singing were often a prelude to group revelations in the Kirtland Temple. Like Jesus and other "faith-healers," Joseph found that some individuals who believed in his message might be restored to health. In some cases, hysterical symptoms of psychosomatic origin could apparently be relieved. Whether such people were permanently cured, we have no way of knowing, yet news of cures led to still other conversions. Did Joseph now take himself so seriously as a prophet that he did not relax his role even in front of his wife, Emma? He was, people said, "bewitching and winning."

He established early hegemony. Although the church he built had a bureaucratic structure and there was widespread lay participation in running its affairs and preaching the new gospel, during his lifetime Joseph was its absolute master. He constantly had to resort to new revelations and commandments from God—"Thus saith the prophet!"—to set things in order. He usually invented these to deal with the practical issues of the day that developed within the church. Interestingly Joseph was following in the footsteps of Moses and Mohammed, who convinced their followers that they were receiving constant communications from God.

Ezra Booth, an early Mormon convert, confesses how at first "the magic charm of delusion and falsehood" had "wrapped its sable mantle around him." But he soon came to the conclusion that Joseph's revelations were not of divine origin but were simply human solutions to the problems they faced. The prophecies of Joseph often failed, they contradicted earlier ones, and they bore the marks of human weakness and wickedness, sometimes masking Joseph's own self-interest. Yet Joseph by now was called the prophet, seer, revelator, and translator, and when he spoke "by the Spirit," it was "received as coming directly from the mouth of the Lord." When he said that something must be so, it was accepted without con-

26. Vernon Holley, *Book of Mormon Authorship: A Closer Look* (Ogden, Utah: Zenus Publication, 1983).

troversy. For example, two elders of the church argued about whether a bucket of water would become heavier by putting a live fish in it. After awhile Joseph decided in the negative: "I know by the spirit, that it will be no heavier." Booth noted that a person could by an actual experiment clearly decide on his own whether the Prophet was influenced by a true or false spirit. Joseph was adamant that no one in the church except him could receive commandments or revelations for the whole church. On one occasion, according to Booth, a woman turned up in Kirtland, claiming to be a prophetess. She ingratiated herself with some of the elders and was welcomed by Rigdon. Joseph viewed her as an "encroachment" upon his sacred domain, and she was ejected from their midst.[27]

I will not recount here the entire subsequent history of Joseph Smith and the church he built but will provide only an outline. It is truly a dramatic and unbelievable story. The early church demanded utmost dedication and commitment from its members. At Kirtland, Joseph was given an important new revelation, which became known as the "Law of Consecration." This meant that goods and properties were to be consecrated to the church, and a form of communal living was established. This conveniently provided funding for the church and was a constant source of new wealth.

Kirtland was not to be the promised land or the new Zion, however, and so the early Mormons again moved westward and established communities in Independence and Far West, Missouri, which was then on the western frontier of the United States. Meanwhile, dissension broke out in the ranks of the church, and Joseph was first accused of adultery. In 1832, a mob dragged Joseph from a home in which he was living and tarred and feathered him. Sidney Rigdon received the same treatment. Building the new church against such odds was no easy task. Nevertheless, the church continued to grow, attracting a constant flow of new believers. Among them was Brigham Young, a young convert who was enthusiastic about the doctrines being proclaimed. Young, according to Ernest Taves, had considerable administrative talent, which was an important asset if the church was to develop efficiently.[28] Young also "spoke in tongues," which others interpreted as a sign he was "filled with the spirit of the Holy Ghost."

By 1833, incredibly, there were more than a thousand Mormons living in Missouri. Hardworking, preaching a strange and apparently blasphemous new gospel, clannish in their ways, even fanatic, they were considered dangerous by other settlers. Again warfare broke out. The lieutenant-governor of the state, Lilburn Boggs, later governor, led a militia against the Mormons. There was bloodshed on both sides, as the Mormons

27. Howe, *Mormonism Unveiled*, pp. 171, 216
28. Taves, *Trouble Enough*, p. 81.

armed and defended themselves. Everywhere there were vociferous critics and detractors. Philastus Hurlburt, whom I have already quoted, had been excommunicated from the church. He became a bitter foe, digging into Joseph's past in Palmyra and publishing damning testimony about him. He accused him of being a fraud and charlatan and said that the Book of Mormon was a forgery. Joseph brought suit against him, was successful, and Hurlburt was ordered to cease and desist for six months and to pay $300 in fines and fees. The people of the church shrugged off all criticism and charges, which they attributed to the "doings of the devil."

The detailed criticisms and skeptical doubts about Joseph Smith's veracity and honesty in his own day did not affect the people who flocked to his banner, nor did they dampen the fervor in his behalf. It may even be the case that the surest way to enhance the growth of a new revolutionary or religious movement is to attack it. This brings it to the center of the stage; and the attacks and vilifications seem to attract the disgruntled, the dispossessed, and the disenchanted to the new faith. Although attacks may arouse suspicion and antipathy, even hatred in the larger population, they may also arouse sympathy and enlist support in others. For those human beings already deeply attached to the cause, vilification may only help to reinforce their siege mentality; and it produces a drawing tighter of the hard core of believers. *We* are the chosen people. *They* hate us and seek to destroy us. *We* will fight with even greater dedication and frenzy for our beliefs. This apparently is what happened with the beleaguered and harassed followers of Smith.

Meanwhile, Joseph's stature had been elevated for those who accepted his mission as divinely ordained. No longer was Joseph the simple money-digger nor even the translator of the Golden Bible. He now became the commander-in-chief of battalions of men, women, and children who were willing to do his every bidding. One can but wonder what happened to Joseph's already inflated ego when he realized that he was able to control the destiny of thousands of human beings and that he was able to give meaning to their lives. All of this had emerged from fantasies concocted in his own imagination, which were now accepted as the gospel truth by the members of the church he founded.

Did Joseph, like others, need constant approval? Did he crave self-reinforcement? Did this drive him on? The psychosexual motives of a person are highly complex and difficult to fathom. Quite early, rumors began to circulate that he was an adulterer; indeed, he had an excessive interest in the ladies and even in other men's wives. It seems that Joseph had affair after affair. Although these were usually discreet, there were constant complaints from aggrieved husbands and brothers about Joseph's indiscretions. We will return to the subject of polygamy shortly, for it became a burning issue within the church.

The early Mormon saga was fraught with conflict and danger. Constantly persecuted and expelled because of their heretical views, the main body of Mormons moved from Missouri, where they were unwanted, to Illinois, where there were already numerous followers. The Mormons were at first welcomed to the state, where they flocked to receive protection. Here they planned to build their Zion; and in 1839 they changed the name of Commerce, a small town in Illinois, to Nauvoo, which according to Joseph was the Hebrew word for "beautiful plantation." The Mormons, indefatigable in their labors, set about reconstructing their city, following Joseph's architectural plans. As the city grew and prospered, Mormon missionaries in Great Britain proved to be extremely effective, and there was an influx of new converts, who were dispatched to the United States through the efforts of Brigham Young, using a special emigrants' fund he established to help pay for voyages. During this period Smith developed a closed theocracy, replete with a militia, and he named himself lieutenant-general of the Nauvoo Legion. He had grandiose political ambitions.

It was also during this time that the church's structure was further elaborated and consolidated. Fawn Brodie describes strange new religious practices that were introduced and sanctified by Joseph.[29] Many of these were similar to those practiced by the Freemasons, from whom Joseph no doubt borrowed. Secret rites held within the Temple involved the washing and anointing of parts of the body. Moreover, Mormons were commanded to always wear a church garment (a suit of long underwear) with Masonic symbols cut into it, as a protection against evil. And there were other Masonic symbols and ceremonies, such as the Mason's apron.

Secret Mormon Temple rites transformed the church into a mystery cult. These were related to Joseph's newly hatched theories of the afterlife. After death, he said, a person's soul was transported to the world of the spirits, where it awaited judgment day. However, only those souls who had affiliated with the Mormon church could attain the highest level within the celestial kingdom. However, souls in the past, who had not lived long enough to receive the Mormon gospel, could be liberated from the world of the spirits and reach the highest level by the intercession of a proxy form of baptism. Accordingly, any Mormon could "seal" his dead relatives or friends, or the great figures of history, simply by baptizing them by proxy.

In Salt Lake City today there are genealogical records numbering hundreds of millions of dead persons. Many of these have apparently been saved (without their permission), by post mortem baptisms. This ritual may seem odd to the non-Mormon, but perhaps no more so than the

---

29. Brodie, *No Man Knows My History,* pp. 278-83.

communion service of the Roman Catholic church, where the blood and flesh of Christ are consumed, or the Mosaic dietary customs. If given the sanction of a prophet, the faithful will obediently perform strange practices for countless generations in the conviction that they are following the commandments of the diety.

Mormonism had an allure for settlers on the prairies and frontiers of America, perhaps because it was an indigenous religion, relating the Old and New Testaments to the virgin continent. Not only did it have a place for the red man, but it also taught a new promised land (out West), and alleged that the original Garden of Eden had been located near Independence, Missouri. All of the ingredients were homespun, the language of the religion was camp-fire style, and its prophet was a native-born American. How beautifully it blended with the environment!

Polygamy

It was during the Nauvoo period that the issue of polygamy came to a head. Rumors that the leading Mormons practiced plural marriage broke into the open. Fawn Brodie has provided a list of Smith's wives; she recorded forty-nine women who were "sealed to the prophet," many of whom were currently married to other men.[30] Brodie says that her list is incomplete and that there were many more who were not sealed by a ceremony. Marriages could be performed either "for time" (this life) or "sealed for eternity." The marriages of these women to Joseph were not simply in this life but lasted throughout eternity. If his wives were married to others, this was merely "for time." A large number of Joseph's marriages permitted sexual relations; others were merely "sealed" to him for the afterlife. As Joseph's power and influence grew, he seemed to take new wives with complete abandon. Other church leaders, such as Brigham Young and Heber Kimball, likewise took dozens of wives.

In the Book of Mormon, polygamy had been expressly forbidden. There is some intimation, however, that as early as 1831 Joseph had a revelation foreshadowing plural marriage. After many years of denying charges of polygamy, Joseph finally came forth with an explicit revelation in 1843 justifying plural marriages on biblical grounds. It was no doubt concocted to legitimize his way of life. After all, if Abraham, Solomon, David, and the ancient patriarchs had many wives, why not Joseph Smith and his brethren? His first wife, Emma, witnessed many of the affairs he engaged in with other women, and she was even forced to be present during the special sealing ceremonies. She objected strenuously to Joseph's

30. Ibid., pp. 457-88.

philandering. Joseph replied through the voice of revelation, commanding Emma to "receive all those that have been given unto my servant Joseph" and "to cleave unto my servant Joseph, and to no one else."[31] The penalty for disobedience, he admonished her, was that she would be destroyed. Joseph then proclaimed a new law:

> If any man espouse a virgin, and desire to espouse another, and the first give her consent, and if he espouse the second, and they are virgins, and have vowed to no other man, then he is justified; he cannot commit adultery . . . and if he have ten virgins given unto him by this law, he cannot commit adultery, for they belong to him.[32]

This revelation was kept secret from the public and even withheld from those who lived in Nauvoo, though it led to a schism when the subject was openly broached within the high council of church leaders. Critics were aware of the doctrine of polygamy, and they charged that the Mormon church encouraged promiscuous sexuality. This issue became a bitter bone of contention with the broader public, many of whom hated Joseph Smith and his followers with a ferocious intensity. To demonstrate the low reputation of the Mormons, Ernest Taves quotes a Protestant minister, W. M. King, who wrote in 1842: "I presume Nauvoo is as perfect a sink of debauchery and every species of abomination as ever were Sodom and Nineveh."[33] The Mormons were accused of every moral crime. Anti-Mormonism paralleled anti-Semitism. Considered to be haughty and proud, staunchly dedicated to their religious commitment, practicing strange customs, and holding bizarre beliefs, they irritated non-Mormons. When word that they practiced polygamy leaked out, this proved to be explosive; it may even have been the key factor that precipitated the death of Joseph Smith.

Many forces no doubt contributed to Joseph's premature demise. He had gathered thousands of supporters, was commander of an army, and sought political power in the states in which the Mormons were strong; he even declared himself a candidate for the presidency in 1844. He had courted political leaders by promising them black voters, but when he declared his candidacy, their political protection evaporated. By then, his critics were charging that he had become a law unto himself: his followers did his bidding blindly, and he had imposed a tyrannical rule that threatened to spread to the adjoining states. Some thought that a despotic new religious empire would be forged by Joseph Smith with himself as emperor-pope. For example, Joseph wished to separate Nauvoo from the

---

31. Ibid., pp. 340-41; also see *Doctrines and Covenants,* Sect. 132.
32. Ibid.
33. Taves, *Trouble Enough,* p. 169.

legal system of the state of Illinois and the federal government. Hence, those who were responsible for applying the laws of the United States sought to incarcerate and punish him and to disband the army under his control.

An incident that especially aroused public ire was Smith's destruction of an opposition newspaper in Nauvoo. The newspaper was founded by William Law, a dissident Mormon, who rejected the doctrine of polygamy and thought it improper for Joseph to be so heavily involved in real estate. After being excommunicated, Law dedicated himself to publishing attacks on Smith. Smith retaliated by destroying his news plant. Public opinion turned against Joseph. He was charged with suppressing the freedom of the press and flouting the First Amendment. Taves notes that in his last years there were signs that Joseph was delusional, out of touch with reality.

The last days were rapidly approaching. Joseph and his brother Hyrum, who was always at his side, were taken to Carthage, Illinois, in 1844 to await trial on a variety of charges. An unruly mob broke into the jail and shot both Joseph and Hyrum. What a shock to the Mormon community when their bodies were solemnly returned to Nauvoo. With the death of its leader and prophet, many Mormons became disconsolate and left the church. It was the enormous ability of Brigham Young, the St. Paul of Mormonism, that saved the church from collapsing. There are, of course, differences within this analogy. Paul wrote doctrine; Brigham administered. Paul was ascetic; Brigham indulged in women and land. Nonetheless, both men helped give their respective churches new foundations. There was some dispute about Joseph's success, but Brigham prevailed. Mormonism was able to continue, for a new element had been introduced: Joseph had been murdered; he was a martyr, like Christ, and he had died for the Mormon cause. The chosen people must now escape still further west to build a new Zion. Smith was like Moses, who did not make the promised land, but his faithful children would.

A short time after Joseph's death, the Mormons again pulled up stakes and transplanted themselves to the Utah territory, where they thought they would be free to start anew. But there were many problems in building the church and constant battles with the federal government. Brigham Young ruled the church with an iron hand. He officially proclaimed polygamy in 1852, and that issue proved to be a festering one with the rest of America. The church leaders were forced to abandon it in 1890, unable to withstand the full weight of the federal system. The Reorganized Church, a much smaller and dissident group, did not practice polygamy, and hence did not suffer the same persecution.

Today the Mormon church is a powerful institution of great wealth and influence, still growing, sending missionaries to all parts of the globe (like the early Christians), and recruiting new believers to the faith. The church has long since been transformed from a radical, nonconformist cult everywhere despised to a powerful conservative institution, a defender of a status quo and the socioeconomic establishment. In the relatively short period of 150 years it showed a process of birth, development, and maturation similar to that of ancient religions, whose origins are buried in an uncertain past but whose prophets are nevertheless revered because of their alleged revelations from God.

Today we may ask, with some assurance that we can give a reasonably definitive response: Was Joseph Smith an authentic prophet? Or was he merely a human being who used "divine revelation" to suit his own purposes? Did Smith receive a revelation in the form of the Book of Mormon? There is little to support this assertion. The preponderance of evidence suggests that the Golden Bible was Joseph Smith's creation. The devout Mormon will, of course, insist that Joseph was a true prophet of God and that the Book of Mormon was a product of revelation. It may be impossible to convince him of the contrary, especially when faith, custom, authority, and tradition dictate otherwise and especially since the religion of Mormonism, whatever its intellectual foundation, has developed a strong new *ethnic* tradition. It has been inculcated by numerous fathers and grandfathers into their young. As a belief-system it not only sets forth a creed and dogma held by a group of closely related people; it also defines their way of life, their self-image as a chosen people, preexistent with God and destined for a special afterlife. This especially applies to the American Mormons centered in Utah and neighboring states, if not to the newly converted in other parts of the world who will no doubt bequeath their new ethnic heritage to future generations.

Today, for the critically minded skeptic (whether Mormon or non-Mormon), Joseph Smith's veracity must be thrown into doubt. Two further discoveries that he "translated"—the Book of Abraham and the Kinderhook plates—provide additional confirmation of the hypothesis that Smith, though seemingly convinced of his prophetic role, nevertheless practiced deception.

The Book of Abraham

In 1835, Joseph Smith was visited by Michael Chandler in Kirtland, Ohio. Chandler was exhibiting throughout the country four Egyptian mummies along with several Egyptian papyri. Learning of Smith's reputation, he sought his aid in translating the ancient documents. Joseph became so interested in the findings that he purchased them. He declared that the papyri contained the writings of the Hebrew prophets Abraham and Joseph. (The general public in 1837 was unaware of the fact that Champollion had by now succeeded in unraveling Egyptian hieroglyphics.) Smith resorted to the same method he had used earlier. Claiming inspiration from God, he "translated" the papyri into the *Book of Abraham,* published in 1842. (He never translated the papyri of Joseph.)

The *Book of Abraham* (part of the *Pearl of Great Price*) is perhaps the most unfortunate publication in the entire corpus of Mormon literature. It provided an account of the creation of the cosmos and Joseph's speculations about the planets and the stars. It also contained reflections on the origins of the black man. According to the *Book of Abraham,* they were descendants of the "loins of Ham" of the blood of the Canaanites. From Ham, it said, "sprang that race which preserved the curse in the land."[34] The blacks allegedly were destined to be the servants of other races. Pharaoh, king of Egypt, was the son of Egyptus, daughter of Ham. All Egyptians had inherited black skin, and with this curse they did not have the right to be ordained into the Mormon priesthood. These views, published before the Civil War, supported the position of white slaveholders, who sought a biblical justification for the lower status of Negroes.[35]

The point of introducing the *Book of Abraham* is that, unlike the golden plates, which had mysteriously disappeared after translation, the mummies and the papyri that Joseph allegedly translated were exhibited in Kirtland and Nauvoo. At the time that Joseph published the *Book of Abraham,* the science of Egyptology was still in its relative infancy. But it was developing rapidly. After Joseph's death, in 1860, the Egyptian facsimiles printed in the *Book of Abraham* were translated by Egyptologists; they differed completely from Joseph's version. Scholars showed that the papyri were ordinary funeral scrolls taken from the Egyptian *Book of Breathings,* a revised form of the *Book of the Dead,* commonly buried with the dead and found in thousands of Egyptian tombs of the period; they represented the gods Maat, Osiris, and Isis.

Later, the scholar Arthur Mace of the Metropolitan Museum of Art described Smith's translation as "a farrago of nonsense from beginning to end." And Dr. W. M. Flinders Petrie of London University stated: "It may safely be said that there is not a single word that is true in these interpretations."[36]

In 1967, eleven fragments of the papyri turned up at the Metropolitan

---

34. *Book of Abraham* 1:21-24.
35. Only after well over a century had passed, on June 9, 1978, did the president of the Mormon church, Spencer W. Kimball, revise this racial doctrine and announce a new revelation. Thenceforth black males could hold the priesthood and be ordained in the Mormon church.
36. Brodie, *No Man Knows My History,* p. 175.

Museum of Art. Scholars again verified the earlier scholarly analyses. Indeed a translation of the papyri was published in 1968 in the independent Mormon journal *Dialogue*. It showed that Joseph Smith had concocted his inspired translation from beginning to end and that it bore no relation to the original.

A similar disconfirmation of Joseph Smith's prophetic abilities occurred with his partial translation of the so-called Kinderhook plates. Here a deliberate trap was set and Joseph fell for the bait. Three men living in Kinderhook, Illinois, claimed that in April 1843 they had discovered in an old earth mound six bell-shaped brass plates, which were engraved in hieroglyphics. Two Mormons were present when the corroded plates were dug up and carefully cleansed. Upon receiving them in Nauvoo, Joseph Smith immediately declared that the plates were genuine and proceeded to translate part of them. He said that they contained the history of a person whose bones lay nearby and who was "a descendant of Ham, through the loins of Pharaoh, king of Egypt."[37] Smith never published a full translation of the plates. Perhaps he was now somewhat more cautious, for the *New York Herald* had reported on December 28, 1842, that the ancient Egyptian language had finally been deciphered and that a grammar had been published in England. Many years later, it was admitted by one of the three men that the Kinderhook plates were a complete hoax, that they were cut out of copper, etched with acid and nitric oxide, and buried in the mound in order to trap Joseph Smith. Joseph, true to form, had again fabricated a "translation."[38]

William Miller and doomsday prophecy

During the days of Joseph Smith another bizarre, apocalyptic movement began to develop. Its origin was in the spiritual climate of New York state and neighboring New England, under the direct inspiration of William Miller, a farmer, an avid fundamentalist Baptist, and a diligent student of the Bible. Miller's reading of the Bible convinced him that the end of the world was imminent. His reasoning was rather contrived, but it was based on his taking the Bible as the literal word of God, and especially focusing on the Book of Daniel in the Old Testament and the Book of Revelation in the New Testament, which allegedly prophesied the last days, Armageddon, and the Second Coming of Christ.

There have been many efforts to draw prophecies from the Bible. The

37. Ibid., p. 291.
38. See George D. Smith, "Joseph Smith and the Book of Mormon," p. 23. A letter from William Fugate, June 30, 1879.

Millerite movement is a classic case of the often tragic character of such predictions. The books of the Bible were written over many centuries, and they reflected the idealized, historical account of the Jews and their yearnings for national identity and liberation from a foreign yolk. To read into the Bible more than that and to attempt to interpret it so that it becomes a plan for the entire future of humanity is a deceptive snare for the unwary. It is always presumptuous of us to read our own times into earlier documents, which express other political, religious, or social interests, or to believe that what is happening now or is about to happen was previsioned long ago by some ancient prophet.

A good illustration of this can be seen with the continued fascination with Nostradamus, a 16th-century French astrologer. Nostradamus composed four-line verses (quatrains), which are arranged in groups of one hundred, appropriately known as centuries. His quatrains were supposed to be used to predict the future, but they are so vague and general that they do not clearly say what is to occur. Indeed, they could not be understood until they were interpreted retrospectively after an event had happened. Thus the latter-day interpreter is left to his own devices to fit the facts to the prediction. But which facts apply to what period? The game is open to ad hoc prophecy **and** a wide range of permissive interpretations, so that virtually anything can be made to fit.

Let us take for example a famous quatrain, Century I, verse 60. Nostradamus writes:

> An emperor shall be born near Italy
> Who shall be sold to the Empire at high price
> They shall say, from the people he disputed with,
> That he is less a prince than a butcher.

To whom does this apply? Napoleon, in the late 18th and early 19th centuries, fits the prediction, since he was born in Corsica and became emperor of France, and the armies he led contributed to the slaughter of great numbers of people. But it also applies to 20th-century Hitler, who was born in Austria (near Italy), became an absolute dictator and butchered millions. It can likewise be applied to Ferdinand II (1578-1637), king of Bohemia and Holy Roman emperor, who ruled during the bloody Thirty Years War. The interpreter is free to choose whatever he wishes after the fact. This is the difficulty in going back and finding mysterious prophecies in earlier writers. Nostradamus is being widely appealed to in the 20th century to prophesy grave forebodings and a massive war.

Is the universe fulfilling a plan? Is the future fixed such that some prophetic minds can peer into it? Claims have been made on both paranormal and religious grounds about this ability. But precognitive clairvoyance has never been empirically demonstrated in the laboratory; and the fact that it has religious embroidery does not make it any more true.

The Bible has been and still is being read in order to prophesy the future course of history. Most specific Bible prophecies fail, and they can be disconfirmed easily if they do not occur, unless they are taken so generally that they can be stretched and interpreted to apply to whatever is wanted in the future or the past. Jesus' own prophecy—that his generation would not taste death and would live to see the end of the world and his return to save them—did not occur in his own time. But it did not dishearten his disciples. In virtually every age, including our own, Bible prophets utter grave new forecasts of doomsday for the world and of salvation for true believers. There is a fatalism in the view that we can do nothing but passively await the unfolding of foreordained events and hope that we will be spared the terrible destruction awaiting sinners and nonbelievers.

Perhaps the most pointed illustration of the misuse of biblical prophecy is the saga of William Miller. It borders on the ridiculous, and yet he was taken seriously in his own day. Unfortunately for Miller and his followers, known as Millerites, his predictions were framed in such a specific form that they could be readily proved false. As we shall see, however, this did not weaken the next generation's faith; they only redoubled their conviction that the end of the world was near.

Miller announced that in 1818, after poring intensely over the Bible, he came to the firm conclusion that in twenty-five years our present state of affairs would be wound up. At that time all the pride, power, pomp, vanity, and wickedness of the world would come to an end and the Kingdom of the Messiah would be established. The Old Testament passages he used as the basis for his inference were in the Book of Daniel. According to that, Daniel had dreams and visions, sometimes when in bed, or once when he had fallen, or when he was sick. In Chapter 8 there is an account of an elaborate vision. One saint speaks to another, asking how long impiety will cause desolation and the land to be trodden down. In response, Daniel says:

> And he said unto me, unto two thousand and three hundred days; then shall the sanctuary be cleansed (8:14—KJV).

Incredible as it may seem, Miller interpreted *days* to mean "years" (as many others of his time did). Miller was convinced that many events in the Bible had been predicted to occur within a specific time: the flood, 120 years; Abraham's descendants' sojourn in Egypt, 400 years; the time in the wilderness, 40 years; the exile, 70 years; etc. Similarly, for the 2,300 years which would mean the end of the world. He estimated that Daniel's prophecy was made about 457 B.C.E. This he identified with the command-

ment of Artaxerxes, which was mentioned in Ezra. Ergo, the prophecy
would be fulfilled 2,300 years later in 1843! Inasmuch as there had been
many changes in the calendar, Miller could not say with certitude the
exact day or year in which the prophecy would be fulfilled, but he thought
that it would be sometime between March 21, 1843, and March 21, 1844.
He interpreted the phrase "the sanctuary would be cleansed" to mean that
the earth would be purged of sinful wickedness by fire and destruction.
Judgment Day, the Second Advent, or the Second Coming of Christ
would be ushered in. At the fateful hour, the sky would open up to reveal
the heavenly host, the dead would rise from their graves, the believers
would be taken up to heaven in the form of "the rapture," and the sinners
would be cast into the hell they so rightly deserved.[39]

Miller began to espouse his theories in 1831; he published and lec-
tured on biblical prophecy and carried on a vigorous propaganda campaign. Large crowds flocked to hear him. Enthusiasm gained momentum
and the Millerite movement began to grow rapidly. Great numbers of
people left the established churches and converted to his apocalyptic
doctrines. But there was widespread rejection of Miller's prophecies by
critics: scientific skeptics found them absurd, and other ministers thought
they were overly simplistic, fanatical, and deficient even on biblical
grounds. Nevertheless, by 1839 the movement has been transformed from
a small rural phenomena to a mass movement with large urban churches.
Estimates of the number of adherents run anywhere from fifty thousand
to hundreds of thousands. In 1842 and 1843, as the end-days supposedly
were approaching, interest in Miller's prophecies continued to grow. If the
cataclysmic end of the world was really in store for everyone, what should
one do to prepare for it, people asked in fear.

In November 1833 there was a bright meteor shower—the Leonid
shower, so-called because meteorites were seen so close to the constellation
Leo. This shower, incidentally, appears periodically. A bright comet appeared in 1843. To the credulous, who are untrained in astronomy, these
were taken as divine omens that portended the imminent end of the world.
In 1843 a rumor swept the Millerite group that D-day would be April 23,
1843, even though Miller and his church leaders never officially accepted
this date. The day passed without any untoward event. Similarly, at the
end of 1843 there was great anticipation of and consternation about what
might ensue. When again nothing happened by March 21, 1844, the
Millerite movement faced a real crisis; opponents mocked them. In anticipation of the end of the world, many true believers had sold their worldly

    39. See Daniel Cohen, *Waiting for the Apocalypse* (Buffalo, N.Y.: Prometheus Books,
1983), pp. 7-34. See also Elmer T. Clark, *Small Sects in America* (Nashville: Abingdon
Press, 1947).

goods and awaited the end in high emotional expectation. Farmers even refused to plant their crops. Since they would not need any money, they gave away their funds and discharged their employees, all in preparation for the appointed day when the Lord would arrive to deliver them. Miller and his followers were dejected when again nothing happened. The faithful had banked everything on the hope that the world would be destroyed and they would be saved. Miller still kept insisting that "the day of the Lord" was near, perhaps in a few more months.

Early in 1844 another Millerite, Samuel S. Snow, declared that D-day would be in the fall of 1844, October 22nd to be specific. Most of the Millerite leaders now focused on that day, and even Miller eventually came to the same conclusion. Finally committed to October 22, 1844, the Millerites again made preparations to meet their maker. There were camp meetings and intensified missionary activities; stores, businesses, and farms ere again shut down. An interesting sidelight to the events was the fact that Miller reaped a considerable sum of money by selling his followers white "ascension robes." Miller insisted that he was simply doing God's work by providing an abundant supply of the recommended attire for the faithful to meet their maker. Many Millerites, dressed in their white robes, made their way to hillsides to await the coming of the Lord. There was much excitement, crying, and shouting, all in anticipation of the great event. Several sought to leap into the air and take off like birds. "One man put on turkey wings, got up on a tree and prayed that the Lord would take him. He tried to fly, fell and broke his arm."[40] Many were deeply frightened at what might happen to them. Twelve o'clock midnight passed and again nothing happened. This time their hopes were truly shattered. People wept openly; for the great expectation was not fulfilled. Miller confessed his mistake about the date and showed surprise, chagrin, and disappointment. But he still held steadfast in his faith that the Second Coming would occur shortly. He died a grief-stricken and virtually broken man. Most of his followers had by then disappeared. His prediction has been decisively disconfirmed.

What are we to make of this prophecy? First, it should be pointed out that many biblical scholars are uncertain as to who the author of the Book of Daniel was. Second, they are uncertain of the date the book was written. Some scholars think it was written during the Babylonian captivity of the Jews, about the 5th or 6th century B.C.E. If so, Daniel spent his career at the court of the Mesopotamian rulers, seeking to remain faithful to Jewish law, even in the Persian court. In this historical context, Daniel was predicting, at best, events that would occur within his own lifetime, and he was probably mourning of the loss of his brethren's faith

40. Cohen, *Waiting for the Apocalypse*, p. 31.

and hoping for a revival of Judaism. Others have said that the Book of Daniel was set in the second century B.C.E. There is some question as to whether parts of it are canonical or apocryphal. Daniel may even be a fictional character. Thus Miller even erred in specifying a date on which Daniel's prophecy was first made.

Third, even if one takes the passages at face value and assumes that they described what happened to a real person, what are we to say about the dreams and visions of a sick man? The prophecy is very subjective. How do we know that Daniel had a vision or dream? Perhaps it was only his imagination at work. Is it not perilous to predict the destruction of the entire world at some future time based on an individual's visionary fantasies? Fourth, these prophecies were stretched out of all proportion to apply to the nineteenth century, and we have seen that they were false.

As we approach the end of the second millennium after Jesus, people are again prone to apply doomsday forecasts to the near future. Today some fundamentalist prophets are preaching that we are living in "the last days," and they view earthquake tremors, wars, and rumors of war, as signs of the impending apocalyptic disaster. The last great battle of Armageddon is approaching, we are warned. Indeed, *this* generation, many of them insist, is the *last* generation, and this was all foretold in the Old and New Testaments.

In *The Late Great Planet Earth,* Hal Lindsey claims that Armageddon is around the corner.[41] According to biblical prophecies, seven years of terrible tribulation will soon befall mankind. This period is about to begin because the Jewish people, after their long Diaspora, have finally returned to their ancient homeland in Palestine, which they left after the destruction of Jerusalem in C.E. 70. Next, Lindsey says, the Jews will rebuild the Temple in Jerusalem. Then a whole series of terrible events will trigger the final Armageddon. A world war will ensue. Israel will be invaded from all sides: by a confederation of nations from the north (said to be led by the Russians), by the Arab nations, and by a great power from the East (said to be the Chinese), with an army of two hundred milllion soldiers. During the period leading up to these cataclysmic events, there will also emerge a ten-nation confederation—this was the old Roman empire, now the European Common Market—headed by an anti-Christ preaching a new religion. These years will witness the greatest devastation mankind has ever seen. Valleys will flow with blood and cities will be destroyed by torrents of fire and brimstone. This, it is said, represents a thermonuclear war, the most terrifying holocaust of all time. But then Christ will return to rescue in rapture all true believers. Christ will reign for a thousand

---

41. Hal Lindsey, *The Late Great Planet Earth* (Grand Rapids, Mich.: Zondervan, 1970).

years and eventually bring into being his final kingdom, which will last throughout all eternity.

We may ask the following questions: Is the Bible reliable as the basis for prophecy about the future? Why should its predictions be viewed as applying to this generation? Why should the Bible be taken literally? I would, of course, deny the very possibility of prophecy. The true Christian has waited almost 2,000 years. He can always extend into the future his hopes for the Second Coming. Such a general statement—that there will be a war of Armageddon and a Second Coming—is impossible to confirm, for it is always being pushed into the future. Not yet, say the true believers, but soon.

Ellen G. White: Inspired prophet

Revelations and visions from on high

The Millerite movement, however, did not die. One would have thought that a specific prediction that failed would have put an end to the matter. But the transcendental temptation was so strong that the faithful were not to be daunted. Those who followed the Millerites and believed that the Bible was the inspired word of God needed some explanation as to why Jesus did not come. This they managed to manufacture. In doing so, they repeated a familiar psychomythological pattern: paradoxically, it is often out of failure that a religious movement gains new momentum. The post-Millerites reasoned as follows: It is our fault, not God's, that the end of the world and the salvation of man did not occur; it is we who have misread God's real intention.

The leader in this new movement was Ellen G. White. William Miller never claimed to have revelations, but Ellen White did. Her many visions led to the founding of the Seventh-Day Adventist church. This church, like the Mormon church, had its roots in the nineteenth century; and it has grown rapidly until it now has over four million members. Like the Mormons, it claims to have a divinely inspired message received by Mrs. White. I will not describe all of the features of the faith, nor will I trace the historical growth of the church; rather I wish to focus on its founder's crucial claims that rested on revelation. I submit that her case illustrates the same syndrome that I have been delineating concerning the foundations of other religions. Ellen White is a false prophetess, claiming to receive visions and revelations from on high and founding a church in the name of God. Yet it has now been fairly well documented that her much heralded prophetic powers were based on a physical and emotional disability and were not received from some divine being but were borrowed from other sources.

Ellen White and her fellow Adventists attempted to overcome the tremendous psychological letdown of the failed prophecy by offering a new explanation for "the great disappointment" of October 22, 1844. They would not accept the fact that their calculation of when Christ would appear was mistaken. Instead they reasoned that they must have been mistaken only in depicting the kind of event that was prophesied.

Ellen Harmon, a seventeen-year-old girl from Gorham, Maine, began to have frequent visions. These visions offered explanations for the apparent failure. In 1846 she married James White, another avid Adventist, and assumed his name. At first they promulgated the "shut door" thesis, namely, the belief that on October 22, 1844, the chance of salvation had been terminated for the wicked world. Only those who had been faithful would be saved. This was a rather harsh doctrine, for it closed the door for all others to be saved. Why bother, if everything had already been decided? This thesis is reminiscent of the Calvinist theory of predestination. Such a harsh thesis had to be modified, if a church was to be built; hence it was transformed into the "sanctuary doctrine." The explanation offered for the great disappointment was that on the fateful day in 1844 Christ *had* indeed interposed himself, but he had moved from one part of the heavenly sanctuary to the "most holy part" of the sanctuary. Beginning on October 22, 1844, Christ began the intensive work of "investigative judgment"; that is, he began to examine the records of all his professed followers throughout history and in the present in order to determine who merited eternal salvation.

Let us outline the details of the emergence of this new theological notion. Ellen Harmon was born on November 26, 1827. She dropped out of school at a young age. Her family was intensely religious. In 1840 when she was thirteen years old, she heard William Miller preach about the imminent end of the world. Frightened, she returned home and spent all night in tearful prayer. She continued in the same anguished state for months after undergoing a conversion experience. In 1842 she again heard Miller preach and was again terrified by his claims. She recounts how condemnations rang in her ears day and night, that she feared "she would lose her reason," and that she was overcome by hysterical despair. She reports that she often remained in prayer all night groaning and trembling. She further states that she was taken to heaven, where she met Jesus and felt relieved. She often attended prayer meetings, would fall unconscious, and would remain in such a state all night. She maintained that during such experiences the "spirit of God" took control of her. She and her family accepted Miller's teachings and were wrapped up in the spiritual fervor of the day. The young girl showed all of the signs of emotional instability. And she suffered terribly along with her fellow believers at the great disappointment. It was a short time after the failure, in December 1844, that she reported having visions, and she had them virtually every day. The early Adventists at first regarded these visions as hallucinations caused by her feeble physical condition and influenced by those around her. Only later did they begin to attribute these visions to a divine source. Notice the analogy with Mohammed's psychic pathology.

Mrs. White continued to elaborate the "sanctuary doctrine" throughout her career, supporting its truth by new revelatory visions. This explanation as to why Christ had not come was a post hoc rationalization if there ever was one. Miller's interpretation of biblical prophecy had predicted Christ's return on a specific date. The fact that it did not occur was clear empirical evidence for its falsity. Mrs. White attempted to ignore that rational conclusion by maintaining that something did occur, namely, that Christ *had* intervened, that a process of investigative inquiry had begun, that one day Armageddon would still ensue, and that the true faithful would be saved.

The objections to this speculative rationalization are so abundant that were it not for the fact that the thesis was based on the authority of religious faith—that one person had a revelation from God—it would have been rejected out of hand. To illustrate, let us suppose that someone had predicted that a destructive earthquake would occur in Chicago on April 10, 1950, and that in fear of this he moved all of his belongings, sold his property, and got his friends to do the same. The event did not occur. Was he wrong? No, he says, there was an "indiscernible" shift in the underlying geological structure on that day, and he insists a terrible earthquake will still occur. It *may* occur in the future, but how does he know that there was an "indiscernible" rearrangement if it is not discernible by others? If he answers "because I had a vision about it," one would laugh at him. Yet this is precisely what Mrs. White had done.

Thus, we have to raise the question anew: Did a self-proclaimed prophetess, Ellen White, really have such visions about God's investigative judgment? Had God opened up to her "the precious rays of light shining from the throne?" We have only her testimony to the occurrence. (It is estimated that she wrote over twenty-five million words in her lifetime presenting her revelatory visions.) If one accepts Moses, Jesus, Mohammed, and the Old Testament prophets, why not a nineteenth-century prophetess with latter-day revelations? Many Seventh-Day Adventists take her word as divinely inspired and hence authoritative.

The beginnings of Ellen White's career revolved around the shut-door thesis. According to D. M. Canright, a Seventh-Day Adventist who knew the Whites intimately and later defected, this thesis was first suggested by O. R. L. Crosier, an Adventist whom Ellen knew. She accepted the thesis because she said she had a vision attesting to its veracity. According to the thesis, Moses had built a tabernacle or sanctuary, which had two rooms or apartments. The first room was the "holy place" and the second, the "most holy place." Ordinary priests were allowed in the first room; no one was allowed to enter the second room except the high priest once a year when he "cleansed the sanctuary" with a blood offering in order to atone for the sins of the people.

According to the thesis, the earthly sanctuary was similar to what existed in Heaven, and this sanctuary in Heaven is what Daniel 8:14 was referring to. Jesus was the high priest, who ministered in the first room, the "holy place," where he received believers and forgave those who had confessed their sins. On October 22, 1844, he moved to the next room, the "most holy place," and began the task of cleansing the sanctuary. The shut-door doctrine said that probation for sinners was now at an end and that Christ had moved into the "most holy place" to prepare salvation for believers. This shut-door thesis was reinforced by quoting the parable of the ten virgins recorded in Matthew 25:1-13, in which Jesus said that one should always be ready, for one never knows when the Second Coming will occur. The parable tells the story of the ten virgins. Some were wise and some foolish. The wise ones took their lamps with sufficient oil and went to await the bridegroom. Five tarried because the bridegroom had not yet arrived. They went to purchase oil for their lamps. While they were gone the bridegroom appeared and took those who were ready with him to the marriage feast. "The door was shut" says the scriptures, and when the five foolish virgins came later and asked to be admitted, they were not let in by the bridegroom. This was interpreted to mean that those who had waited for the Second Coming on October 22, 1844, did not do so in vain for they would be admitted eventually to the "most holy place" if Christ deemed it available. Jesus had become the advocate for the saints, the true believers. Those who did not believe or who had sinned would never be admitted; the door to salvation was now closed to them.

This doctrine proved to be too harsh, for it did not allow for new converts to be brought into the faith; it tended to promote a passive apathy, since everything had already been decided and there would no longer be probation for sinners. How can one possibly build a church if new converts cannot be let in? This led to a new doctrine that was put forth in 1849. The shut-door thesis had applied to the remnant church, that is, to the first Adventist believers who had been disappointed. The new theory propounded was now called the open-door or sanctuary theory; it would allow people to enter the church as new members. Under the shut-door thesis, the Adventists could not even bring in their own children, who had been born recently or were growing up. To be saved, however, a person must know about the change that Jesus underwent in heaven in 1844, as he moved to the most holy place. The only ones who had knowledge of this were in the Seventh-Day Adventist church. All other Christians, who remained in ignorance of this doctrine, could not hope to achieve salvation. Any prayers that they made were to Jesus in the first room, but the prayers were lost since he was no longer there. This put the Adventists in opposition to all other Christian churches, since they alone had the exclusive key to salvation.

Mrs. White reinforced her divine authority by claiming that she had talked to angels in her visions and that she went up to Heaven to converse with Jesus. Her critics have pointed out, however, that her visions changed over time, which meant that God's revelatory messages were often contradictory. D. M. Canright denied her claims to divine revelation and concluded that "her revelations were not from God, but were the unreliable products of autosuggestion and an abnormal state of mind. . . . Her professed revelations were simply the products of her own mind reflecting the teachings of those around her."[42] Canright also maintained that these visions were most likely "the results of her early misfortune, nervous disease, and a complication of hysteria, epilepsy, catalepsy, and ecstasy. That she may have honestly believed in them herself does not alter the fact."[43] Canright tells of other hysterical Adventist women who had visions, who were rejected by the church and eventually abandoned them. When Mrs. White's so-called revelations were accepted by those around her as divine, did this only encourage her further in her misapprehensions? Canright is mystified that any sane person would accept these theories simply on the say so of Mrs. White.

Moreover, critics of the Adventist church have charged that the earlier writings of Mrs. White, outlining and defending the shut-door policy, were suppressed. Although they had been published, they were censored by the church because they contradicted later doctrines. Canright shows, for example, that a vision revealing that "all the wicked world which God had rejected" was lost forever and would not be saved was deleted when her "early writings" were later republished. Here she was arguing that there was no sympathy for the ungodly, nor must we pray for them in the hopes that they would be saved. It is only natural that a person's views would change over a period of years, but this is difficult to reconcile if one claims that each point of view is the infallible word of God.

42. D. M. Canright, *Life of Mrs. E. G. White* (Cincinnati: Standard Publishing Co.), p. 25.
43. Ibid., p. 170.

Mrs. White constantly borrowed doctrines from those about her and promulgated them as her own. Her husband came up with a scheme to have each person make a pledge to the church each year based on his net worth; this was called "systematic benevolence." Mrs. White then divulged in her *Testimonies* that "the plan of Systematic Benevolence is pleasing to God."[44] Canright maintained that Mrs. White rarely initiated a new revelation herself but that men in the movement came to her with suggestions, which she then adopted as her own. They sought to control her, especially her husband.

Can a physiological, as well as a psychological, explanation be given for Ellen White's visions, as Canright suggests? At the age of nine Ellen was dealt a blow to her face which broke her nose and rendered her unconscious, so that she lay in a stupor for three weeks. Did this cause her to suffer visions later? There are reports that she fainted frequently, had heart palpitations and pains, and was violently ill. Many of her visions apparently were preceded by a fainting spell and occurred when she was very ill. She reports that her blackouts were so severe that her friends did not think she would live, and she did not seem to be breathing. She became violently sick, then appeared almost to be dead; only then did she have a vision. This process is apparently repeated throughout her life. Most of her visions occurred in front of other people while she was praying, speaking earnestly, or when ill, and not while alone. She had been diagnosed as subject to hysteria; her experience resembles an epileptic fit. After coming out of the spell she remembered and repeated her thoughts and feelings, and recorded them. A description of the trance-state follows:

For about four or five seconds she seems to drop down like a person in a swoon, or one having lost his strength; she then seems to be instantly filled with superhuman strength, sometimes rising at once to her feet and walking around the room. There are frequent movements of the hands and arms, pointing to the right or left her head turns. All these movements are made in a most graceful manner. In whatever position the hand or arm may be placed, it is impossible for anyone to move it. Her eyes are always open, but she does not wink; her head is raised and she is looking upwards, not with a vacant stare but with a pleasant expression, only differing from the normal in that she appears to be looking intently at some distant object. She does not breathe, yet her pulse beats regularly.[45]

Mrs. White's visions were interpreted by uneducated people as being from God since they were ignorant of medicine and psychology. Can her "transcendental experience" be given a natural explanation? Was it a result of a

44. Ellen G. White, *Testimonies for the Church*, pp. 190-91.
45. Canright, *Life of Mrs. E. G. White*, p. 185.

brain disorder that was precipitated by an emotional state? An epileptic seizure usually begins with a loud cry, and after the seizure there is no memory of what occurred. Mrs. White did not begin her trance with convulsions; her physiological state is more akin to an ecstatic state, where the body is immobile and breathing is reduced markedly. Quoting a Dr. Wood, Canright says that "both ecstasy and catalepsy can co-exist in the same person and may alternate." The strong, direct character of her visions apparently decreased as she got older; she often stated that her revelations came from a voice or someone speaking to her, reporting that she was "instructed." Earlier in her life she always said "I saw."

The Adventist church accepts both the Bible and the writings and visions of Mrs. White as the "word" of God. On February 7, 1871, the General Conference of the church stated that "we reaffirm our abiding confidence in the Testimonies of Sister White to the Church, as the teachings of the Spirit of God."[46]

In 1914, church publications declared that Mrs. White expressed the "Spirit of Prophecy" and that Mrs. White was "a prophet of God."[47] Many even go so far as G. A. Irwin, for many years president of the General Conference, to claim that the "Spirit of Prophecy is the only infallible interpreter" of the Bible and that it is "Christ through this agency giving the real meaning of his own words."[48]

Mrs. White supported this interpretation of her work by saying, "In ancient times God spoke to men through the mouths of prophets and apostles. In these days he speaks to them by the Testimonies of his Spirit,"[49] and also, "It is God, and not an erring mortal that has spoken."[50] This allegedly applies to *all* of her writings! They are not her words but God's. "I am presenting to you that which the Lord has presented to me. I do not write one article in the paper expressing merely my own ideas"; rather they are "the precious rays of light shining from the throne."[51]

Was Ellen White a plagiarist?

It now seems abundantly clear that what Ellen White alleged to be the result of divine revelations was often the crudest form of plagiarism. Many of her writings can be shown to have been lifted in toto from other

---

46. A statement by the General Conference Committee, May 1906, Nov. 10, 1906.
47. *Seventh-Day Adventist Year Book,* 1914, p. 253.
48. G. A. Irwin, *The Mark of the Beast* (Washington, D.C.: RHPA, 1911), p. 1.
49. Ellen White, *Testimonies,* vol. 4, p. 148; vol. 5, p. 661.
50. *Testimonies,* vol. 3, p. 257.
51. *Testimonies,* vol. 5, pp. 63, 67.

authors. Scholars have carefully compared her writings with others and found that she borrowed on an enormous scale, without quotation marks and without footnotes, maintaining all along that they were "the word of God" given to her in a vision or trance.

A charitable observer might say that Mrs. White had read various writings, that they had entered her subconscious, and that they returned in a trance state. If this is the case, then they are not of divine origin. Ellen and her husband, James, repeatedly denied that she read any books or that she used books as sources for her "inspirations." She says, "Although I am as dependent upon the Spirit of the Lord in writing my views as I am in receiving them, yet the words I employ in describing what I have seen are my own, unless they be those spoken to me by an angel, which I always include in marks of quotation."[52] Her husband, who collaborated and edited her writings, insisted that her words were *not* contained in books by others.

A second inference that might be drawn is that Mrs. White consciously borrowed from other writers without giving credit—the most blatant form of plagiarism. This charge has been leveled by some recent defectors from the Adventist church, notably Walter T. Rea, Ronald L. Numbers, and Douglas Hackleman, and it has caused quite a stir.[53] We now have the tools that enable us to test the claims of a prophet or prophetess. After her death, it was discovered that Ellen White had an extensive library. When we compare her writings with those of others, it is clear that many of her books, magazine articles, letters, testimonies, and diaries, written over a seventy-year period, were full of plagiarisms. She frequently began a statement or passage with "I saw" or "I was shown," implying that she had had a vision. Yet what followed were generally passages taken from other writers. Even when Christ or an "angelic guide" spoke to her in a vision, their "words" came from other authors. Yet Ellen White repeatedly insisted that the passages contained in her visions were "not of human production" and that her "views were written independent of books or of the opinions of others."[54]

Douglas Hackleman points out that Ellen White was sometimes confronted by the similarities between what she claimed to be a divine utterance and the writings of others, but that she repeatedly denied that these influenced her. In a letter written on July 13, 1847, to Joseph Bates, who noted the similarities between her "shut-door thesis" and that proposed

52. Douglas Hackleman, "Ellen White's Habit," *Free Inquiry* 4, no. 4 (Fall 1984):17; quoted from Ellen White, *Selected Messages*, vol. 1.
53. See Walter T. Rea, *The White Lie* (Turlock, Calif.: M & R Publications, 1982) and Ronald L. Numbers, *Prophetess of Health: A Study of Ellen G. White* (New York: Harper and Row, 1976).
54. Ellen White, Manuscript 7, 1867.

by another, she denied that she had read a paper delineating this thesis, insisting that she "took no interest in reading," for it made her "nervous."[55] Many years later another Seventh-Day Adventist, J. N. Anderson, noted the remarkable similarity beween John Milton's epic poem *Paradise Lost,* depicting man's fall from divine grace, and her own "great controversy" vision that culminated in her book *The Great Controversy* in 1858.[56] When confronted with this similarity, she denied ever having read Milton; yet recent researchers have shown the striking parallels between the two and the fact that there were little or no differences between *Paradise Lost* and Ellen White's vision. A charitable—or sarcastic—critic might suggest that both Milton and White received the same vision from God "independently." But Milton never claimed divine revelation; it was his artistic and creative imagination that produced the great epic. Hackleman also shows that there were strong similarities between *The Great Controversy* and a similar book published by G. H. L. Hastings.[57] Hastings' book had been reviewed in the *Adventist Review* and *Sabbath Herald,* published by her husband, and so one could say that Ellen White was probably familiar with it. Incidentally, to support the thesis that Ellen White was a plagiarist, Canright reports that an Adventist woman once saw Mrs. White copying from a book in her lap and that when someone came in the room she covered the book until he or she left and then resumed her work.

Various other charges made about Ellen White: that she made large sums of money by selling her books; that she revised her works, which means that they could not be the inerrant word of God; that her writings were edited, corrected, and compiled by members of her staff, etc. All of this would be natural, except for one who claims to be a "prophetess of the Lord." Critics have also pointed out that Ellen White made a number of predictions based on her "divine visions" that did not come true. For example, during the Civil War she prophesied that the war would be a failure and that slavery would not be abolished. She also advocated vegetarianism and forbade the eating of meat. Yet it has been said that she herself was a secret meat-eater all of her life. All of this no doubt points to the fact that Ellen White was all too human, not a messenger of God, but a fallible person.

The moral of the story is that we are close enough to White's life to see her idiosyncrasies and frailities, whereas the prophets and saints of ancient times who have been deified by later generations of believers are

55. Hackleman, "Ellen White's Habit," p. 18.
56. Ellen G. White, *The Great Controversy Between Christ and His Angels, and Satan and His Angels* (1858).
57. G H. L. Hastings, *The Great Controversy Between God and Man, Its Origin, Progress, and End.*

beyond the reach of our historical binoculars. What they did has receded into the past beyond our scrutiny; that they behaved in ways not dissimilar to Ellen White is not beyond the realm of plausibility.

False Prophets

One possible inference that can be drawn by a believer in orthodox religions after reading this chapter on the lesser prophets is that they are all "false prophets" masquerading as divinely anointed ones. But then should the same inference be used to strengthen a belief in the prophets of the Old and New Testaments? On the contrary: If the credentials of modern prophets are questionable, why not apply the same standards of critical skepticism to the ancient ones? Of course, the historical record is spotty and much of it has been lost; but this is all the more reason why we should doubt the older prophets' claims. We have seen similar processes: self-proclaimed prophets or prophetesses claim to speak for God and a gullible multitude eventually accepts them. Later an authoritarian church must be built if his or her word is to survive. If it happened in the cases of Joseph Smith and Ellen White, can we not wonder if this is how Judaism, Christianity, and Islam began?

I do not believe we can say that *all* prophets, mediums, and psychics are or were conscious frauds. Many individuals may have genuinely believed that they had special powers; because of this sincerely held conviction, they were able to convince others. A psychotic may be so convinced of the reality of his or her delusional system that there is no crack in the armor. Actually, however, different psychological characteristics can be applied: (1) In some cases we have clear-cut fraud; (2) in others, partial fraud appears in times of crises or of waning faith; (3) in still others, the person may believe or half-believe in his or her divine powers, but this may be due to a serious emotional or delusional disorder. Many prophets and prophetesses who have appeared in history were schizophrenic or disturbed personalities; they truly believed that they had visions of God. Yet no one believed them, because they lacked the ability and talent to convince others of their authenticity.

There are any number of other individuals who have founded religious movements that survived their deaths. In the late nineteenth century Mary Baker Eddy founded the Christian Science church, which was based on her reading of the Bible. Her devout followers believe that her every word is divinely inspired revelation. Christian Science denies the existence of sickness and considers it to be an error of the mind. Christian Science practitioners offer prayer as an alternative to medicine; the church is based on biblical readings and the revealed word of Mary Baker Eddy.

Charles Taze Russell, a haberdasher from Allegheny, Pennsylvania, founded Jehovah's Witnesses. He maintained that the millennium began in 1873, the apostles were raised in 1878, and the end of the world would occur in 1914. Russell preached that he was the angel referred to in Ezra, and was the seventh messenger of the church. Since his death the leaders of his movement have prophesied the end of the world and still continue to attract avid followers. Armageddon will happen, they allege, as soon as the work of the Witnesses is completed, and this will be followed by the millennium period in which sinners will have a second chance at salvation. Their by-word is the statement "millions now living will never die."

In the twentieth century other prophets have come forth claiming divine revelation; the Reverend Moon, for example, implying that he had a revelation as a young man in Korea, went on to create the Unification church. The soil fertilized by the Bible is so rich that self-proclaimed prophets are ever ready to crop up. Most of these growths fall by the wayside. Some, however, are able to plant their roots deep in subsequent generations, and their influence grows and develops. Believers are willing to renounce all their critical faculties in the hope of obtaining the peace of transcendence. In extreme cases people have even followed their prophet to their death, as befell hundreds of people who obeyed the Reverend James Jones in Guyana and committed suicide. Jones had used all sorts of paranormal tricks to convince his followers of his divinity.

Beginning as a sect or a cult and rejected by the majority in its own day, the institution founded in the name of the prophet survives in the hearts of his followers. Eventually it may develop into a church, and loyalties are nourished, much as in a family or a nation. The prophet is accepted by later generations as a "messenger of God." Since large sections of the population, in the past and in the present, believe in divine prophecy, one should not be surprised by this recurrent phenomena. It will persist until the entire basis for prophecy and revelation is questioned by skeptical rationality and scientific investigation. Yet to do so freely and openly always risks hostility, enmity, and suppression at the hands of the believers, who, when they gain sufficient political control, usually condemn as heretics those who dare question the foundations of their own unquestioned faith.

Conclusion: The argument from revelation reappraised

The argument from revelation is central to many religions. Did God enter into human history at various times in the past? Did he reveal himself to men and women especially chosen by him? Did God convey messages to these individuals and were his messages accurately transmitted to us?

The claim that there is a revelatory God is an empirical claim; we can examine its factual content in order to determine the strength of the evidence and the probability of its truth.

As we examine the "logic" of the argument from revelation, let us ask how much evidence is required before we would be warranted in accepting the claim that divine revelation is true. The argument from revelation is a quasi-inductive argument; it generally takes the following form. (1) Some divine being—Jehovah, Allah, God the Father, etc.—has manifested himself to especially appointed individuals (a prophet, disciple, or saint) and communicated unique messages through them to us. God has made his presence and intentions known either directly or by means of an emissary —an angel, the Holy Ghost, or even his son Jesus. (2) The messages conveyed are in the form of truths, sayings, commandments, or prohibitions. These are prescribed as obligatory and are supposed to govern our beliefs and moral conduct. (3) The reality of God's revelations is, however, based upon some human testimony. It is claimed that some prophet has witnessed the divine presence and/or received information about God's intentions. God's revelatory manifestations are relatively rare phenomena Since divine revelations occur so infrequently, our knowledge of them is largely historical. (4) Nevertheless, on the basis of these alleged revelations, it is inferred that God exists. The argument from revelation is thus used to demonstrate that some divine being exists and that he has made his intentions known to humanity.

We have recounted various cases of revelation accepted as sacred by different religious traditions, as reported in texts associated with Moses, Jesus, Paul, Mohammed, Joseph Smith, Ellen White. Should we accept them as divinely inspired? Are the later revelations from God superior to earlier ones, or vice versa? Many of the revelations in the New Testament supplant those in the Old Testament, and are held by Christians to modify or replace them. The Koranic revelations, Muslims maintain, are supposed to supplement and supersede those of Judaism and Christianity. The Mormon revelations supplement the Old and New Testaments (though Mormons reject Islam) but are rejected by other Christian churches.

Is there any explanation for the contradictory character of the revelations? Since God revealed himself at different times in history, perhaps he had to send different messages. But which should be our ultimate guide? Perhaps we should compare the revelations and accept only those that predominate? For example, Moses, Mohammed, and Joseph Smith all practiced polygamy and maintained that it was divinely ordained. Does this mean that Christianity's defense of monogamy as the only legitimate form of marriage is wrong? Whose revelations are genuine and whose are specious? It is unsettling to have God talking in so many different tongues and in so many different ways, and to have alternative systems of priesthood following divergent revelatory traditions and providing different sets of rules and prohibitions. Which of these express the *authentic* word of God and which do not? Should we accept the Old Testament, the Koran, the Book of Mormon, and also Ellen White, Mary Baker Eddy, the Reverend Moon, and all others who have claimed to have had divine revelations? Or should we be selective and argue, as virtually all separate religious systems do, that only *some* of the revelations are true? Christians adamantly reject Mohammed as a false prophet who denied that Jesus was the son of God. Thus, the internal contradictions within the entire body of alleged revelatory experiences are difficult to reconcile.

I submit that *none* of the claims to revelation should be accepted as veridical for another reason, and that is because they egregiously violate certain elementary canons of common sense and inductive evidence as the latter are used in ordinary life and in science. Unfortunately, belief in a revelatory tradition is often considered to be immune to critical inquiry. A person's faith is usually based upon his or her ethnic background (the exception being conversion), and one usually accepts the faith of his forefathers and rejects those of other groups. Thus it is often difficult to examine the foundations of religious faith objectively. Nevertheless, a number of critical issues can and should be raised.

Several key questions come to mind. First, *did* the revelation in fact occur as reported? (1) How many people witnessed the event and can attest to its having occurred? As far as we know Moses, Mohammed, and Smith were either alone when they had their experiences or underwent them subjectively. (2) Who first recorded the report of the revelation? Has anything of the original account been lost? Was it transmitted by an oral tradition? Was it distorted by propagandists in favor of the faith? The Old and New Testaments and the Koran underwent some transformation, and whether the surviving account is accurate has been questioned by critical scholars. (3) Most important in evaluating a revelatory claim is the question: Were objective and impartial observers present, persons who are able to confirm the occurrence of the revelation? Is there any independent physical or circumstantial evidence in support of this claim? In none of the preceding instances is this latter condition satisfied, though those near Mohammed and Ellen G. White sometimes observed them to go into a trance.

What constitutes adequate corroborative eyewitness testimony? There is, of course, firsthand testimony based upon direct observation. But if a prophet is alone when he undergoes an experience, then we can question whether or not he is accurate in portraying what transpired. We can ask, was he truthful? Was he a disturbed personality? Did he suffer hallucinations? Did he believe in it himself? It is clear that reports of bizarre, purely private experiences are to be taken with great caution. People have seen everything from pink elephants to the tooth fairy, and we surely cannot accept their accounts simply because they say they saw them. Now, it is important that some secondhand testimony be available to support a revelatory claim. This independent corroboration can be of two sorts. (1) There is a weaker kind, where some observers were actually present when someone claimed to have heard voices, seen a vision, or had a visitation, even though they may not have had the same experience themselves. At the very least they can testify that the prophet in question believed that he was having an experience. In Acts, it is said that those present on the road to Damascus with Paul were left speechless and heard the voice, but they did not see anything. We have no independent confirmation that this was indeed the case, since the observers were not named, nor is their testimony recorded. Moreover, it is not claimed that they saw what Paul saw. (2) There is a stronger kind of testimony, however, where other individuals are present and actually hear and see the same phenomena. In none of the preceding cases is even minimal testimony available.

Another question that can be raised about revelation is this: *How* are we to *interpret* the alleged experience, even if we assume that the experience occurred? There are at least two possible explanations: (1) the revelatory experience was divine in origin and was a unique message being conveyed by some deity or by his messenger to the person who received it, or (2) there are alternative naturalistic causal explanations for what occurred. What might some of these natural explanations be? Perhaps the person who had the revelatory experience (if it was an entirely private and subjective phenomena) was schizophrenic, or delusional, or a psychotic. Perhaps he or she confused a hallucinatory experience with reality. We know from the abundant psychiatric literature that millions of people suffer delusional states, many of whom claim to hear voices or see visions, which they are convinced are genuine. Perhaps the prophet had an epileptic seizure. It is difficult to conduct post hoc medical diagnoses, but plausible and rational interpretations of "revelatory states" are possible.

Another explanation for some of the behavior of so-called religious prophets who claim to have extraordinary powers is that some may have skillfully used the arts of magic and conjuring to deceive their followers. Their claims to revelation were designed to arouse awe and mystery. If ancient emperors could claim divine authority for their hegemony, why not the prophets?

The claim to divine revelation has powerful political implications. A leader is better able to direct armies and do battle with enemies if those who follow believe devoutly that he is divinely appointed for the task. This provides a powerful psychological motive for a religious prophet to resort to revelation whenever it suits his fancy: this commandment does not come from me alone, he insists, but from God. What a way to strike terror into the hearts of simple folk! What an effective way to exact allegiance and sacrifice for the holy cause!

There may have been a mixture of causes and motives to explain the resort to revelation by the prophets: psychiatric disorders, the desire for political power, economic gain, or adulation. Even if at first a prophet has a vague and confused belief that he has a divinely appointed mission, if it is in time accepted by other people, this may only induce him to continue his efforts to persuade others. And having once succeeded in convincing them that he is special, he may continue his deception. Eventually his power, influence, and fame may grow to such an extent that the temptation to continue the deception that he received messages from God may be too strong to resist. The revelatory impulses may feed on themselves, and in convincing others that he has a special connection to the deity, he may come to believe or half-believe that his powers are divinely ordained. This explanation for the behavior of religious prophets seems far more applicable to the data than any fanciful notion that God came down and talked to a limited number of prophets in history.

Revelations in large measure are buried by the sands of history, and thus resist definitive empirical analysis. I have only suggested a few possible alternative explanations. But let us ask: How should we deal with a revelatory claimant who might appear today? Let us suppose that we encounter an individual who proclaims that he spoke to God and that the latter gave him special commandments we are obliged to follow. (I have occasionally encountered such individuals on street corners; perhaps you have too). What should we say about such a person? Is he mad? Does he belong in a mental institution? Is he a danger to himself or to others? Should we humor him as we would a harmless crank? If he has a group of disciples who follow him around, we might wonder if they are a threat to our security or safety. What is his motive, we ask. Is he a charlatan? Perhaps we should consider that he might be genuine. Perhaps the Lord did send him.

There are certain tests of his legtimacy that common sense would have us apply. If we have the opportunity, we might cautiously inquire: (1) Was the man or woman alone when he or she spoke to God, or were others present? If alone, we probably would laugh at the gullibility of his followers. (2) Can anyone corroborate his startling claims? Were there any other eyewitnesses, and are they his confederates? If so, perhaps they are insincere. Are any of these witnesses reliable? Are they careful in their observations, objective in their judgments, trained to detect chicanery, fraud, or delusions? (3) Can perfectly normal explanations be given for the alleged revelatory experiences? Is the "prophet" hallucinating? Was it a ruse or a joke? Were there hidden microphones and loudspeakers that would account for his hearing voices? Is he taking drugs?

Common sense bids us exercise reasonable caution. Let us develop this illustration and in so doing suppose that the "prophet" insists that he is sincere and commands you to give him all your worldly goods, to leave your family, sacrifice everything, including your life, for him and his cause. What would you do? Would you be prepared to follow his bidding? Or would you be annoyed by his audacious presumptuousness?

Surprisingly, none of the classical cases of revelation satisfies the demands of reasonable common sense. We found that the claims made to divine revelation were uncorroborated.

Moses was alone when he claimed to have encountered God. At the very least we might ask: Is the Pentateuch the accurate word of God? Has it been garbled and embellished by later generations? According to the Bible, Moses was able to convince the ancient Israelites that he was the prophet of the Lord, and he had revelations throughout his life—which included slaughtering his enemies with abandon, all in the name of God. If the biblical rendition is accurate (and we have no way of knowing that it is), Moses forbade anyone to enter the innermost sacred Tent of the Presence or to climb the Mountain of God (except Joshua), from whence he brought forth the Ten Commandments. It is apparent that the momentous encounter of Moses with Yahweh on Mt. Sinai, on which so much religious faith has been focused, was unwitnessed. Why should we accept the argument for divine revelation in this case as trustworthy?

None of the authors of the New Testament knew Jesus directly. The synoptic Gospels are based on contradictory hearsay accounts and an oral tradition later written down and elaborated upon by propagandists for a new faith. The claims of miraculous events and cures allegedly performed by Jesus have not been verified by independent or impartial observers. Similarly, in regard to Paul's experience on the road to Damascus. No one but him encountered Jesus. No one can independently corroborate the truth of his claim.

Mohammed was also alone when he first encountered the angel Gabriel. Afterward, he ran home to tell his wife. Though she accepted his account, he had difficulty at first in convincing his neighbors of his divine mission, and they mocked him. It was only after he conquered a nearby city that he was able to come back and vanquish his own people. Yet it is on Mohammed's initial and repeated ecstatic trance experiences that the entire Islamic faith is grounded.

Joseph Smith had revelations from the angel Moroni, son of Mormon, on a hill near his home and earlier in a grove across the road. But these revelations are unsupported.

There is a fourth criterion we should apply to all eyewitness reports, especially when they are appealed to in support of extraordinary claims that run contrary to normal experiences based upon past experience: extraordinary claims require extraordinarily strong evidence. Human eyewitness reports are often confused. This is particularly evident when a situation is charged with excitement and emotion. Where there is a strong will-to-believe there is a tendendcy to leap in and color the facts to suit the individuals involved. Where a powerful charismatic personality is present, his ability to influence and persuade may be greatly enhanced.

I think that we should be extremely cautious of all the claims made in the so-called sacred books. They have not been adequately corroborated. The evidence adduced in their behalf is based upon hearsay, or often the vision of revelation has no more to support it than a report that someone has had a dream. There was a tendency for the ancient mind to accept such reports at face value without exhaustive investigation. Accounts of revelations were written by dedicated proponents who believed in them and attempted to persuade others to do so as well. They were not written by neutral, objective, or dispassionate reporters, who attempted to ferret out the facts. Thus we should be extremely wary of the claims made on behalf of revelation.

XI: Does God exist?
Deity and impermanence

Sacred books are expressions of human longings to fathom the unknown; they represent an effort to postulate something that transcends human existence and is beyond this world. Perhaps Moses, Jesus, Mohammed, and the lesser prophets—however, fraudulent, self-deceived, or mistaken they were—nevertheless had some intimations of an "ultimate reality." Although Moses did not encounter God "face to face"—this is an anthropomorphic conception of deity—we may ask: Is there some being or reality over and beyond the mundane world, and did Moses have some inkling of its hidden presence? Did Christ too have an awareness of the depths of being when he spoke of the "kingdom of heaven" or of his "Father in heaven?" Did Mohammed, in his reverence for Allah, have a sense of the majesty of the eternal? Are these prophets' revelations only metaphorical or symbolic expressions, fleeting and inadequate as they are, of some ultimate noumenal character to existence? Are we ordinary mortals imprisoned in our world of experience, limited by our sense organs, which can only perceive what is impressed upon them? Are we restricted by our logical framework which compels us to hammer out conceptual distinctions and to draw rational inferences that miss the essence of ultimate reality? Similarly, for our efforts to manipulate things and to test them experimentally by using methods of prediction and control. Are noumena immune to normal perceptions, inferences, or predictions, and the normal modes of verification and validation? That is the familiar protest of religious mystics and prophets. I have already raised all sorts of objections to their postulation of indefinable divine entities and their alleged revelations, which are based upon uncorroborated testimony.

Still, if we reject the mythological and historical elements of religion as just so much human posturing, is there nonetheless something not in this world but outside of it, as its source and ground? May we call this being "divine," "noumenal," or "sacred"? Is this, in the last analysis, the insight that religious experience has captured, however inchoately, uncertainly, or indefinably it is expressed? Is this primordial religious intuition a genuine impulse insofar as it can grasp a glimmer of the

284

transcendental? Is the error only that the prophets and mystics have tried to give it human form, whereas it is inexpressible in language and unidentifiable in form?

I am introducing the age-old question. And I ask: What can we say that is new today, given the many attempts through the millennia to fathom the nature of the deity? Is he knowable? Can we validate, or prove his existence? What is the extent and reach of human knowledge? Is the theist correct in his conviction that some such supernatural being exists, however incomprehensive it may be to our ordinary categories of understanding? Should we adopt the stance of the atheist, denying any truth to the theistic claim, insisting that it is meaningless and without rational or empirical foundation? Or should we instead assume the position of the agnostic, maintaining that God's existence is still an open question and that we are not able either to affirm or deny it?

Theistic efforts to plumb the depths of existence usually begin with the "mystery of being"; then they work back to a necessary being who is the first cause, artful designer, or creator of the universe. The familiar question asked by Heidegger is this: Why should there be something rather than nothing? How can we account for existence in general? The theist responds that there is some being who is responsible for reality. Otherwise, how can there be order, regularity, and causal laws in the universe? Again, the theist postulates an ordering being, who allegedly is responsible for the order encountered.

Theologians and philosophers who are bewildered by the great mystery of being seek to settle their quandary by introducing an eternal principle that is outside the universe and yet the ultimate cause of it. The immediate objection the skeptic raises to such a possibility is whether it makes any sense to "explain" the universe by attributing it to a God outside of it. He asks: Can we define such an unknowable "God"? Whether or not God exists, it is difficult to know what we are talking about until we can specify to what *God* refers.

The classical tradition uses *God* to designate an "omnipotent," "omniscient" and "omnibenevolent" creator, immanent though transcendent, a personal being who has some relation to human persons. Do we have any ground for believing that such a being exists? The claim appears unwarranted on evidential grounds We have already dealt with various efforts to vindicate belief in such a God by appeals to custom, emotion, authority, faith, mysticism, or revelation—all of which we found illegitimate. What about philosophical and theological efforts to prove the existence of God by demonstrative arguments? Because of unclarity as to what we are talking about, perhaps we will never be able to resolve this problem or get beyond it. Perhaps the only sensible position is that of the agnostic, who maintains that the **God** question has its back broken on the meaning question, floundering within the realm of murky unintelligibility, unable to say for whom or what we are seeking except in the most vague or incoherent way.

Why disorder and chaos?

Nevertheless, I propose to reexamine the question of whether God exists. But I will do so by raising another ontological question equally puzzling to human consciousness: not, why do things exist in general, or why there is order in the universe, but why do things *change* and why is there *disorder*? We can directly observe that things undergo a constant process of transformation; in both the physical and the biological realms they come into being and pass away. Why should there be transience, impermanence, evolution, growth, degeneration, and decay? Why do not things last forever? Why at the height or the full power of an organism is it sometimes overcome and destroyed? A flower is brought to fruition in glorious bloom, but it withers and dies. A young colt is born, quickly learns to run, and grows into a handsome mature stallion, but senescence sets in and it eventually dies. A resplendent rainbow flashes for a few moments and is no more. A volcano may roar for decades or centuries but then becomes quiescent. Tens of thousands of species have had their day in the sun on the planet earth and countless generations of their type are spawned, but many species are extinguished, only to be uncovered by paleontologists and geologists probing a rock strata. A meteor shower glows through the heavens for millennia, eventually encounters a distant planet, and is burned up upon entering its atmosphere. Astronomers tell us of the birth and death of distant suns and galaxies. The universe manifests a continuous process of change. You cannot step into the same river twice, observed Heraclitus.

In the life of men and women there is a deep-felt yearning for permanence, especially when confronted by the contingent character of the human enterprise. Uncertainty and ambiguity cloud the human condition. The future not only is not fixed but also it is often without rhyme or reason; the unexpected and bizarre intrude. A child is struck down by an incurable disease; a tornado decimates a town, killing hundreds of innocent people including the priest and his entire congregation huddled in prayer. A freak automobile accident unexpectedly paralyzes an Olympic athlete. A nation is defeated in war and destroyed—as was ancient Carthage, which disappeared from the map. The best-laid plans often go astray. The universe, from that perspective, looks chaotic, unpredictable, indeterminate. That may only be from the human perspective, we are told, for there *must* be a deeper unity to all things. But that notion only expresses a human supposition, a pious hope. For the brute facts of nature point to the existence of polarity; order and disorder exist side by side.

In some human cultures change is more apparent than in others. Contemporary culture has the benefits of comparative historical inquiry. Archaeologists and anthropologists have unearthed the ruins of impressive cities of the past, which give eloquent testimony to the shattered hopes and dreams they embodied. And so we cannot have any illusions about our own finitude. Human beings vainly seek to immortalize their heroes and geniuses and even confer sainthood on some: Beethoven and Pindar, da Vinci and Pericles, Lincoln and Plato, Joan of Arc and St. Francis, Buddha and Jesus. A few are elevated to divinity, and the beauty and wisdom of their moral insights are taken as immortal and divine. Yet it should be clear to our civilization, if not to those of the past, that all civilizations come onto the stage of history, have their day in the sun, and eventually are replaced or disappear. Did Rome decline because of the advent of Christianity, the invasion of the Visigoths and Huns, or because of lead poisoning caused by the lead vessels in which Romans cooked and ate their food?

The drama of human history reveals the frailty of human existence, of individual personalities, and of the cultural artifacts they created. How rapidly time passes. It was only yesterday that we were young, dreaming and imagining a great future, and now we grow older. As we look back, we remember the people we knew and loved who are no longer alive—our relatives and friends, the people we worked and played with. The flood of time will engulf us all. Beauty and power are of short duration; the moments of both enjoyment and melancholy pass quickly from the scene. What are we left with for now but our memories of the precious days of our years, the joys and achievements, the sorrows and pain? We cannot live fixated upon our pasts, for we are always overtaken by the urgencies of the present and the opportunities and demands of the beckoning future.

Some societies are fairly stable, and they change slowly. Constant efforts are being made, especially by the favored classes, to solidify their institutions and erect their monuments in stone. Yet the houses we build, the art objects we purchase, and the monuments we help to erect will decay like the gravestones of yesteryear, and they will eventually lie strewn in ancient cemeteries, uncared for and unkempt. All human efforts will fail in the last analysis, as storms and decay, rust and erosion eventually defeat whatever barriers we may seek to raise against the tides of time. The lines on our faces can be smoothed by the best efforts of cosmeticians and facial surgeons, but new lines will form. Preventing change is not possible, either for communities or individuals. The ever-present facts we face are the precarious character of the human condition, the volatility of desires, the provisional nature of all human plans and institutions.

Some human beings seek to deny the onrush of change. The illusory religious quest is nourished by the hunger for longevity and permanence. People wish to prolong the pleasures of existence, extend the beauties of experience, capture forever the eloquence of memorable events. There is the continuing quest for eternity, a flight to the timeless, in which we imagine there will be a release from the tyranny of time and in which we can perpetrate forms that express perfect order and nobility. It is that quest that Plato, Moses, Jesus, Augustine, and Mohammed attempted to enshrine. Theologians tell us that there is a deeper order to all things, and that even though there is flux and impermanence, there is underlying harmony. But is not this view only a supposition, introduced to help calm anxiety, overcome our dread of nothingness, and sooth the pain of death? Can the hope for things unseen be sustained by the evidence?

Looking at the life-world, we see what appears to be disorder, irregularity, deviation, and variation. The religionist points to the regularities, periodicities, and causalities uncovered. We do find a strong basis for the belief that there is order and structure discernible in nature. But at the same time, a pluralistic universe emerges. For although we may find harmonies we also find disharmonies; there is regularity but also irregularity; stability but also variability. Given certain causal conditions, we can predict that specifiable consequences will most likely occur. But this is a form of hypothetical, not absolute, necessity. Granted, there are recurrent seasonal regularities, periodic continuities, pulsating rhythms, and these may be described and plotted. But there are also discontinuities, fluctuations, dislocations, shifts, turns, breaks, divergences, and deviations from expected norms. John Dewey has described ours as a "precarious universe."

One has to guard against anthropomorphic categories. Why should we read into nature our own uncertainties and indeterminacies? How can we say that order has collapsed or is limited, or that there is chaos, when whatever happens seems to follow lawlike causal patterns? Does the fact that we are baffled by unpredictable and bizarre events mean that nature is equivocal about what will happen? From the standpoint of physics and astronomy, it is difficult to read disorder into the universe, for the cosmos appears to behave in terms of immutable and universal laws.

Surely one cannot argue easily for a strictly deterministic universe today, given the developments within quantum mechanics and the significance of evolution as a key principle in interpreting the universe.

Moreover, the question is raised: Even if one admits to the evident facts of change, are there not laws governing the processes of the universe, are these not fixed, and do they not underlie the change? Is there permanence under the impermanence? Parmenides, Plato, and others thought so; and they postulated an intellectual ground: timeless forms and the *principle* of change is itself beyond change. Doesn't reason lead us directly to that truth? But, if so, how can one explain participation of the particulars in the universal, Plato asked. The Platonists had great difficulty explaining change; there was even a tendency to deny its reality. Indeed, the world of flux became only a secondary or lesser order of being, existing only because of the essences that govern change. Yet change is an empirical given. The desk that I am now writing on is already peeling and chipped, and it will need a new coat of shellac. Some day it will end up in the fireplace or junk heap, as will this paper and ink—and even myself. Though objects persist and have a material structure, they will erode in time. How can we account for the processes of change? One cannot deny the reality of change, for it appears to be omnipresent. Change is not a human invention, but a cosmic fact, applying to all forms of life.

One of the most revelatory experiences of my life occurred with my wife in Kruger National Park, a game preserve on the South African–Mozambique border. The huge park was established to permit wildlife to thrive in its natural form without being decimated by encroaching human civilization. In the park are thousands of diverse species of animals, birds, insects, reptiles, trees, grasses, and flowering plants. They live, as nearly as possible, in their natural state, largely untrammeled by human intrusion. There is some human presence, although people are not permitted to hunt, camp, or leave their vehicles except at designated campsites.

We took a microbus from a campsite into the interior in order to observe the game in its natural habitat. The countryside teems with various hues of greenery; one could see thickets of thatching grasses, red-top grasses, and the high-growing tamboekie grass. There are all kinds of trees, from the awkward-looking apple-leaf and prickly-thorn trees to the imposing baobab and the beautiful fan-palm trees. Gorgeous flowering plants of many colors are visible everywhere. Throughout the terrain, in the open fields and under shade trees, rise numerous termite mounds. The animals that live in the park are wondrous: there are tens of thousands of graceful impalas, stenbok, grysbok, dainty klipspringers and antelope, larger kudu, grazing herds of African buffalos, giraffes, zebras, and elephants, troops of monkeys and baboons, and hippos, rhinos, and wild boar. Most intriguing are the predators, such as leopards, cheetahs, and lions, followed by the spotted hyenas and black-backed jackals. There are exotic birds: purple-crested loeries, turtledoves, bee-eaters, oxpeckers, kingfishers, hornbills, and sandpipers; not to be outdone are the occasional, awesome eagles, the fearsome vultures, and fast-moving ostrich. Parking for several hours by a lake, we observed crocodiles, lizards, tortoises, and terrapins on the banks and in the water. And, of course, the insect world was abundant with wasps, termites, bees, beetles, flies, centipedes, millipedes, and scorpions of various kinds.

Two things overwhelmed me. First, there was the utter silence in the bush. The quiet was deafening, save for an occasional bird chirping. Second was the all-pervading sense of fear infusing the lushly beautiful panorama. Animals everywhere seemed to tread lightly, whether the delicate fawn or the ponderous elephant, and all were ever alert to danger. At the slightest rustle of a leaf, ears would prick up. The male, leading his harem of females, whether deer or cattle, was ever alert, ready to protect them; at the first sign of danger they would gallop off in mad haste.

The struggle for survival is starkly apparent: males of many species are ready to do battle with other males for the favors of prized females. A graphic sculpture at the entrance to the park depicts two massive antelope bucks, their horns interlocked in a struggle to the death. A band of baboons are lorded over by a dominant older male, until he is eventually challenged and defeated by a younger male. Mothers are protective of their grazing offspring, defending them and sprinting from predators who might strike at any moment. We watched from afar on a high plain, as a pride of lions gave chase and then killed a kudu, feasting on it while blood gushed from the animal. At a quiet lake, a bird would suddenly appear, swoop over the water, dive and catch a fish, which it would take to a nearby shore and swallow. I saw two handsome jackals softly stalking nearby as they picked up the scent from a recent kill. One of them gazed at me directly and then silently moved away.

In recent years, Africa had suffered a terrible drought. A large number of the animals in Kruger Park died from the lack of water and food. Visitors were appalled by the thousands of animal carcasses that were found. The animals were not watered or fed, nor were their carcasses removed; the park management tried not to interfere in the natural state of the preserve. They did intervene in two ways, however: (1) They systematically burned the veld grasses in order to maintain their luxuriant growth for grazing animals and to keep the underbrush clear (natural fires caused by lightning are a common occurrence in the bush). (2) Some herds, such as the elephant, were culled to prevent them from proliferating too rapidly and thereby destroying the bush. Elephants tend to overturn and devour vast numbers of trees in order to suck and nibble their roots.

We may ask: In what sense does this scene imply the presence of "order?" What we witnessed was continuing conflict and a struggle to survive against threatening adversities in the environment. Natural cruelties seemed to be present everywhere, at least from the standpoint of individuals and/or species who are vanquished and consumed by others. Many males were unable to mate, or infants and old adults were eaten by predators, and the entire land was decimated by a drought. Our guide told us with a mixture of horror and sorrow that the night before a sleeping tourist in one of the campsites had his entire foot bitten off by a hyena, who stealthily entered his cottage. Does sorrow and tragedy also apply to the animal, reptile, bird, or even insect world? Monogamous geese mourn the death of their mates for a long period of time. Chimpanzees, similarly, seem to suffer the loss of their companions. Curiously, the theist denies that the struggle in the natural world spells "disorder" or "evil," and seeks to read a deeper "meaning" in the ecosystem. But what about the price paid by those individuals and species who are sacrificed to this "harmony"?

Why should there be chaos, conflict, disorder, the death of individuals and/or species, the constant erosion and destruction of various forms of life? To deny that this exists is to ignore the observable facts; to seek to explain them away is to fly in the face of the obvious. But this is precisely what the religious consciousness does. Overwhelmed by suffering and the struggle for survival, the religious person seeks to overcome them by postulating some hidden, transcendent order. He finds beauty and majesty in the natural scene—and it *is* present—but he is unable to account for the ugly and loathesome aspects and proceeds to explain them away. I do not mean to overload my view of nature with pessimism. It is true that life preys upon life in order to survive. But the universe is not as Schopenhauer depicted it—completely hopeless—for life does have its virtues and pleasures. Darwinian hypotheses provide an explanation for what occurs. Chance factors intervene: the struggle for survival, adaptation to the environment, mutations, and selective reproduction. In the process of evolution, all individual members die and entire species may disappear. This demonstrates that internal disorder, conflict, rupture, and dissonance exist in nature. There are failures in nature and there are fluke occurrences and accidents. Individual lines of causal systems interact and collide. Nature shows the existence of anomalies, unexpected or unforeseen eventualities. From the standpoint of both the individual and the species, life is unstable and threatening.

It has been argued that to claim there is disorder or chaos in nature is to impose a moral category upon it, and that this masks our own human bias. But if this is the case, then the same considerations must apply to any effort to read order or harmony into nature or to infer that the total scene manifests goodness and perfection. We should be neutral and not permit any moral categories to be imposed on nature. Even if we leave out moral judgments, there seems to be some basis for the view that there is both disorder and harmony in nature. Although every individual shares with others of his species common characteristics and duplicates the morphology of the species, still individuation, variation and deviation exist. Although heterosexuality is the norm for sexual behavior, a certain percentage of some species manifest bisexual or homosexual tendencies.

The degradation of the systems of nature may also be seen from the standpoint of the second law of thermodynamics which was introduced in physics in the nineteenth century. The first law of thermodynamics states that within any isolated volume of space the quantity of energy remains constant. According to the second law, the quantity of that energy—i.e., how much work can be done with it—runs steadily downhill. The German physicist Rudolf Causius showed that heat flows spontaneously from hotter to cooler regions, and as useful energy declines, entropy increases. The usefulness of thermal energy derives from the orderly division of heat and cold in distinct regions. The movement of heat erodes order; as this process continues structures break down. The second entropy law was reformulated to show that arrangements of all kinds, not simply tempera-ture, tend to degenerate into disorder. According to this theory, given enough time, the temperature in the universe would everywhere equalize and all motion cease. The universe would become a lifeless void without activity or motion. The great exception to this are living things, which indicate a favoring of order over disorder, the stubborn persistence of life in the face of degeneration. In nature we find isolated systems in various states of equilibrium and homeostasis—the functioning of the human body, a balanced ecosystem, the rotation of the planets around a sun. In time, however, each of these systems will erode and will eventually be upset by new forces. Entropy overwhelms, as it were, the orderly systems that exist. Order and chaos thus coexist in nature.

Is there an underlying order?

The appeal to a hidden order underlying change and diversity is invoked not simply by theologians but also by some natural scientists, who likewise believe that there is a cosmic order. Even Einstein thought that it was a "miracle" and an "eternal mystery" as to why the physical world we encounter has an intelligible order. A priori, it would have been far more probable, he suggested, that the world should have been unintelligible and even chaotic than intelligible and orderly. We find in physics and astrono-my simple and beautiful mathematical and causal laws that govern the cosmos; and this puzzled many scientists. Lewis Feuer maintains that Einstein even approached the belief that a personal intelligence might underlie the cosmic order. He quotes from a letter that Einstein wrote to a friend in March of 1952:

> You find it curious that I regard the intelligibility of the world (in the measure that we are authorized to speak of such an intelligibility) as a miracle or an eternal mystery. Well *a priori* one should expect that the world would be rendered lawful only to the extent that we intervene with our ordering intelligence. It would be a kind of order like the alphabetical order of the words of a language. The kind of order, on the contrary, created, for example, by Newton's theory of gravitation, is of an altogether different character. Even if the axioms of the theory are set by men, the success of such an endeavor presupposes in the objective world, a high degree of order that we were *a priori* in no way authorized to expect. This is the "miracle" that is strengthened more and more with the development of our knowledge.
>
> Here is where the weak point is found in the positivists and dedicated atheists who regard themselves fortunate because they feel that not only have they, with entire success, deprived the world of gods, but also "despoiled" it of miracles. The curious thing is that we have to content ourselves with recognizing the "miracle" without having a legitimate way of going beyond it. . . .[1]

According to Feuer, when we use the term *miracle,* we mean that something departs from known laws and cannot be explained by them. Einstein is using the term *miracle* metaphorically, because he thinks that it applies to the very existence of law itself. Presumably, there are an infinite number of possible universes. Why this one, evincing such an elegant order?

Appealing to the authority of Einstein is no doubt powerful. He has been called upon to support Spinoza's theory of God against theism, and he has been quoted to defend atheism. The major question I would raise is the premise that the *entire* universe manifests a perfect or elegant order. Physics and astronomy are impressive sciences, which explain large portions of the physical cosmos, but they do not explain the entire universe. Nor have they yet presented a unified theory for the physical universe, let alone the universe of life.

Some writers are likewise mystified by Heidegger's question: Why should there be something rather than nothing in the universe? But is such a question meaningful when it is phrased in so general a form? The universe we encounter contains *particular* causal systems; and we can try

1. Albert Einstein, *Lettres á Maurice Solovine* (Paris: Gauthier-Villars, 1956), pp. 114–15. Quoted in Lewis Feuer, "Noumenalism and Einstein's Argument for the Existence of God," *Inquiry* (Norway) 26:251–85.

to explain the origins and genesis of individual substances or systems by references to the relevant science. To ask why there should be *something,* if taken in a universal or general sense, is purely formal and is empty of content. To ask why there should not be *nothing* in the universe is vacuous. The concept of absolute and universal nothingness is hard for us to imagine. We might conceive of empty space without any matter in it, perhaps even the idea of an absolute universal vacuum. But this implies some sort of container or place where all particles of mass or energy are absent. The idea of nothingness implies at least that some mind is observing, or that space is not totally devoid of everything. The linguistic reference only makes sense when we refer to a specific *thing* or *things* being absent or present.

The phrase "why should there be something?" makes sense if it emphasizes the *some* interpreted concretely. This form of the question merits a response. The question makes sense if we interpret it by reference to particulars. We can, at least in principle, account for and explain the origin of a *specific* entity or system. Thus we can try to explain the development of a specific glacial formation in the Northern Hemisphere or the remains of a volcanic eruption on the surface of Mars; or we can try to trace the line of descent of a particular species, say the horse, or the emergence of a social system, say in ancient Crete, or the birth and career of an individual, say Winston Churchill. Why there should be *this thing* here and now thus requires a historical and causal reconstruction in which we trace the antecedents as far as we can by specifying its efficient causes. And here we also draw upon common similarities with other things like it, its essential or formal cause governing things of its type (perhaps as a concrete universal). Thus the specific is explained by reference to general laws, hypothetically stated, applicable to all things of its kind. But to obtain an explanation of the *totality of all things* is not easy, unless we specify the content.

Similarly, the question that I have raised—"Why should there be impermanence in general?"—is beyond the range of easy resolution. And my response is that processes of change are discovered; they are the data given in observation. We attempt to account for the observed changes in specific systems. To seek to explain the totality of change in all systems is an ambitious but perhaps misdirected effort. The parts of the universe, things changing and manifesting order and disorder, may very well be eternal in the sense that they always existed; there may never have been a time when they did not. In any case, are we entitled to introduce a being outside of the universe to explain being, order, and impermanence within it? That everything within the universe has an origin does not mean that the totality has an origin, such that there was nothing before it. The question usually raised is not "Why should there be something?" but

"Why should there be *everything?*" In other words, there is an effort to explain the totality of all things in the universe.

The cosmological argument

Historically, in the cosmological argument, the existence of God is postulated as a ground for the entire cosmic scene. The question I wish to raise now is whether the total universe has an origin. Does it need a "ground?" There is a tendency to overgeneralize by assuming that there is one universe, when there may very well be many universes. Much of this tendency is due to the assumption that everything is interconnected with everything else. This implies a deterministic model in which the same causal laws are held to apply everywhere. This uniformity-of-nature principle may be viewed as a methodological rule governing scientific inquiry, not as a universal law of nature; as a useful pragmatic presupposition, not as a tested generalization about the entire universe. We do not know with any certainty that the total concatenation of events are interconnected so that whatever happens in one part reverberates and/or influences whatever happens in all other parts. On the contrary, the universe appears to be an open system in which there are various universes, including individual systems and subsystems. These systems have some gravitational and other physical relationships to all other systems. But they are also relatively isolated by distance and time, and the evolutionary processes that transpired in one may not have been repeated in another.

For example, that life has evolved on the planet earth in manifold forms is obvious. Organic life, as far as we can tell, has not evolved in the same way on Venus, and it is unlikely to have evolved on Mercury due to the proximity of the sun and extreme temperatures. Thus not all possible causal systems come into operation in all parts of the universe. It is possible, perhaps even likely, that life exists on other planets in our galaxy and/or in other distant planetary systems in other galaxies. But what forms life may have assumed elsewhere we do not know, though if what has happened on the earth is repeated elsewhere there is an extremely wide range of possible variations depending upon the unique conditions existing in another planetary system. That other intelligent civilizations have developed is, of course, a possibility that has excited the imaginations of both science-fiction writers and serious scientists. The inference I draw is simply that we do not know with any certainty that the various systems and subsystems discovered are interrelated in a grand causal network, nor do we know whether everything has a uniform structure.

Theologians, metaphysicians, and some scientists have believed that it does. Standing on earth and viewing the heavenly bodies on a clear night gives the impression that the starry vault is one and we are part of it. The geocentric theory was dislodged by modern science, which undermined both the Ptolemaic astrological and astronomical systems. Modern-day astronomy, applying high-powered visual and radio telescopes and using space vehicles, have extended the range of the universe enormously, far beyond what the naked eye can see. We are only a small minor planet in an expanding universe. It is obvious that what happens in outer space, however, influences our past, present, and future on this planet. Solar energy, solar and lunar tides, gravitational and cosmic radiations from far distant stars have had a continuing effect on earth and the life-forms that have evolved. Not only are there continuing influences but there have also been catastrophic events. Of late, it is fashionable to speculate that the dinosaurs became extinct because a meteor shower killed most of earth's vegetation. It may have even been the case that organic matter did not emerge from the primordial soup on our planet when certain elements were combined some two or three billion years ago, but rather that organic molecules were transported on meteors from distant galaxies and once here began to evolve into their present forms. Fred Hoyle and Francis Crick have suggested that possible scenario. We do not know at present which account of the origin of life is accurate or whether both are true.

There are large gaps in our knowledge; but there may also be large gaps in the universes around us, such that there may be separate lines of causality or distinct series of processes unrelated to each other except minimally. The conception I am presenting is similar to William James' pluralistic universe. It is populated by stars and galaxies, suns and planets, quasars and quarks, electrons and charms, flowers and birds, mountain ranges and glaciers, chimpanzees and human beings, architecture and poetry, social systems and civilizations—in short, a universe of variety and complexity, containing diverse individuals and systems coexisting at the same time. It is a universe which is difficult to reduce to single set of formulas or to unify into a general theory, in which everything is explained by reference to a few ultimate principles. Indeed, there may be no ultimate principles that govern all things in the same way. Pluralistic diversity may be characteristic of nature, which resists any easy reduction to universal sameness or structural order; similarities and dissimilarities, order and disorder, reversible and irreversible processes, dynamic and thermodynamic systems, universality and individuality, periodicity and concreteness may all be present. In this universe there is genuine novelty, uniqueness, inventiveness, emergence, and qualitative difference.

There does appear to be a tendency for the universe to take on habits, as Peirce suggested. Nature displays general regularities, but these may not be universal or complete. Chance, contingency, and multiformity (rather than uniformity) may also be pervasive and irreducible traits. The universe indeed appears to manifest dissonance and discordance as well as constancy and symmetry, which suggests that it is an open system, not one closed, fixed, or pervading unified system. This at least describes what we encounter and know at the present stage of human knowledge. This is deceptive, we are told by theologians and philosophers, for there is a hidden order, in terms of which the pluralistic diversity can be explained. Elegant and uniform causal laws are said to underlie the heteromorphism and heterogeneity. Surely the history of science demonstrates the continuing discovery of deeper, formerly unknown causal relationships. Democritean materialistic explanations locate the essential structure of being on the micro level; here we discover molecular, atomic, subatomic, and genetic structures underlying the changes that take place on the macro level. Others find order on still other levels of analysis.

There is in all of this a tendency to leap beyond unified explanations in one or more sciences to a uniform explanation from all of them. There is the conviction that there is an ultimate unifying principle for all of existence. A number of assumptions are made: (1) that the universe has a common unity and is one; (2) that the entire cosmos displays a basic order and structure; (3) that the underlying structure involves ordering principles, which are universal or near universal in character; and (4) that this can be transformed into a religious article of faith. This ordering principle is divine; it is both the ground of being and the object of piety.

Aristotle, in his formulation of the cosmological argument, postulated a first cause—unmoved mover and necessary being—in order to explain the existence of causality, motion, and contingency. This being was itself uncaused and noncontingent. As a first principle it coexisted eternally with matter and form. It did not create the universe of matter ex nihilo or give matter a form. Moreover, Aristotelian physics required fifty-five unmoved movers, a plurality of cosmological principles. But these did not imply a personal, theistic god. It was left to St. Thomas Aquinas to modify the argument so that God was not only the cosmological ground but also the creator of the universe, the omnipotent, omniscient person of the Judeo-Christian tradition.

The cosmological argument has a weaker and a stronger form. In the weaker form it was introduced to explain causality, motion, contingency. It sought to do so by (1) denying the intelligibility of an infinite regress and (2) by postulating a principle, or being itself, immune to causality, change, motion, or contingency. Kant, Hume, and other skeptical critics have shown that this notion of a first cause makes little sense when extended and applied to the universe in general. For the idea of cause implies constant conjunction, a hypothetical relationship: whenever *a* is discovered, *b* occurs. Our knowledge of causal relations implies that the observer has been able to observe that whenever certain antecedent conditions are present certain effects result. The cosmological argument, however, goes outside the universe and postulates a totally unobservable infinite and transcendent being. The concept of cause is here being misused; it has only a metaphorical or analogical signification. "God" is a purely speculative idea lacking any identifiable empirical content. Kant's famous dictum that "concepts without percepts are empty" (and conversely "percepts without concepts are blind") is relevant here; for both the meaning of the term *God* is unclear and the empiricist criterion is violated. This does not deny that scientists often introduce hypothetical constructs (for example, subatomic particles or genes) before they are empirically observed. The scientist notes an experimental consequence that enables him to infer the presence of an unobserved entity. Particle traces were eventually observed, for example, by means of an electron microscope or traces on a photographic plate taken in a cloud chamber. Not knowing the cause of a disease, researchers postulate a virus, with the expectation that it may eventually be identified. God, as a transcendent being, however, is not so readily identifiable or discernible—other than by means of scriptural revelation.

Aquinas rejected the concept of an infinite regress as being "unintelligible," because he could not understand how motion or cause could be possible without it. He maintained that unless there were an intermediate cause there would be no end cause; but there would be no intermediate cause unless there were a first cause, which is itself uncaused. But the idea of an infinite series of causes is not any less intelligible than postulating an uncaused cause whose essence is infinite and unknowable. Remaining within the universe seems more intelligible than going outside of it. The theistic explanation is deceptive; it provides little help. The most pointed objection raised by the skeptical naturalist is to ask, "What is the cause of the first cause?" The theist retorts that this question is illegitimate, being unknowable. There is no going behind or beyond God, he says. But if this question is illegitimate, so is the quest for the transcendent ground of the universe. We do not advance our understanding by invoking a divine force, an uncaused, unchanging, unknowable entity to account for what is observed. The universe involves matter and energy, space and time, order and stability, change and process; it is not necessary to go outside or around it in order to "explain" it. One might simply say that we do not at present know the origin of the entire universe, and there is not sufficient reason to believe that it is divine. The theist replies that atheism is untenable, because even if the theist is unable to prove that God exists, the atheist likewise has not been able to prove that he does not exist. Here the argument *ad ignorantium* is drawn upon. But is one entitled to believe whatever he wants where there is not a proof? As I have already shown in Chapter V, the burden of proof is upon the person who makes a claim; he has to justify his hypothesis. It does not rest on one who denies it to prove the negative. I do not have to disprove that there are no mermaids or flub jubs; the claimant who maintains that there are must first make the case for them.

There is a stronger form that the cosmological argument has taken, especially by the theist. This is the claim that not only is God the first cause—conceived as the sustainer of the universe—but is also its creator. As we have noted, for Aristotle the unmoved movers were physical principles introduced within the cosmos, to lend order and completion to the entire scheme of things. However, for Aristotle, both matter and form were eternal and coexisted with the divine agency. Aquinas' view that God created the universe ex nihilo is a leap required by his acceptance of the Pentateuch story of creation, which dominated his cultural milieu; it was an article of faith, not of proof.

Aristotle's unmoved movers as teleological causes are not doing anything beyond thinking of themselves, thinking of thinking; they have no concern for humankind. Aquinas' God became a personal God of moral commandments, justice, love, and compassion. He is interested in human destiny. Thus the universe is not eternal, only God is; and he has created it out of nothing and governs it in terms of his plan. For Aristotle, since the universe of matter and form is eternal along with God, God is not a creative principle.

The skeptic asks: How do we know that God created the universe? The classical theologian offers no resolution of this problem other than to appeal to his a priori commitment based upon faith. Taking the universe as we find it, however, we might argue for polytheistic cosmological principles, which better explain the variety of systems found. Even if God created the universe, he may now be dead, after his initial creative act. There is no evidence that he persists today and is now "sustaining" the universe. Thus we may say that the cosmological argument does not prove the existence of God to the unbeliever; nor can it develop any conviction about his reality where none is present. The most that can be said in its behalf is that if one already has faith in the existence of God, the cosmological argument may help reinforce that belief-state; though by itself it does not prove what it sets out to prove. At the very best, it provides the believer with some plausibility for his conviction.

Though invalid, the cosmological argument nevertheless continues to persist through history, assuming different forms. Some nontheists have even been persuaded by it. Why should there be a universe, they ask? Why should there be order? Even the eighteenth-century deists, who rejected biblical revelation or any form of a personal God, appealed to the cosmological argument. God created the universe, they said, but after doing so he does not intercede or perform miracles. The universe is like a machine or watch; it keeps ticking away in perfect harmony and obeys strictly mechanistic laws. The deterministic assumption of Newtonian science has been seriously questioned in the twentieth century by quantum physics. Heisenberg's principle suggests that some events are chancelike and that they may occur without clearly definable causes. Thus, the exact moment of the radioactive decay of the nucleus of an atom is unpredictable, even if we have knowledge of the nuclear state. The uncertainty principle states that one cannot predict the position and the velocity of a particle simultaneously. This has been extended by some astronomers from quantum mechanics, on the micro level, to the cosmic level. Whether this principle refers only to our present inability to isolate and determine effects or is inherent in nature has been hotly debated. In contemporary physics the absolute space and time of Newton gave way to relativity concepts.

The inflationary-universe scenario

Nevertheless, some modern-day physicists and astronomers have given a new lease on life to the cosmological argument. This is related to the nebular hypotheses (first suggested by Kant). Noting the Doppler effect—that there is a shift toward the red in the spectroscopic analysis of light—they infer that radiation and matter are traveling at vast speeds away from a center and that they are diffused homogeneously throughout the cosmos. This is taken as evidence for the big-bang hypothesis, namely that there was a cataclysmic explosion some eighteen to twenty billion years ago and that our universe is a result of that cosmic event. At first, this would seem to lend some credence to the first-cause argument. According to this theory, the universe was initially concentrated into a point of infinite density, a fairly small fireball at a temperature of approximately one trillion degrees centigrade. At some point this ball exploded; and the universe has been expanding ever since. This implies that the universe is spatially finite.

Some physicists have interpreted this initial big bang as creation ex nihilo, for matter and energy and the physical laws, even space and time as we know them, came into existence at that time. St. Augustine's comment that "the world was created with time and not in time," according to

Paul Davies, a theoretical physicist, was a "remarkably prophetic pronouncement." Davies has argued that contemporary physics provides some evidence for the classical view.[2] According to the general theory of relativity, singularity was present: the big bang began from a condition of infinite compression, at which space-time was infinitely warped. Under such conditions, concepts of before or after and causality are difficult to fathom and may not be relevant, for time and causality apply to the physical universe as we know it. Davies further argues that the entire physical universe, including space-time could have come into existence "spontaneously" and "uncaused" as a sort of "quantum fluctuation." He elucidates the idea of a "self-creating universe," using the theory of an "inflationary scenario" whereby many elements came into being in the initial phase following the explosion. Nevertheless, theists have sought to use the big-bang model to lend credence to their belief in a divine being, who created the fireball and exploded it.

Some physicists have suggested another speculative scenario, that of an oscillating universe, in which at some point the expanding universe will collapse back on itself and eventually may even begin the process again, much as an accordion or yo-yo. This implosion may be called the "big-crunch" theory; it depends on whether or not the universe contains enough mass for gravitational pull to bring it back eventually to its original point. If not, then the second law of thermodynamics (entropy) might prevail, and the universe of motion and time will degenerate into a dead, lifeless universe. This is the "big-freeze" scenario.

The British physicist Stephen Hawking[3] maintains that there are three possible scenarios, which he attributes to the Russian physicist Alexander Friedmann: (1) The big-crunch model: at the present we do not know whether or not there is a sufficient quantity of matter for gravitational pull to exert itself. (2) The galaxies are moving so rapidly that gravitation can never stop them and the universe continues to expand. (3) The galaxies are moving apart at a critical rate sufficient to avoid recollapse.

In one sense we have reached the limits of our knowledge of the physical universe: for the laws of physical matter, space, and time apply to what happened after the initial fireball explosion, and it is difficult to say what existed or occurred before the big bang or will occur after any future possible big crunch, even whether space-time and the same general

2. Paul Davies, *God and the New Physics* (London: J. M. Dent and Sons, 1983); "God and the New Physics," *The New Scientist,* June 23, 1983; "The Inflationary Universe," *The Sciences* 23, no. 2 (March/April, 1983).

3. Stephen Hawking, "The Edge of Spacetime," in *Universe,* ed. W. J. Kaufman and W. H. Freeman (1985). Also *New Scientist,* Aug. 16, 1984.

laws of nature have or will apply. Here, one may have entered the realm of pure speculation—in terms of present-day science—which it is difficult to predict or confirm experimentally. This does not deny that experimental theories may be introduced within physics and astronomy, and one cannot preclude the possibility that a priori future developments in science will resolve this impasse.

Interestingly, Davies has suggested that the universe may be uncreated and might be unexplained without God. The miniuniverse might have exploded and inflated in an instant until it reached its present size. The tremendous energy at that moment powered the cosmos and provided all of the radiation and matter in existence today. Surprisingly, our telescopes, if powerful enough, provide a picture of the expanding universe and of its temporal history; and mathematical calculations enable us to make these inferences. All of this is occurring in terms of the known laws of physics, which are held to be sufficient to account for what happened since the explosion. Our physical universe thus can be understood in terms of the laws of science and mathematics. Whether there are any boundary conditions to the universe of science is an open question. If there are not, then we need not go outside of the universe. If there are, then we can't understand it yet; perhaps we never will, some say, in terms of the present categories of science.

Some physicists, however, go beyond the cosmological argument by other means. Davies, for example, introduces a variation of the design argument. He believes that we can infer the existence of God, but this follows from his awe of the existence of the laws of nature. Like Einstein, he holds that these laws have such symmetry and harmony that they would seem to point to some divine intelligence. "The laws of physics dovetail together with such exquisite consistency and coherence that the impression of design is overwhelming."[4] God, for Davies, is a mathematician, and theoretical physics is reduced to logic and mathematics. There is no need for God as a creator in the traditional sense; God is, as it were, "above nature sustaining all of existence." God here is timeless and the mathematical order subsists in some Platonic sense. Did the laws of science predate the big bang (or final crunch) in some timeless realm?

Mario Bunge believes that those who speculate in this fashion have abandoned the physical sciences and entered the realm of theology. According to Bunge, Davies is engaged in religion, not science. Like Jeans and Eddington a generation ago, he has set himself the task "to debunk materialism and replace it with a spiritualistic" philosophy of nature. The same point is reiterated by physicist Mendel Sachs, who maintains that

4. Davies, "God and the New Physics," *op cit.*

faith has intervened and that what we have is a metaphysical and speculative rather than a scientific interpretation of nature.[5]

Hawking believes that it may be possible to resolve these cosmological problems within science. According to him, it makes some sense to
talk about singularities in the universe; this refers to places where the
curvature of space-time ceases to have any meaning—the big bang, the
big crunch, even black holes (where massive, burnt-out stars collapse and
allow little radiation to escape). This means also that we might be able to
ascertain what happened before the big bang; and the laws of science
would enable us to determine the initial state of the universe at the
singularity. There would be no boundary conditions to our knowledge of
the universe. Perhaps the singularity may be "smeared out," and space and
time together may form a closed, four-dimensional surface without an
edge or boundary. If this were the case, the universe would be completely
"self-contained" and not require any boundary conditions. If such were
the case, the laws of physics would not break down and there would not
be a prior state in which the laws of science do not apply. Moreover, we
would not need some supernatural being or force to create and/or sustain
the universe.

Some theists still raise the question: How are we to account for the
symmetry and elegance of the laws of science? They reply by reducing the
cosmological argument to the teleological argument: namely, they maintain that the universe points to the presence of some ordering and pur-
poseful intelligence. Does God exist in some form as a necessary timeless
being responsible for the design of the intellectual and causal laws?

The argument from design

The design argument is a very popular effort to demonstrate the reality of
God. Even the Bible presents a version of it: "The heavens declare the
glory of God; and the firmament sheweth his handywork" (Ps. 19:1—
KJV). The most influential philosophic form of the argument goes back
to Aristotle and Aquinas, though it became popular among the deists.
Thus there are two major versions: the first is related to Greek teleological
science adopted by Christian theologians, which has been abandoned
with the modern scientific rejection of final causes; the second is the
design argument, which has been defended even by mechanists, and is
independent of teleological explanations.

5. Mario Bunge and Mendel Sachs, "God and the Physicists," a review of Davies' *God
and the New Physics, Free Inquiry* 4, no. 2 (Spring 1984).

It was Aristotle who was largely responsible for formulating the classical teleological argument. In his *Physics* and *Metaphysics* he introduced four kinds of causes used in explanations: (1) the material cause: when we ask what a thing is, its underlying stuff; (2) the essential cause: why it is what it is, its essence or form; (3) the efficient cause: how a thing came to be what it was, the processes of change and motion it underwent; and (4) the final cause: when we seek to understand its purpose, end, or goal.[6]

The four causes were central to Aristotle's explanations of processes found in nature. His science was biologically oriented, and it was based on the use of functional and teleological explanations. Aristotle drew an analogy between the things that happen "by art" in human creation and those that happen "by nature"; and he maintains that much the same as there are means-end relationships in the production of human artifacts, so there are means-end relationships discernable in natural processes.[7]

For example, a human architect constructs a public sports arena; each part—seats, open stage, vestibules, and doors—have a function in the total arena. He has created the arena to satisfy man's ends and the uses to which we put it. Similarly, Aristotle finds that functional adaptation applies to nature: the purpose of the eye is to see and of the ear to hear; and these organs serve the broader purpose of survival and reproduction and the higher functions by members of a species. Here potentialities are actualized in processes of growth and development as ends are realized; this applies to all the normal members of a species and helps to define its fixed essence or form. Why do we have teeth? asked Aristotle,

6. "*Cause* means: (1) That from which as immanent material, a thing comes into being, e.g., the bronze is the cause of the statue and the silver of the saucer, and so are the classes which include these. (2) The form or pattern, i.e., the definition of the essence, and the classes which include this (e.g., the ratio 2:1 and number in general are causes of the octave), and the parts included in the definition. (3) That from which the change or the resting from change first begins, e.g., the adviser is a cause of the action, and the father a cause of the child, and in general the maker a cause of the thing made and the change-producing of the changing. (4) The end, i.e., that for the sake of which a thing is, e.g., health is the cause of walking. For 'Why does one walk?' we say: 'that one may be healthy'; and in speaking thus we think we have given the cause. The same is true of all the means that intervene before the end. . . ." (*Metaphysics,* Book 5, chap. 2).

7. "Of things that exist, some exist by nature, some from other causes. 'By nature' the animals and their parts exist, and the plants and the simple bodies (earth, fire, air, water)—for we say that these and the like exist 'by nature.'" All the things mentioned present a feature in which they differ from things which are *not* constituted by nature. Each of them has *within itself* a principle of motion and stationariness (in respect of place, or of growth and decrease, or by way of alteration). On the other hand, a bed and a coat and anything else of that sort, *qua* receiving these designations—i.e., insofar as they are products of art—have no innate impulse to change. But insofar as they happen to be composed of stone **or** of earth or of a mixture of the two, they *do* have such an impulse, and just to that extent—which seems to indicate that *nature is a source or cause of being moved and of being at rest in that to which it belongs primarily,* in virtue of itself and not in virtue of a concomitant attribute" (*Physics,* Book 2, no. 1).

and he replies, to help us to chew our food. Aristotle, however, extended this analogically to natural processes. He sought to account for why many or most things in nature happened "invariably," "always" or "for the most part." These processes and substances, he thought, had natural ends: a river seeks its delta, a stone when thrown up will come down, the rains fall so that corn may grow, so that we may eat it. But attributing these processes to teleological purposes appears to be fallacious to modern physical scientists; for what seemed to operate were material and efficient causes, given the formal or lawlike structures, without the need for reading in final causes in nature. In other words, vitalistic entelechies were of no help in physics; and Galilean-Newtonian science overthrew any vestiges of natural teleology as it applied to the physical world. It was not until Darwin, however, that teleological explanations were also abandoned in biology and the life sciences. For in the first place species were not fixed but evolved over long periods; moreover, great numbers were now extinct. Instead of finding purpose in the biological world, there were instead other explanations: chance mutations and differential reproduction, adaptation, and the survival of the fittest. There seemed to be no discernible teleological order in the biological realm.

It is interesting that Aristotle introduced teleological explanations within the context of his scientific and metaphysical perspective. Nevertheless, Aristotle did not use the teleological argument to maintain that a divine mind had designed the world. Although the world manifested order, it did not have design and was not fulfilling a preconceived plan. Nor, as we have seen, did Aristotle claim that God had created the universe.

Aquinas again used Aristotle in his version of the teleological argument. In his fifth effort to prove this existence of God, he argued:

> The fifth way is taken from the governance of the world. We see that things which lack intelligence, such as natural bodies, act for an end, and this is evident from their acting always, or nearly always, in the same way, so as to obtain the best result. Hence it is plain that not fortuitously, but designedly, do they achieve their end. Now whatever lacks intelligence cannot move towards an end, unless it be directed by some being endowed with knowledge and intelligence; as the arrow is shot to its mark by the archer. Therefore, some intelligent being exists by whom all natural things are directed to their end; and this being we call God.[8]

If, however, we dispense with teleological explanations in the physical and biological sciences (they are still useful in part as intentional and motive explanations, helping us to interpret deliberate human conduct), then there is little basis for this argument. Things in nature do not move

8. Thomas Aquinas, *Summa Theologica,* Part 1.

to fulfill ends. This is a kind of animistic invoking of purpose: the rock or lake has no end per se. Therefore, there is no basis for injecting an intelligent designer to account for it.

It was in the post-Newtonian period that deists abandoned teleology, yet retained the design argument. It was also based on the analogy between the artifacts of man and the ordered systems found within nature. In his *Dialogues Concerning Natural Religion,* Hume has Cleanthes state and defend the design argument:

> I shall briefly explain how I conceive this matter. Look round the world: contemplate the whole and every part of it: you will find it to be nothing but one great machine, subdivided into an infinite number of lesser machines, which again admit of subdivisions, to a degree beyond what human senses and faculties can trace and explain. All these various machines, and even their most minute parts, are adjusted to each other with an accuracy, which ravishes into admiration all men, who have ever contemplated them. The curious adapting of means to ends, throughout all nature, resembles exactly, though it much exceeds, the productions of human contrivance; of human design, thought, wisdom, and intelligence. Since therefore the effects resemble each other, we are led to infer, by all the rules of analogy, that the causes also resemble; and that the Author of Nature is somewhat similar to the mind of men; though possessed of much larger faculties proportioned to the grandeur of the work, which he has executed. By this argument *a posteriori,* and by this argument alone, do we prove at once the existence of a Deity, and his similarity to human mind and intelligence.[9]

Hume then has Philo meticulously examine this argument and submit it to a devastating critique, which we may summarize as follows. It is difficult to prove anything from analogy. Analogies do not demonstrate the truth of a hypothesis; they can only suggest its possibility, not its proof. Analogical reasoning is often easy to refute; one can stretch the analogy so that it boomerangs. Actually many men have manufactured the objects and products that we use. Are there not then many gods who produced the things found in nature? Human designs are often imperfect and inadequate; many houses may be bungled and botched. Were many universes similarly bungled before this imperfect one was created? Why simply magnify human intelligence and attribute it to the divine being? What about other human qualities: pride, jealousy, vindictiveness? If man possesses these properties, does not the divine being? Is this the explanation for evil in the world? May we not infer that God is a malevolent being?

There is a tendency for the design argument to reduce everything to an anthropomorphic view of the universe; it is an effort of man to create

9. David Hume, *Dialogues Concerning Natural Religion.*

God in his own image and to read into nature his own traits. Moses surely invoked divine retribution to punish people who disobeyed Moses' commandments: God's qualities reflected human tendencies, though on a greater scale that approached omnipotence. Similarly, teleological and design theologians and philosophers invoked intelligence in the universe at large, extending it beyond human reason so that it becomes omniscient. Both efforts reinforced human qualities in idealized form. But one cannot reason selectively; we need to take the analogy all the way—even at the risk of its boomeranging. Man is not only a creator but a destroyer at times, expressing spite and employing terror to vanquish his enemies. Is the divine being destructive, and can this explain the big bang, the big crunch, or the big freeze on the cosmic scale? No, the theist retorts, this would be ridiculous. Surely it is, but so is the effort to reduce the entire universe to the qualities of one species on one planet. Why not an analogy with other species or forms of life? Why is the universe not like a vegetable? From a small seed, enormous expanding and inflationary processes develop and the organism grows larger, until senescence, decline, and decay sets in. The universe began with a loud cry; will it end with barely a whimper? All of this is on the level of poetic metaphor; it is not an established truth claim. Instead of reading human intelligence into nature, why not infer that living processes manifest degrees of order as reflections of nature, and that human intelligence is similarly a reflection of natural order, without turning the universe into the image of man? What conceit and pride is at work to endow the creator with the approved-of qualities of human personality!

There is still another defect in both the cosmological and design arguments: that is in attempting to reason back from what is observed in nature to the existence of a divine being. Regretfully, the theist selects only some empirical facts and chooses to ignore others. He observes order, regularity, and causal law in nature and so he infers that someone has designed the system of nature. But order and design are not equivalent. It is possible to recognize degrees of order in nature and yet not have an orderer, causal laws and not a lawgiver, regularities and not a regulator. If my analysis in the earlier part of this chapter is accurate and if disorder exists, then why not attribute the cause to a divine disorderer, a destructive or demonic being? Classical theologians have indeed done so insofar as they attribute disorder to Satan or some other demonic force, and insofar as they maintain that there is a cosmic struggle between the forces of good and the forces of evil, the children of darkness and the children of light. This, of course, raises the classical problem of evil. Is evil real in the universe or is it only apparent? And if it exists, is God ultimately responsible for it or Satan? If God is all-powerful, why does he permit Satan to exist?

Actually, the theist is faced with a double paradox, though there are two ruses he employs to deny them: (1) the paradox of disorder, and (2) the paradox of evil.

(1) The theist may deny the existence of disorder, chaos, chance, and indeterminancy and insist upon a perfectly ordered and harmonious universe. But this ignores the abundant evidence of the struggle for survival in the life-world, of conflict in the human world, and of the interaction of systems in the physical world—the origin and death of stars, of meteor showers, volcanoes, etc.

(2) Turning to the problem of evil, a rationalization process sets in. The existence of evil is only apparent, we are assured. But what of the destructive forces of tornadoes or rampaging floods, or the extinction of entire species. We have all witnessed pain, sorrow, anguish, suffering, perfidy, duplicity, cruelty, and eventual death. Indeed, all individuals and species have to struggle to survive, and like men and women, they enjoy and suffer the vicissitudes of life until they eventually die. Ah, responds the theist, there must be a *deeper* order. But this is only an article of faith, not of proof; to selectively emphasize order and ignore disorder is to beg the question. He may claim that there are laws of nature, but he must also admit the existence of concrete particulars and allow that systems interact and there is apparent disorder and chaos on the level of observation.

If the theist will grant that there is some disorder, then we may inquire whether his God is not limited, impotent, finite, unable to organize the huge universe and to arrange all the furniture of Heaven and Hell in some kind of perfect symmetry. Is the universe expanding so fast and is it so complex that it is difficult for God to be everywhere at the same time? Does this explain the creaks, groans, and breakdowns in the cosmic scheme? God is mighty and powerful, but not all-powerful. Perhaps he needs all the help he can get in running the conglomerate; perhaps he needs our help in our small corner of the universe to help get rid of some of the confusion. If so, God is not omnipotent and perhaps not even omniscient. The uncertainty principle may even govern his actions, and he, like us, may be unable to predict both the position and velocity of bodies, save in general probabilistic terms. If we are at times uncertain, why not God? Thus, the theist is faced with the insuperable problem of reconciling a pluralistic universe with a monotheistic being. Is not God responsible for impermanence and change? Why does he not put things in perfect harmony? Why did he have to create the universe in the first place, if he is going to allow it to expand and run down like an old steam engine or a rusty watch? Why did he create the marvelous mess? Perhaps the universe is not taking on habit or order, but disorder and disunity, and in an ultimate breakdown all of its parts, including us, will eventually be destroyed. Evil, says the theist, in a new effort to explain it, is necessary if we are to understand or appreciate goodness; and it occurs only when there is a defect or lack in a natural process (nonbeing); it disappears when there is full realization or fulfillment of being. This neo-Thomistic argument applies only to a fixed and finite universe and is not appropriate to the Darwinian or astronomical models we have discussed.

In any case, if God exists, then is he not responsible for the universe as we find it? Perhaps he is not a malevolent being but is impotent, and though opposed to evil he is unable to prevent it. Hume remarked that if we could create or govern the universe we perhaps might do a better job. We would attempt to mitigate much of the unnecessary sorrow and pain, for example, by getting rid of drought in the deserts or floods in the rainy areas or by diminishing the destructive fury of gales and hurricanes. Indeed, human history demonstrates continuing efforts on our part to conquer, by means of intelligence, disease, accidents, and other forces that naturally denude or destroy us.

For the theist to attempt to resolve the problem of evil by denying its existence is a copout, given the great amount of misery and sorrow discernible, particularly in the life-world. The theist believes that we are reading our own narrow moral interpretations into nature and that from the total scheme of things, all will work out well in the end and is good. Voltaire sets up for ridicule the naïve Candide, who insists that this is "the best of all possible worlds." But what is the meaning of such a cosmic good? The Buddhist also believes that there is a kind of perfection in the total order of things, but he looks beyond conflict to a state of nirvana and nonbeing. The pious hope—that being, when ultimately fulfilled, will be perfect—only begs the question. And we may ask, since all things in time disappear, why is not all disorder and decay, nonbeing and nothingness part of the ultimate ground? Although a thing may come into being for a period of time—an atom, planet, meteor or sun, a lilac, elm tree, dolphin, or human being—eventually they will all be reduced to ashes and dust, and the material particles which compose them will be rearranged, transformed, or taken up elsewhere. Perhaps there is a deeper chaos within the flux of things? The theist has simply allowed his wishful thinking to intervene: all will work out in the end. In order to secure the human condition and to provide some ultimate mooring for human aspirations, he postulates the existence of eternal souls, which are able to escape the temporal flux.

Still another ploy of the theist is to admit the existence of *some* evil—that which affects the affairs of men—but to deny the existence of evil in nature. Natural evil cannot have any reality, given the theists' fixation on a beneficient God; only human evil exists, and this he attributes not to God, but to human sin. By means of mental gymnastics he removes the responsibility for evil from God and places it on man, and this instills a sense of shame and guilt.

The orthodox Judeo-Christian explanation for human evil is continued; the wonder is that it has satisfied so many believers. The doctrine of original sin is supposed to resolve the paradox of evil. God created man in his own image, capable of free choice. Man thus is responsible for the sins of omission and commission. The fall of Adam and Eve in the Garden of Eden from divine grace has been taken both literally and metaphorically. The literal interpretation of the fall held by fundamentalists is based on pure myth. The metaphorical interpretation advanced by liberal theologians indicates that God's first moral commandment is that his creatures should not eat of the tree of the knowledge of good and evil and that their first obligation is to be obedient to divine commandment. Human beings suffer death and a fall from divine grace only because of disobedience: human sin is the cause. Man can do nothing on his own; his entire moral enterprise depends on submitting to the divine will. Thus insofar as evil exists, it is due to human failure, for which we will eventually be punished.

This rationalization may satisfy the faithful, but it leaves too many questions unanswered. Why did God, who created man in his own image, not also create us so that we could share with him perfect virtue and beneficence? Why did God endow us with a double nature, and then punish us for transgression, when he could have prevented us from committing sin from the very start? If God is omniscient and knows that we will commit sin, why does he allow us to do so, when he could easily have prevented us from sinning? Why should he punish the wicked here and in the afterlife, when he could have made us like him, capable of doing only good? And what about the innocent and helpless? Why punish them? Why did he permit the Temple of Jerusalem to be destroyed or the sack of Rome? Because of the people's wickedness? Why did he permit the slaughter of the Huguenots, the massacre of the Armenians, the genocide of the Cambodians, Nazi tyranny, Stalin's infamous Gulag? How could a just and merciful deity have permitted these awful things to have happened? Perhaps some of the adults tortured or massacred were wicked, but then why innocent children?

Many Roman Catholics and fundamentalists rail against abortion because they insist that fetuses are innocent human beings. Why then does God permit huge numbers of miscarriages to occur? Why does he allow an infant to die of spina bifida or some other incurable disease, suffering and wracked with great pain? Is this man's fault? We are against it and try to mitigate suffering and to do what we can to prevent it. Why does not God do the same? Why do the innocent often die young and the wicked enjoy the fruits of power and wealth in this life? Will they suffer in the next? What assurances do we have? Job, in the Old Testament, raised for the religious consciousness such disquieting questions. He concluded that we could not fathom the ultimate purposes of the divine mind but needed faith that justice would right things in the end. But the biblical view of evil doesn't resolve the paradox; it only exacerbates it.

Does God have morally sufficient reasons for not preventing evil? On the basis of ethical reflection, one can argue that it is wrong for a moral being who can prevent evil not to do so. If we see that an innocent blind person is about to step off a curb onto the street and be killed by an onrushing car, should we not do what we can to warn him of the danger by crying, "Watch out!" If we can prevent a person's death, yet stand by indifferently and allow evil to happen, would we ourselves not be morally deficient? Although we surely are not responsible for preventing or miti- gating *every* evil in the world, don't we have a prima facie duty to attempt to forestall those that we can. And we must give a sufficiently good reason for not acting. Does God have a morally good reason for allowing an innocent child to suffer an early death in great pain? Perhaps he is punish- ing the child's parents for sins they have committed. But this is not a morally justifiable reason. Perhaps he is punishing the child for the sins of his forefathers even back to the third or fourth generation. What a convo- luted form of collective guilt—to burden a child with suffering if he or she is not guilty. Or perhaps the child is guilty and God is punishing the child for what he or she did in a previous life—even though the child has no memory of his prior existence, as Lucretia Borgia, a Nazi stormtrooper, or some other sinful person. This later rationalization requires the postula- tion of the doctrine of reincarnation in order to explain why God is not sufficiently moved by compassion or mercy to prevent such terrible things from happening. Many people have pointed out that there is a logical incompatibility between the traditional conception of a benevolent deity and the presence of evil.

The conclusion we may draw is that, given the facts of evil in nature, it is difficult to maintain that God is benevolent, omnipotent, and omnis- cient. With these unpleasant facts in mind, may we not with some reason attribute ignoble traits to him who has created the universe? The argument from analogy or design, if followed consistently and not selectively, can only lead the theist to draw an unpleasant conclusion about the character of the divinity. Using his methodology, there is an escape hatch: he is faced with an insuppressible paradox of reconciliation. The theist replies that the ways of God are inscrutable to mere human understanding. One must have faith, he insists, in divine salvation and the ultimate conquest of evil. But this only skirts the issue and does not absolve God from guilt.

Using the analogical and teleological arguments, we can likewise infer that the divine being is responsible for impermanence and disorder, evil and wickedness, nothingness and death; the ultimate chaos that may await us is the big crunch or the big freeze. Let us call this the "demonological argument." Using the same process of reasoning, we can reason back from nature to the existence of an all-powerful demon, as the creator, sustainer, and eventual destroyer of the universe. God would be the Destructive Force writ large in the Cosmos!

The ontological argument: What is God?

The theist persistently denies evil and impermanence and postulates goodness and order, which become the ultimate defining characteristic of the deity. This leads us to still another classical approach to God: the effort to discover a timeless being above and beyond the world, untrammeled by concreteness, particularity, impermanence, or evil.

Plato and Augustine are especially responsible for this argument, though it was St. Anselm who gave an influential formulation to it in the eleventh century. Later, Descartes, Spinoza, and other rationalists appealed to the ontological proof. Even today, philosophers such as Alvin Plantagina and Norman Malcolm have repeated essentially this same argument. The ontological argument is so simple that it is deceptive. It is based upon formal reasoning that alleges that the existence of God is true by reason of analysis of the meaning of the concept. It runs as follows: I have an idea of "a being of which none greater can be conceived." If I submit this idea to analysis, I find that existence must be part of its essential meaning. If by reduction *ad absurdum*, I assume that he does not exist, then he would not be "a being of which none greater can be conceived!" For if he does not exist, then his greatness would be diminished. Therefore, the idea of God necessarily entails his existence.

An objection can be made to this argument as follows: the very fact that someone has such an idea does not entail the existence of the idea independently. Thus, if I have an idea of a unicorn, this does not necessitate its existence. The ontologist responds that we can use the ontological argument only to demonstrate the necessity of God's existence. Since existence is not part of the defining essence of the unicorn, we are not entitled to infer its existence. Existence, he asserts, is essential to our very understanding of the concept of God, much the same as three-sidedness is essential to the idea of a triangle. If I had an idea of a triangle that was not three-sided, it would contradict the very idea of triangularity. Similarly, God's essential defining property is that he exists. Ergo, God exists.

My response to the ontological argument is, first, that the neo-Platonists did impart existence to unicorns, triangles, and other universal concepts, and they thought that these subsisted in an ideal realm of pure forms. Perhaps one can make the same inference about God, who would then have the same ontological status as other ideal essences. But one can raise serious doubts about Platonic realism. It is a mere postulate. How can we claim that what is present in our minds subsists or exists in reality? One move that has been made is to reject Platonic realism but to insist that existence applies *only* to the idea of God. To which I respond: surely not everyone has the *same* idea of "a being of which none greater can be conceived." The idea is not intrinsic or innate; it is not implanted in us by some deity (this is without evidence), but is only the product of a specific culture. Indeed, the idea of an infinite being is beyond comprehension: we may have only a glimmer of what such a thing is. It is a vague notion that cannot be defined easily. The ontologist is begging the question by maintaining that existence is essential to the idea he has of God. To claim that existence is intrinsic to the idea of God is to assume what he wishes to prove. We are dealing here with a mere presumption; we have no way of knowing whether the idea of God has any extrinsic, existential import. The sentence "God exists" is analytic, not synthetic. It is an incomplete and truncated statement; for *exists* is not a predicate but is only the copula. We cannot reason about God until we say something about his properties and indicate in what sense he exists.

Another variation on the same argument is to begin with the idea of "a perfect being." Again, the ontologist maintains that he has such an idea. Upon analysis he infers that the "perfect being" must exist. For if he did not exist, he would be less perfect than I, who does exist. It would be contradictory to say that the perfect being did not exist, for he would then be imperfect. Therefore, he must of necessity exist.

The speciousness of this argument is readily apparent, and it is also culturally biased. The Buddhist, for example, maintains that ultimate perfection is a state of sheer nothingness or nonbeing, where an entity is released from the struggle for survival and persists over and beyond the world of flux. God, rather than existing as the fulfillment of being, might exist on the level of pure nonbeing. Thus we might argue that the idea of a perfect being entails his nonexistence. How can the ontologist prove that the preceding argument is fallacious? The point is, we are dealing with the realm of speculative fancy, with pure concepts devoid of all empirical content. There is no way of knowing whether one system of axioms and postulates, arbitrarily selected, has any more claim to existence than any other.

Indeed, using the same mode of inference, we might reason as follows:

I have an idea (pure supposition) of an omnipotent, maximally evil, satanic being. If he is omnipotent and maximally evil, then he must exist. For if he did not exist, he would not be all-powerful and that would be a contradiction. If he is maximally evil, that means that he is able to inflict destructive chaos, disorder, degradation, and entropy on the universe, beginning with the big bang and ending with a big crunch or big freeze. To deny that he is capable of this would contradict the idea of his being omnipowerful and maximally evil. Therefore, he must exist. If one rejects this argument, then one must reject the ontological argument, which assumes the omnipotent being to be all-perfect. Whether there is a perfect being, embodying the qualities of Buddha, Christ, or Socrates on an extended cosmic scale or a maximally malevolent being with the extreme qualities of Satan, Hitler, Himmler, and Stalin has never been proven, though one can speculate about the one as much as the other. In either case, we are engaging in a deceptive ballet of bloodless categories by inferring from the use of concepts in our language something about the nature of ultimate reality.

The meaning of the concept of God

We have in this chapter raised the question whether the existence of God can be demonstrated. We have examined arguments introduced by theologians and philosophers and have found them invalid or unconvincing. A far more basic task perhaps is to clarify the meaning of the concept of God. Unfortunately, traditional notions of God are notoriously vague and ambiguous. They are analogical extensions of human qualities that have been extrapolated to the universe at large: omnipotence is inferred from witnessing human power, omnibenevolence from admiring human virtue, a divine creator from observing human creative arts. What does it mean to say that a divine being possesses such extended human capacities? Does God have bodily form? Does he have a soul? Is he a person? The Greek gods on Mount Olympus were invested with human powers on a magnified scale. Does God assume human form, and does his omniscience, benevolence, and creativity follow from his biological structure? Christianity, the most anthropomorphic and mythological of present-day religions claims that God took human form in the guise of Jesus, his Son. And Moses conversed "face to face" with God. Such conceptions of the deity are expressions of human conceit writ large. They demonstrate that human pride is all too ready to attribute to the vastness of nature its own slender powers. To believe that God is a person like us is, of all human vanities, the greatest self-deception. God has become the idealized projection of human need, the craving of the heart and the mind, seeking to complete the universe, bending it to suit human desires. Alas, there is no evidence to sustain this most presumptuous of human claims, fashioned from impudent imagination. The human wish is made father not only to the fact but to the entire universe as well.

Theists admit, however, that they cannot define God, that he transcends all the categories of experience and understanding. God thus is not simply an immanent but is a transcendent being. He cannot be narrowly described or characterized by human distinctions. Although a theist may demonstrate *that* some such being must exist, he cannot say what or who he is or know anything about his essence, except what has been revealed in Scripture. Karl Barth maintains that philosophic or scientific reason is unable to grasp that which is ultimately "unfathomable," "ineffable," and "inconceivable." If this is the case, then those who attempt to demonstrate that God exists are totally confused about what they are doing. Their conception of God is largely vacuous and God becomes an *x* factor, for which almost anything can be (and has been) substituted. If one rejects the Old and New Testament, the Koran, and other so-called sacred books and thus denudes God of any identifiable human properties, then God becomes a purely formal and empty concept. Since we cannot define the concept, or give it referential form, it has no designated or descriptive meaning; it is literally a concept without any cognitive signification.

The *x* factor, however, should not be viewed simply as an intellectual construct; it may be given other forms of meaning. Religious language has other functions than merely conveying information about the world. The idea of God does point to the fact that there are things in the universe we do not understand and that transcend our present experience and knowledge, and so this is a quasi-descriptive claim that no one need dispute. But this only masks a deeper function of the idea of God.

"God," in my judgment, is primarily a projection of the human longing for completeness. The idea of the deity is an evocation of our emotional desire for permanence. It gives vent to our passion for eternity, our wish to transcend chaos, danger, impermanence, and adversity. God is man's hope for an ideal world beyond this vale of suffering. Fixated on the father image, we look for someone over and beyond, guiding us safely through the sea of adversity, a being who, having created all things, will ultimately rescue us from nothingness and death. God-language thus has expressive meaning. It provides a distinctive moral component, for it recommends how we ought to live and how we ought to behave in the face of the universe. It postulates an external godhead and demands our devotion to it. Its primary function is normative. In praying to and sanctifying "unseen things" that we cannot comprehend, and in commanding faith, religious language is imperative. Growing out of the primordial scream and yearning for immortality, it constructs a system of moral commandments and rules. God refers to how we ought to live. It is the moral code made sacred.

What is left? Natural piety

Is the idea of God totally worthless and meaningless? Is it pure illusion? If God has no referent in the universe, should the idea of divinity be entirely abandoned? I submit that there is one sense, and one sense alone, in which the idea of God has some minimal meaning for man. He is a figment of human imagination and contrivance, expressed creatively in religion, art, morality, and poetry. God does not exist. He is not a separate person or a being, an all-powerful, all-knowing, benevolent creator of the universe. These are all anthropocentric misattributions. To worship such a being is mistaken. Prayer is only a psychological reaction to the troubles we encounter and an expression of hope that we will overcome them. No one is on the other side of the transaction to hear our supplications, no one with whom we can plea-bargain for salvation. Human beings do not have a cosmic purpose, nor are they fulfilling a divine plan. There is no drama of history in which God's purposes can be discerned. The human species will probably disappear someday. In spite of these stark realities, can something of the religious attitude be preserved? Is there still some room for religiosity, some rationale for cultivating natural piety? Yes, I believe there is. As a skeptical atheist, I submit the following:

1. The universe we observe is vast in its spatial and temporal dimensions. Human beings are relatively insignificant from the standpoint of the total cosmos.

2. We should cultivate an appreciation for the magnificent splendor of nature and the order and regularities we discover within it. Likewise, we should note, with wonder and dismay, the spectacles of conflict and chaos everywhere in nature, the polarities of life and death, order and disorder, stability and precariousness.

3. We have learned many things about nature. Science continuously expands our knowledge of the intelligible universe. But many things in the universe remain beyond our present understanding, transcending the present boundaries of knowledge.

4. The cosmic scene is rich in a variety and plurality of organic and inorganic processes. We can do no less than appreciate its great magnificence, from heavenly bodies viewed in astronomy to the intricate arrangements of its minute parts observed in physics and microbiology.

5. From this cosmic scene can emerge some intellectual appreciation of and sense of awe at the sheer grandeur of the natural world and some humility about the brute character of the individual substances and systems we encounter and their causal interrelations.

6. Nevertheless one must be dubious about the so-called "supernatural realm." The skeptic is not unmindful of the vastness of the cosmos and the fact that these distant reaches are still unexplored and little understood. If the supernatural refers only to that which we do not presently understand, it does not mean that that part of the universe is independent of nature. The skeptic does not wish to graft our transient human hopes onto the totality of nature or to maintain conceitedly that the universe was created for man and that man is the image of some creator. Although there is no evidence for any of these claims, they continue to persist in human culture. Rooted in faith and conviction, stimulating religious fervor and belief, God in general remains the myth *extraordinaire,* which persists in every age and comes back to haunt us no matter how brilliant the critique or how decisive the refutation.

Does the idea of God have its roots in something psychological or biological? Is it extrarational? Does the transcendental temptation fixate in God because of a deeper need in man that it seeks to satisfy? These are questions we shall subsequently have to address.

Part Three

SCIENCE AND THE PARANORMAL

XII: Scientists, spiritualists, and mediums

The resort to deceit and fraud that appears in religion is clearly evident in the spiritualist movement that developed in the nineteenth century and continued into the twentieth, especially in the United States, Great Britain, and Europe. It is possible for us to submit spiritualist claims to careful scrutiny, and in one sense, it is a far easier task to analyze them than religious claims. For, with few exceptions, the spiritualists did not claim revelatory or prophetic powers and hence did not develop powerful churches willing to deify them, to defend their message against disbelievers, or to suppress those that threatened their authority. Yet psychological and epistemic processes similar to those at work in religion are evident in the spiritualist movement. The spiritualists *did* claim that some individuals possessed special paranormal powers, which enabled them, among other things, to communicate with discarnate spirits. And so here we have a new form of flirting with the transcendental.

The basic premise of spiritualism is that human personalities survive death and are able to communicate through specially selected persons. It is claimed that this spirit world manifests itself in many ways: by mental communication, such as telepathy, visions, and apparitions; and by physical phenomena, such as poltergeists, levitations, table tappings, ectoplasmic emissions, materializations, the playing of musical instruments, and automatic writings. The most convenient method of communion is a special séance, a darkened room where a medium would have the discarnate spirit communicate through him or her, by mental or physical manifestations.

The spiritualist movement is religious or quasi-religious in motivation and function. For it points to a hidden realm, transcending this world. And it gives the promise of an afterlife: hope for survival was one of its main inspirations. There soon developed two camps: (1) believers, who accepted the existence of another dimension of reality and the ability of certain mediums to come into contact with it, and (2) skeptics and disbelievers, who rejected the entire phenomenon as self-deceptive nonsense. Into this battleground emerged science. For at long last, there seemed to be some empirical method to carefully weigh the evidence pro and con and to test supernormal claims. All of this illustrates, incidentally, that dallying with transcendental revelations, spiritualistic, or paranormal phenomena is as tempting to highly educated and otherwise sophisticated individuals, including those committed to scientific methodology, as it is to the great unlettered.

I do not intend to provide a full history of the birth and development of spiritualism, as it came to be known, but shall touch only on certain aspects of the movement for the light that it sheds on our general inquiry—the nature of the religious quest for the transcendental. Nor can I deal with all of the many spiritualists and mediums who flourished.

Belief in spiritualism is common in human history. The Bible is full of reports of supernatural phenomena, of visitations from dead persons and apparitions during both dream and waking states. And this continues throughout history. In the third century, for example, the skeptical philosopher, Porphyry, investigated the alleged movement of inanimate objects that were untouched, the levitation of mediums, the apparitions of spirits, and other unusual manifestations. Similar accounts of spiritualistic phenomena are found in the lives of the saints, the accounts of witch trials, and reports of poltergeists and apparitions in haunted houses and castles. In the eighteenth century, the Swedish mystic and religious philosopher Emanuel Swedenborg gave much impetus to spiritualism when he claimed to possess clairvoyant and precognitive powers.

The Fox sisters: Who's that rapping on my floor?

Modern spiritualism began in 1848, however, when the members of a poor and isolated family living in Hydesville, a small town in upstate New York not far from where Joseph Smith got his start, believed that their house was haunted by the spirit of a man alleged to have been murdered and buried in the cellar. This incredible tale was circulated by the Fox family and their two young daughters, Margaretta (age 11, later called Margaret) and her younger sister Catherine (age 9, called Kate), and their much older sister, Leah.[1]

The story of the Fox sisters reads like fiction; the only realities are the sisters' gumption in perpetrating a fraud and the public's unbelievable receptivity in swallowing it whole. It all began when the Fox family heard strange rappings and noises in the house. The mother was prone to belief in spiritualistic forces. On March 13, 1848, after Kate and Margaret had gone upstairs to bed, their parents heard strange rappings downstairs.

---

1. There are conflicting accounts of the ages of the girls. Arthur Conan Doyle says that they were 14 and 11 when the hauntings began. The Fox sisters themselves, later in life, made themselves younger by claiming that they were only 8 and 6. I have used the ages provided by the *Encyclopaedia Britannica* (11th ed.).

They rushed upstairs to be greeted by knockings and rappings throughout the room. When Kate snapped her fingers, rappings would follow. The disembodied rapper counted to ten and even gave the ages of the children. When Mrs. Fox asked, "Is this a disembodied spirit that has taken possession of my dear children?" the answer was a sharp rap, which was taken to be an affirmative. Mrs. Fox ran to tell the neighbors, and they flocked to the house. One neighbor, William Duesler, reported that one evening twelve to fourteen people were in the house too frightened to enter the bedroom. Duesler went in, heard Mrs. Fox ask questions, and the rappings that responded. Various neighbors asked questions, to which there were replies. The ages of those present, one rap per year, and the number of children in each family were given.

The news of this remarkable occurrence swept the area surrounding Hydesville, and new visitors appeared day after day. The ability of the Fox sisters to interpret the spirit world was enhanced when their older brother, David, suggested that they spell out words. He would repeat the letters of the alphabet and the spirit would rap when the proper letter was reached. Although this was a slow and laborious process, messages were received in this way from the beyond. The careers of the Fox sisters took a dramatic turn when the girls' elder sister, Leah Fox Fish, who lived in Rochester and had heard about the foregoing events, appeared on the scene. She immediately took Margaret and Kate aside and had them undress and show her how they were able to produce the mysterious noises.

Much of this was revealed to the world in a public confession by Margaret Fox, made in New York City and published in the *New York World* some forty years later, October 21, 1888. According to Margaret, the deception first began when the two girls went to bed. They would tie an apple on a string and allow it to bounce on the floor. The mother heard it downstairs and became alarmed. She never imagined that her nice young daughters were capable of deceiving her in such a way. After the mother had called in the neighbors, the girls contrived a way to make the knocking more effective. They felt cornered by what they had already done; self-preservation forced them to keep up the act. The method that they developed was to crack their large toes, allowing the noise to resound against the bedstand, wooden floor, or other wooden object. The effect was eerie, particularly if one believed it emanated from the spirit world. In subsequent months, hundreds of curious people visited the Foxes in Rochester and Hydesville, and many came away convinced believers.

At the behest of Eliah Wilkinson Capron, a newspaper editor, and George Willets, a large hall was rented in Rochester. They evidently saw money-making potential in the phenomenon and they were right, for four hundred people turned up the first night, each paying 25 cents to see the Fox sisters. The audience was treated to the rappings and then heard a lecture given by Capron. A committee of five persons was selected from their midst to investigate the strange doings. Curiously, the rappings were no longer confined to the haunted house in Hydesville but followed the Fox sisters wherever they went; and they were able to invoke any number of departed spirits, not simply the poor soul who was supposedly buried in their cellar.

The public demand for the show was intense. Three performances were scheduled for Rochester. Each night the crowd increased. None of the three appointed committees was able to detect any evidence of fraud. There was audience opposition, however. On the third evening a riot ensued, led by Joseph Bissell and others, who insisted that the whole thing was "humbug" perpetrated on the unwary, in order to fleece them of money. The local newspapers were also highly critical of the Fox sisters and ridiculed the notion of spirit communication. The whole matter soon received national attention, when Horace Greeley's *New York Weekly Tribune* reported on the three meetings after receiving a letter from Capron and Willets. Greeley was one of the most powerful and respected journalists of the time.

The response to the reports in the *Tribune* was immediate. The sisters were invited to New York City, where they held séances in Barnum's Hotel. They were visited by great throngs of people, many of whom hoped to make contact with dead relatives and loved ones. They were not disappointed. Spirit communication was in constant display, and many of those who asked questions of the Fox sisters were stunned at the accurate replies they received.

Horace Greeley became a staunch believer. He said that he was confident of the "perfect integrity and good faith" of the Fox sisters. Indeed, they were invited to stay at his mansion. Other New York newspapers, however, were highly skeptical, and they accused the girls of "jugglery" and "outright fraud." Yet, belief in the Fox sisters grew rapidly. Many distinguished persons were won to the cause, including Judge Edmonds of the New York Supreme Court. Many made pilgrimages to their séances and were duly impressed: James Fenimore Cooper, the famous novelist; George Bancroft, the historian; William Cullen Bryant, the poet; Governor Tallmadge of Wisconsin; General Bullard; and others. Moreover, hundreds of other mediums now appeared everywhere, all able to manifest similar powers; spiritualism swept the country like a prairie fire. Séances were convened all over the United States. The Fox sisters went on a money-making Western tour, and when they left each city, new circles of spiritualists were established. Adherence to spiritualism was now deeply implanted in the popular imagination. Dozens of magazines and newspapers were founded to proclaim the new faith.

George Templeton Strong, a prominent New York attorney, deplored the popularity of spiritualism and found these developments ominous. He had attended several séances but could not figure out how the many things he had witnessed had occurred. Nevertheless, he thought that the spiritualist explanations were unacceptable, and he particularly decried their easy acceptance by otherwise sensible people. "What would I have said six years ago," he wrote in 1855 in disbelief, "to anybody who could have predicted . . . that hundreds of thousands of people in this country would believe themselves able to communicate daily with the ghosts of their grandfathers?" He observed this included "ex-judges of the Supreme Court, senators, clergymen, and professors of physical science," who were lecturing and writing books touting spiritualism.[2] At that time millions of Americans apparently accepted spiritualism, and it is estimated that some 30,000 mediums were conducting séances.

From the earliest days the critical exposés of the Fox sisters were available to the public. But these had little effect on those who wished to believe. Thus, Dr. E. P. Langworthy, a Rochester doctor, reported on February 2, 1850, in the *New York Excelsior* that his investigation indicated that the rappings invariably came from under the girls' feet or from objects, such as tables or doors, with which they were in contact. He concluded that the noises were produced by Margaret and Kate themselves. Others came to a similar conclusion. Rev. John M. Austin wrote to the *New York Tribune* that the knocking could be produced by cracking toe joints. And in December of 1850, a Rev. Dr. Potts demonstrated to an audience in Corinthian Hall, Rochester, that he could replicate the raps in this manner. In 1850, the Rev. Charles Chauncey Burr and his brother published a book, *Knocks for the Knocking,* in which they maintained that they had investigated the Fox sisters and other mediums in five states; they attributed the phenomenon to "fraud" and "delusion." Burr said that he could produce the rappings in seventeen ways, his favorite being by toe cracking.

In February 1851, a committee of three doctors from the University of Buffalo, Austin Flint, Charles A. Lee, and C. B. Coventry, investigated Margaret and Leah, who by now was also able to produce raps. They concluded that the raps were voluntarily produced by the sisters. They had observed the sisters' facial expression and inferred that the sounds were produced by cracking bone joints, probably the knees. The semidislocation of the bone joints they inferred caused the distinct jarrings of doors, tables, and other objects with which they had contact. Using a

---

2. Allan Nevins and Milton H. Thomas, eds., *The Diary of George Templeton Strong* (New York: Macmillan, 1952), vol. 2, pp. 244-45.

controlled test, the doctors had Margaret and Leah sit on a couch and extend their legs so that their heels rested on cushions. The sisters sat in this manner for fifty minutes, urging the spirits to communicate. But there were no rappings—until they were again able to put their feet on the floor. Leah admitted that the spirits did not manifest themselves when their feet were on cushions. But she offered the lame excuse that her friendly spirits had retired when they saw "the harsh conditions imposed by their persecutors," and she could do nothing to detain them. This is the pretext often offered by psychics: that skeptics prevented the expected effect from happening.

Another decisive indication that skullduggery was afoot occurred when Mrs. Norman Culver, David Fox's sister-in-law, admitted in 1851 that the girls had confessed to her how they produced the rappings. Mrs. Culver gave her testimony in the form of a deposition. "The girls have been a great deal in my house," she said, "and for about two years I was a sincere believer in the rapping." She went on to say that she eventually began to suspect that they were deceiving people. She offered to assist Kate, who then "revealed to me the secret. The raps were produced with the toes. All the toes were used. After nearly a week's practice with Catherine [Kate] showing me how, I could produce them perfectly myself."[3] Margaret also told her that when people insisted on observing her feet and toes, she could produce raps with her ankles and knees.

Given these exposés, Rev. Burr predicted that the "reign of these imposters is nearly at an end."[4] How wrong he was! For exposés continued to be offered by critical scientific bodies—but to no avail. A challenge to the ardor of believers often produces the opposite effect; it only strengthens their faith. In 1853, Professors Henry and Page of the Smithsonian Institution and in June 1857 a committee of three Harvard professors concluded that the raps were produced not by the spirit world but by the Fox sisters themselves. In 1884 the Seybert Commission at the University of Pennsylvania attributed the rapping sounds to foot pulsations.

There were, of course, defenders of the Fox sisters. Of particular note was the famous British scientist, Sir William Crookes, who held séances with Kate in London in 1871. His tests of Kate Fox led him to the following conclusion:

> With a full knowledge of the numerous theories which have been started, chiefly in America, to explain these sounds, I have tested them in every way

3. E. W. Capron, *Modern Spiritualism, Its Facts and Fanticisms, Its Consistencies and Contradictions* (Boston, 1855), p. 421.

4. Ernest Isaacs, "The Fox Sisters and American Spiritualism" in *The Occult in America: New Historical Perspectives* (Urbana: University of Illinois Press, 1983), p. 95.

that I could devise, until there has been no escape from the conviction that they were true objective occurrences not produced by trickery or mechanical means.[5]

Was Crookes right? Did the Fox sisters have genuine powers? Were all their critics and detractors mistaken? In time, the Fox sisters became world-famous celebrities, giving séances even in the White House and for Queen Victoria. Large sectors of the public remained enthralled. Yet the mediums had their permanent critics. The Bar Association of New York City decided to do what it could to clear the city of mediums, whom they considered to be con artists. With this in mind, the Academy of Music was engaged for the evening of October 21, 1888, in order to expose the spiritualist movement. Margaret Fox Kane was signed up as the main speaker. What a coup to have one of the leading exponents of spiritualism denounce it. She had begun to blame her sister Leah, with whom there had been a bitter feud, for sucking her and Kate into a grand deception in order to make money. Beginning with toe crackings, Leah had added other spiritualistic attractions to her repertoire, including levitation, the plucking of musical instruments, and "ghosts" made out of luminous paper. Kate Fox Jencken, in an interview published in the *New York Herald,* concurred with Margaret in the need to expose spiritualists. She insisted that "spiritualism is a humbug from beginning to end. . . . the biggest humbug of the century."[6]

With Kate in the audience at the Academy of Music cheering her on, Margaret began:

> There is no such thing as a spirit manifestation. That I have been mainly instrumental in perpetrating the fraud of spiritualism upon a too-confiding public many of you already know. It is the greatest sorrow of my life . . . When I began this deception, I was too young to know right from wrong.[7]

Doctors in the audience were invited onto the stage to observe a demonstration. Margaret came forward in her stocking feet and mounted a low pine table. The audience listened raptly. One could hear a pin drop. Margaret snapped her big toe against the top of the table, and loud distinct rappings were heard reverberating throughout the auditorium. Here was the woman who helped launch the spiritualist movement now contributing to its exposure. Margaret was highly excited and agitated;

5. William Crookes, *Researches in the Phenomena of Spiritualism* (London: Burns and Oates, 1874), p. 88.

6. Joseph F. Rinn, *Searchlight on Psychical Research* (London: Ridder and Co., 1954), p. 57.

7. Ibid., p. 60.

she danced on the stage, clapped her hands, and exclaimed: "Spiritualism is a fraud from beginning to end! It's a trick. There's no truth in it!"[8]

Could spiritualism survive this crushing exposé? Other spiritualists later denied that Margaret Fox had ever made a confession; they said that the Fox sisters were alcoholics (which was true), that they were paid to make the exposé, and that their testimony was unreliable.

After the 1888 meeting both sisters took to the road, giving numerous demonstrations wherever they went. Their elder sister, now Mrs. Leah Underhill, angrily denied their entire story. The spiritualists attempted to have Margaret's children taken from her. A year later Margaret recanted her confession, saying that she had been misled by unscrupulous men. In desperate financial straits, she found that there was far less money in the debunking business than she had anticipated. Arthur Conan Doyle, a dedicated proponent of spiritualism, was typical of those who accepted this retraction. He said that she had been led astray. Moreover, Doyle postulated that "ectoplasmic rods" had protruded from her feet in some way and were responsible for the raps, but he maintained that this was still a "psychic force."[9] Histories of spiritualism written by sympathetic paranormalists even today consider the Fox sisters to have been genuine. They either fail to mention their confession of fraud or discount it.[10]

Nevertheless, the Fox sisters revealed other tricks of the trade. They said that at a séance they could watch intently the facial movements of a questioner, as he or she read off the alphabet, for clues to the correct letter. The technique is known as "muscle reading" and is widely employed by spiritualists. They also revealed how they were able to make rapping sounds appear to come from different parts of the room. By placing their feet at the bottom of a table or door, they could make the sound seem to emanate from the top. If they wished to make the sound seem to come from a distant wall, they simply made the raps louder and looked earnestly at a spot on the wall, whereupon people fell for the clever suggestion. By putting a toe on the footboard of a bed, they could make the sound reverberate throughout the room.[11] They were not above using accomplices when it suited their purposes; they instructed someone to rap in the cellar or under the stage at the signaled time.

Much of this is detailed in a book entitled *The Death Blow to Spiritualism*.[12] The sisters contributed a preface authorizing the book and

---

8. Ibid., p. 61.
9. Arthur Conan Doyle, *The History of Spiritualism* (London: George H. Doran, 1926), p. 113.
10. For example, see the incredible rationalizations by Brian Inglis, *Natural and Supernatural* (London: Hodder and Stoughton, 1977).
11. Ruth Brandon, *The Spiritualists* (Buffalo: Prometheus Books, 1984), p. 7.
12. Reuben B. Davenport, *The Death Blow to Spiritualism: Being the True Story of the Fox Sisters* (New York, 1888).

testifying to its veracity. But although the book may have momentarily weakened the spiritualist movement, it did not cause its demise. After the death of the Fox sisters, other mediums and spiritualists took up the cudgels and led the unwary into a further quest for discarnate spirits.

Other famous mediums, many of them controversial, appeared in the late nineteenth century and continued their performances well into the twentieth century. Many were tested by committees of distinguished scientists, some of whom found them to be blatant frauds; other testers thought that some of them possessed spiritualistic powers. A curious impasse developed. There were those who insisted that once a medium was caught cheating all of his or her other manifestations were to be discounted. Proponents of a spiritualistic, paranormal universe maintained, however—and still do—that although some mediums or psychics may be tainted by obvious cheating, there is other well-documented evidence for the existence of anomalous phenomena and hence that some mediums are genuine.

Nevertheless, the number of outright cases of chicanery and fraud are overwhelming. There have been mediums galore. For example, Jonathan Koons, a farmer from Dover Village, Ohio, built a log house and locked it tight. But he discovered that reams of spirit messages were coming from its locked interior. He left various musical instruments inside, as well as phosphorous solutions for the spirits to dip their hands in. Word spread quickly that the spirits of departed persons played the instruments. Another medium, Henry Gordon, developed the power of levitation. Much as Christ was thought to have walked on water, he was allegedly raised from the ground by spirit forces and carried to the ceiling of the adjacent room. Since the séance room was always dark, the audience could only imagine his levitating from his cries that he was "going up" or floating in air. There was never a confirmation of his levitation.

A quasi-religious fervor swept large regions of the United States. The Rev. C. Hammond claimed that he had received a spirit communication from Tom Paine. In Columbus, Ohio, a Miss Vinson, a medium, and an elderly Methodist minister gave public sittings. Musical instruments were suspended from the ceiling on strings; when the lights were turned off the spirits played the instruments. One evening two men turned on the lights in the midst of the concert to find the clergyman and the medium playing the instruments. The general public was often able to debunk such blatant chicanery. Why not scientists? In 1857, after criticism that orthodox science was unwilling to examine the claims of spiritualism, three professors at Harvard University—Louis Agassiz, Benjamin Peirce, and Eben Horsford—agreed to test the authenticity of mediumship; and the *Boston Courier* offered five hundred dollars for proof of any genuine spiritualistic ability. Leah Fox, Kate Fox, and four other mediums were brought to Boston to be tested, but they were unable to demonstrate any powers under the stringent controls of the professors.

Other mediums volunteered to be tested by scientists, believing that they could outwit them. Indeed they were often able to do so, for many professors were convinced of spiritualistic phenomena and were easily hoodwinked. Other professors were more skeptical, and mediums fell prey to their exposure. Henry Slade, a famous medium, was caught cheating by John W. Truesdell in 1872 in New York. He was unmasked again, in 1876, by Professor Lankester and Dr. Donkin and was convicted and sentenced to three months in jail, though he was able to flee to Europe before the sentence was carried out. He had been caught substituting a slate upon which a spirit message had been written. He was caught again in Ontario, Canada, in 1882, but he continued to find gullible audiences.

A popular development was the sudden appearance of mysterious spirits who were able to walk about the séance room. Robert Owen, the famous English utopian and socialist, was a believer in spiritualism; he reported in 1860 having seen a luminous female figure walk through the room during a séance led by Leah Fox Underhill. These materializations soon became common. The religious overtones were obvious. Those participating in the séance would begin by singing hymns. In a dimly lit room, in an emotionally charged atmosphere, a variety of spiritualistic phenomena would be manifested.

Many of those who visited a medium did so in the expectation that they would encounter a dead relative or loved one. They entered the séance in high expectation, and were often drained emotionally by the experience. Unscrupulous charlatans preyed on their gullibility. They would sit around a séance table, with the corner of the room curtained off; from this emerged all sorts of "occult" phenomena. Mediums compiled books of likely prospects—people who tended to revisit mediums, and so the mediums often knew something about their clients from previously compiled data. Upon arriving in a strange town, the medium or his accomplice would pay a visit to the local cemetery, taking down names and dates of death. It would be unnerving to an unsuspecting pigeon to be provided with information "that could not have been known from any other source." Because of the sheer audacity of mediums, many were detected, and so the general public was alerted to possible dupery. But true believers, who wished fervently to believe in the afterlife, did so in spite of the evidence to the contrary. The passion for supernatural belief, particularly concerning matters of life and death, is so powerful that the critical defenses against folly can often be broken down, particularly if there is even a glimmer of hope that immortality is real.

D. D. Home: He floats through the air with the greatest of ease

Paranormal believers consider Daniel Dunglas Home to have been the most awesome medium of the nineteenth century. Home performed miraculous feats, which few were able to explain, so that even today spiritualists and paranormalists look back to his career with respect bordering on reverence. Unlike other great gurus, such as Madame Blavatsky, for example, he was a loner; he did not seek to establish a new religion, though he might have. Everything that he did seemed to have been designed to enhance his private fortune. Some miracle makers in history have religions founded in their name, which exalt and deify them. Others simply use their talents for personal gain, enrichment, and the enjoyment of the luxuries of this life. This was apparently the intention of Home. Although he was on occasion tested by scientists, notably Sir Walter Crookes, who vouched for his "remarkable powers," Home was never caught cheating, though there is some evidence that he had engaged in masterful deceptions.

The mediumship of Home illustrates a key principle: if a person is devious and wily enough and if he is prepared to use his talents to confuse and defraud innocent people, it often becomes very difficult to detect chicanery and to convince them that the medium is not authentic. A consummate flim flam artist will never let his guard down, even for those in his immediate circle, recognizing that if those closest to him ever suspect that he is a charlatan, others are more likely to be skeptical also. The Home case is important even though we do not have a massive amount of negative tests, as we do with the Fox sisters or Eusapia Palladino, whom we will treat later. (In one sense, Home's case is similar to those of Jesus and Moses, who may—or may not—have lived so long ago that it is difficult to ferret out the true facts of their lives.) We are close enough in time to Home that we have no reason to doubt that he performed many of the feats attributed to him; indeed, we may even surmise and possibly reconstruct how he operated.

There is always some risk to the fraud thesis, for Home may have been genuine. The phenomena reported to have occurred in his presence may have in fact occurred, and perhaps only a supernatural explanation will suffice. As scientific skeptics, we cannot commit the error of the religious believer: refuse to examine the evidence. We cannot have a closed mind or foreclose fair-minded inquiry. If willful belief tends to cloud the perception and judgment of true believers, who are inclined to accept paranormal matters readily, conversely skepticism can bias the observations and interpretations of unbelievers, who might refuse to accept the existence of anomalous events no matter what the evidence, or who might believe that in all cases fraud must be present. It is clear that in studying the paranormal, objectivity is essential.

Daniel Home was born in Edinburgh in 1833. He claimed descent from the tenth earl of Home, a distinguished Scottish family, and so he assumed the middle name of Dunglas. There is no proof of his aristocratic lineage, and some commentators, such as Trevor H. Hall, have attempted to demonstrate that Home fabricated his lineage in order to make his way into the upper crust of society.[13] While quite young, Home was adopted by an aunt and taken to the United States. In his autobiography, *Incidents in My Life,* Home reports that he had "second sight" or spiritual powers early in life, claiming that he was able to foresee the death of a school chum when he was only thirteen and that he thereafter communicated with his spirit. He also said that, in 1850, he clairvoyantly saw the death of his distant mother; loud raps and moving furniture disturbed his quiet household, so much so that his aunt, a devoutly religious person, turned him out.[14]

From that time onward Home assumed the role of a professional medium, as did so many others following the great success of the Fox sisters. His modus operandi was different from that of other mediums, however, in that he did not charge fees for his séances. Apparently he was able to exist on the largesse of others, who were influenced by his personal charm, frankness, and affection. He tended to gravitate to people of great wealth and social standing, who were so impressed by his spiritual gifts that he was able to live off their hospitality and generosity. His friends in America, where many famous personalities came under his spell, raised money in 1855 to send him to Britain, and he soon acquired the reputation of being a fine medium.

Home said that his mission in life was "a great and holy one"; it was not to make money but to prove the existence of immortality, which will "draw us nearer to God." Nevertheless, his grateful admirers became so enamored of him that they showered him with gifts—as did a widow, Mrs. Jane Lyons, who presented him with 33,000 pounds and other emoluments. Later she changed her mind, declaring that Home had unduly influenced her by implying that her deceased husband had told her to follow his bidding. Spirit messages had bade her adopt Home as her son and to make a will giving him the family arms, name, and property. Mrs. Lyons took Home to court and he was required to make full restitu-

13. Trevor H. Hall, *The Enigma of Daniel Home: Medium or Fraud?* (Buffalo: Prometheus Books, 1984).
14. D. D. Home, *Incidents in My Life* (New York: Carlton, 1864).

tion. Later he married in turn two women with considerable fortunes—one a Russian noblewoman—and these marriages enabled him to live affluently thereafter.

Home gave spiritual advice and séances to a whole host of distinguished personalities; it is largely on their testimony that his reputation is based. Among these were Sir David Brewster, Lord Brougham, Sir Edward Bulwer-Lytton, T. A. Trollope, Viscount Adare (who became the fourth Earl of Dunraven), and Lord Lindsay (later Earl of Crawford). Home also gave séances on the continent before the French, German, Dutch, and Russian crowned heads.

Home's séances were similar to those performed by other mediums of the day. At first there were knocks, raps, and vibrations of the table. Next the séance table levitated, and the hands and clothing of the sitters were touched, presumably by a spirit. More complicated phenomena were then produced: hands and arms and other luminous objects would manifest themselves, and pieces of furniture would suddenly move up to the séance table. But Home added special features to his act. He had an accordion float in the air, or he held it with one hand, and with no visible fingers touching the keys it played a tune—invariably "Home Sweet Home" or "The Last Rose of Summer." Most astonishing to his bewildered guests was the appearance of Home's body as "elongated" and his ability to plunge his hands into the fireplace and carry a piece of glowing coal. Even more remarkable was Home's apparent ability to float bodily above the sitters, sometimes touching them as he passed over their heads.[15] Home's most famous feat was his supposed levitation out of one window and into another over a London street.

As nearly as we can surmise, Home's great reputation was based on the hearsay of those who attended his séances. He was rarely tested by scientists under rigidly controlled conditions. Neither his hands nor feet were held or bound, as mediums were later tested. Home himself decided who was to sit next to him at his séance. At the beginning of the séance, when raps and knocks were heard, the light was fairly decent. But before the more complex manifestations occurred, such as the appearance of spirit hands or the movement of furniture, the lights were usually turned off and only a fireplace, often screened, or an open shutter emitted light. Under such conditions there were few, if any, safeguards against trickery, and it is possible that Home used wires, hooks, pulleys, and other contraptions to achieve his effects, and perhaps even ventriloquism, or other forms of conjuration. Home used suggestion effectively, as in his levitation

15. Accounts of these events are provided by Frank Podmore in *The Newer Spiritualism* (London: T. Fisher Unwin, 1910). See also Podmore's two-volume work, *Modern Spiritualism* (London, 1902).

sequence, when he would declare, "I am now going up." No one saw him rise, yet they believed him. Or he would say, "I am being elongated," when he might have been simply stretching, contorting, or even standing on his toes. Perception in a dim light could always be influenced by suggestive remarks by Home. Some critics postulated that Home caused hallucinations in some of his visitors; others thought he was an outright imposter. Robert and Elizabeth Browning, the noted poets, watched Home at a séance, and Robert Browning later accused him of being a "cheat." Browning even composed a poem to defame Home, calling him "Mr. Sludge, the Medium."

It is said that Home, unlike virtually all other mediums of his day, was never directly caught in the act of outright cheating. Frank Podmore indicates, however, that there is some evidence that he was caught, and he cites a letter from a Mr. Merrifield, written in 1855, giving an account of a sitting conducted by Home that he witnessed. Merrifield reports that the only light in the séance room was that afforded by the dim light of the stars, as the moon had already set. It was eleven o'clock in the evening. Home sat on one side of the table, the participants on the other side. Home's hands were under the table. After some raps, someone reported that they saw a "spirit hand," an object resembling a child's hand with a long wide sleeve attached to it. The object appeared at one or two different places near the medium, either near his foot or his outstretched hand. Merrifield said, "I noticed that the medium's body or shoulder sank or rose in his chair accordingly." Then he says, "I saw the whole connection between the medium's shoulder and arm, and 'the spirit hand' dressed out on the end of his own."[16] He reports that the trick was so plain to his eyes while the company present was so reverential and adoring that he was seized with a strong impulse to laugh.

The journalist Delia Logan reported a similar séance held at the home of a nobleman in London. The nobleman observed Home groping his way through the darkened room, while every now and then luminous hands would appear. According to Logan, the nobleman saw Home place a small bottle upon the mantle. The nobleman thereupon covertly took the bottle and upon later examination discovered that it contained phosphorated oil, which was responsible for the luminosity. Everyone in the room nevertheless had been impressed by Home's "spiritualistic marvels." Unfortunately, the nobleman never confronted Home directly with his discovery.[17]

16. Podmore, *The Newer Spiritualism*, pp. 45-46.
17. John E. Coover, "Metapsychics and the Incredulity of Psychologists," in *The Case For and Against Psychical Belief*, ed. Carl Murchison (Worcester, Mass.: Clark University, 1927), pp. 243-44.

Home was generally very cautious in conducting his séances and thus he never suffered a direct exposure. He always prepared his performance with consummate skill. Much has been made of the fact that Home was tested by Sir William Crookes.[18] The first séance, which took place in 1871, was attended by Crookes and his personal friends. Various phenomena were demonstrated: a partial levitation of Home, the dipping of his hands into the fire, the levitation of the table, the appearance of luminous hands, and the playing of the accordion without any fingers touching the keys. How did Home perform his bag of tricks? Did he use a black string? Did he play a harmonica (as has been suggested), or did he use a music box? Was Crookes taken in by conjuration? It is difficult today to reconstruct the sequence of events or to explain how they occurred.

Three things stand out in the history of psychical research and its romance with science. First, psychic phenomena more readily occur when the test conditions are fairly loose and informal. Second, as test conditions are tightened up, the effect tends to disappear. Third, men of science, if they are predisposed to believe in paranormal or spiritual phenomena, are as easily duped as the ordinary person. Indeed, many conjurers believe that some scientists, who naïvely accept the ethical standards of science and trust the honesty of their test subjects, are often easier to fool than others. Magicians, who have learned the arts of deception, often find it easier to distract the mind of a logical scientist than that of a child. Further, there is no guarantee that, because a scientist is qualified in one field of science, he will be able to bring the same standards of observation and interpretation to other fields, particularly where his subject is a hoaxer dedicated to deceiving the experimenter.

Was this true of William Crookes? He contrived an experiment using a cage in which he placed an accordion. According to Crookes, Home could not touch the keys, yet Crookes reported that the accordion played and even hovered inside the cage. The séance room was, of course, somewhat darkened. We are entitled to wonder if there was sufficient light to determine whether a spirit hand or some material means could account for the musical sound. Various contrivances have been used by other mediums to achieve a similar effect. A concealed loop of catgut to which a hook is attached is used to pull the lower end of the accordion and produce notes. If the séance room is very dark, the medium can use a rubber tube to blow air into the accordion. His lungs take the place of the bellows and air produces some notes. In addition, we might ask if Crookes' perceptive ability was influenced by Home's powers of sugges-

18. This "test" was published by Crookes in 1889, some fifteen years after the séance had occurred, in the *Proceedings of the Society for Psychical Research,* vol. 6, pp. 98-127. See also Podmore, *The Newer Spiritualism,* p. 46.

tion. Did Crookes accurately report what occurred? The fact that Crookes vouched for the Fox sisters when others found them fraudulent and had testified that Florence Cook was able to materialize a spirit form provides considerable ground to question his judgment.

Levitating over London

No doubt, the single most famous spiritualistic feat in recent times was the alleged levitation of Home over a London street. This astounding event has been cited in the annals of spiritualism as a miraculous event; it is not unlike Jesus' walking on water.

Arthur Conan Doyle, who maintained that Home had levitated more than a hundred times before reliable witnesses, gives an account of the most famous of all levitations. It reportedly occurred on December 16, 1868, at Ashley House, in the apartment of Lord Adare. The witnesses were Lord Adare, Lord Lindsay, and Captain Wynne. According to Doyle, Home put himself into a trance state and then "floated out of the bedroom and into the sitting room window, passing seventy feet above the street."[19] After he arrived in the sitting room, Lord Adare was surprised and remarked that he could not understand how Home could have done this, at which point Home allegedly told Adare "to stand a little distance off." He then went through the open space head first quite rapidly, his body being nearly horizontal and apparently rigid. He came in again feet foremost."[20] Doyle insists that the three eyewitnesses were "unimpeachable." Sir William Crookes maintains that "no fact in sacred or profane history is supported by a stronger array of proofs."[21] If one were not to allow the admission of this testimony, Doyle insists, then the possibility of verifying any facts of human testimony must be given up.

The circumstances surrounding this alleged feat, however, are far more cloudy than Doyle allows; they hardly meet the standards of objective impartial inquiry. An account of the actual event was written up by Lord Adare in a book entitled *Experiences in Spiritualism*, which he published privately. Apparently only fifty copies were printed; and most of these were later withdrawn by Adare, who may have had second thoughts. He sent still another description of the strange event to a friend, Sir Francis D. Burand. Lord Lindsay also related his version of the story on two different occasions: before the Committee of the London Dialectical Society, which was interested in investigating spiritualistic phe-

19. Doyle, *History of Spiritualism*, p. 196.
20. Ibid.
21. Crookes, *Researches in the Phenomena of Spiritualism*.

nomena, and in a letter to the *Spiritualist,* a weekly newspaper. The Dialectical Committee later held four controlled séances with Home, but without any significant results. There is also a letter from Wynne to Home testifying to the event.

Trevor Hall has investigated the subject with great care. He has a photograph of the building (which has since been torn down) where the event was said to have occurred, showing the windows and balustrades. Hall demonstrates that it could have been possible for Home to make his way from one balcony to the adjacent one by normal means. Hall further shows that Adare and Lindsay were unreliable witnesses, that they were given to seeing apparitions, and that Adare in particular came under the strong influence of Home's personality. Interestingly, Home and Adare lived together for a period of time and even shared the same bed (which is suggestive of a homosexual relationship), so that Adare's so-called impartiality has hardly been demonstrated. Hall hypothesizes that Adare's state of mind could have been abnormal and prone to Home's suggestive influence.

In any case, on that fateful evening, according to Adare's account, Home began to walk about the room. "He was both elongated and raised in the air. He spoke in a whisper, as though the spirits were arranging something." Adare next reports that Home told them, "Do not be afraid, and on no account leave your places." He then went out into the passage.[22] Home's directions effectively precluded any scrutiny of his subsequent behavior. They are reminiscent of Moses' warning to the children of Israel never to climb the mountain, which Moses and Joshua alone were allowed to do, or of Joseph Smith's warning to his disciples not to look into the box containing the Golden Bible.

According to Lord Adare, Lindsay heard a spirit voice tell him that Home was going out one window and coming into the next. Then, says Adare, "We heard Home go into the next room, heard the window thrown up, and presently Home appeared standing upright outside our window. He opened the window and walked in quite coolly. 'Ah,' he said, 'You were good this time,' referring to our having sat still and not wished to prevent him. He sat down and laughed."[23] Lindsay confirmed that he heard the window go up but indicated that he could not see it as he sat with his back to it. But he said that he "saw his [Home's] shadow on the opposite wall,"[24] and saw him floating out the other window.

But, according to Hall, Home could have left the window in one room, made his way along the ledge of the building, and climbed into the

22. Hall, *Enigma of Daniel Home,* pp. 109-10.
23. Ibid., p. 124.
24. Ibid., p. 125.

next window. He could even have placed a board between the two bal-
conies (which an earlier critic had suggested), which Hall estimates to
have been only 4 feet 2 inches apart (not 7 feet 4 inches as Adare said).
Moreover, the height of the third floor was not 85, 80, or even 70 feet
above the street, as different versions of the incident have alleged, but was
approximately 32 feet. Hall maintains that it was possible for Home to
move from one balcony to the next without much trouble; at least it was
not impossible. Thus, instead of levitating, Home could have climbed out
one window and come in the next while the three men remained seated in
the darkened room.

Home's levitation was not corroborated by anyone in the street below
or in adjacent or opposite apartments. Since the stories of Lindsay and
Adare differ in many details and since the only light in the room was that
of the moon, critics have a legitimate basis for skepticism. It is Hall's view
that Home's principal secret was his ability to influence those who had
contact with him and to suggest supernatural interpretations for his behavior. This explanation seems more plausible than the spiritualistic account; it is more in accordance with David Hume's principle that we
should not accept an event as miraculous unless we have extraordinary
and reliable testimony about it, which is hardly the case in this instance.

Interestingly, the London levitation occurred only a few months after
Home lost his court case with Mrs. Lyons. Did Home need to resuscitate
his reputation in the eyes of those whom he influenced and depended
upon for a livelihood? Shortly afterward, in 1871, Home ceased the active
practice of mediumship and moved to France, where he married a wealthy
woman. Little was heard from him again. Did he become turned off by
his former life, no longer needing the money it afforded to support himself? He published *Lights and Shadows of Spiritualism* in 1877, in which
he exposed the fraudulent practices of other mediums, particularly those
who claimed they were able to materialize departed spirits in the séance
room. Home died in 1886.

Other mediums, however, came forth to proclaim and manifest their
marvelous abilities. They were able to evoke a positive response from
those individuals who cried out for some assurance that the end is not the
grave and that life has some deep spiritual purpose.

The Society for Psychical Research

The intense public interest in spiritualism no doubt was responsible for
helping to stimulate an interest among scientists and philosophers in
psychical research. Part of this scientific interest was motivated by a

genuine desire to prove the existence of paranormal phenomena. Darwin's *Origin of Species,* published in 1859, had dislodged the human species from the center of the universe, and this was viewed with alarm as a blow to spirituality. If it could be demonstrated that human nature had psychic or spiritual dimensions and that these transcended the limits imposed by materialistic science, what a boon this would be to religious faith. One could again have the "right to believe"—and on evidential, scientific grounds. Accordingly, in 1882 the Society for Psychical Research was founded in Great Britain, headed by Henry Sidgwick, the noted English philosopher. An American branch was established two years later under the creative tutelage of philosopher-psychologist William James. Three investigators, Richard Hodgson, Edmund Gurney, and Frederic W. H. Myers, worked assiduously for many years to place the examination of psychical phenomena on a scientific footing. They were able to enlist a number of other distinguished scientists and men of letters in their cause: Sir Oliver Lodge, Sir William Crookes, Charles Richet of France, and Sir W. F. Barrett.

There is considerable evidence that the original founders of the S.P.R. entered into an examination of psychical phenomena not in the spirit of impartial or neutral inquiry but in order to demonstrate the reasonableness of religious belief. It is therefore remarkable to have both Sidgwick and James say twenty-five years after the founding of the S.P.R. that they were surprised that they had made so little progress toward this end and that they found the field so full of fraud. James wound up with only one "white crow"—Mrs. Piper; most of the other mediums were rejected as untrustworthy or unproven. The stated aims of the society were certainly high-minded. The society would make "an organised and systematic attempt to investigate that large group of debatable phenomena designated by such terms as mesmeric, psychical, and spiritualistic." According to the society's organizers: "From the recorded testimony of many competent witnesses, past and present, including observations recently made by scientific men of eminence in various countries, there appears to be, amidst such delusion and deception, an important body of remarkable phenomena, which are *prima facie* inexplicable on any generally recognised hypothesis, and which, if incontestably established, would be of the highest possible value." They go **on** to state that the task of examining such residual phenomena until now has been undertaken by individual efforts but "never hitherto by a scientific society organised on a sufficiently broad basis."[25]

The society focused on the survival-after-death issue, including the

25. *Proceedings of the Society for Psychical Research,* vol. 1.

evaluation of reports of apparitions, trance-states, automatic writing, and the role of mediums in communicating with discarnate spirits. But it also dealt with other paranormal phenomena, such as telepathy, precognition, clairvoyance, and psychokinesis.

Six committees were appointed to take over different parts of this inquiry. Among their tasks would be:

> An examination of the nature and extent of any influence which may be exerted by one mind upon another apart from any generally recognised mode of perception.

> The study of hypnotism, and the forms of so-called mesmeric trance, with its alleged insensibility to pain; clairvoyance, and other allied phenomena. . . .

> A careful investigation of any reports, resting on strong testimony, regarding apparitions at the moment of death, or otherwise, or regarding disturbances in houses reputed to be haunted.

> An inquiry into the various physical phenomena commonly called Spiritualistic; with an attempt to discover their causes and general laws."[26]

The early work of the S.P.R. members was infused with enthusiasm; often they were duped in their investigations by unprincipled charlatans. Especially graphic illustrations of this were the telepathic tests conducted on Douglas Blackburn and G. A. Smith. These tests were cited for many years as evidence for thought transference. It was only after the death of the principal figures in the S.P.R. that Blackburn revealed that he and Smith were frauds. This is all too typical of the history of psychical research: news of a remarkable psychic or medium is reported in the press; the person may then be tested by psychical researchers; his powers are uncritically accepted and heralded as real; after many years and usually only after meticulous investigation is some fatal flaw in the experiment uncovered or fraud demonstrated.

This is what happened with Blackburn and Smith. They were first discovered by Edmund Gurney and F. W. H. Myers and then introduced to other members of the S.P.R., including Mrs. Sidgwick, Alice Johnson, and Frank Podmore. Some members of the S.P.R. became so excited about their alleged abilities that they published an account of the Blackburn-Smith tests in their first volume of the *Proceedings of the S.P.R.* Smith was later hired as a secretary of the S.P.R., and he worked in that capacity for many years.

26. Ibid.

It was some thirty years later that the full story of what had happened was revealed by Blackburn in a story in the *Daily News* in 1911.[27] According to Blackburn, he had been editing a weekly journal called *The Brightonian,* when he had began a campaign to expose fraudulent mediums, for Brighton, a seaside resort, had become the happy hunting ground for mediums of every kind. In 1882, he came across Smith, a 19-year-old youth, who gave an exceptionally versatile psychic performance. The two struck up a friendship, and they entered into a compact, said Blackburn, to "show up" some of the professors of occultism, who were then holding forth unchallenged. Thus they began their own thought-reading act. Their exhibition was described enthusiastically in *Light,* a spiritualist magazine. According to the article, Smith was able, by strong concentration, to read in an uncanny fashion the thoughts and mental images of Blackburn. It was on the strength of this article that the S.P.R. contacted Blackburn and Smith. Blackburn describes these learned gentlemen as "superior types of spiritualistic cranks." He reports incredulously that they accepted the results of private séances without hesitation and without taking reasonable precautions. According to Blackburn, both he and Smith were highly amused; but they were determined to show "how utterly incompetent were these 'scientific investigators.'" Their plan was to bamboozle them thoroughly.

They employed a rather simple set of signals, such as the jingling of a pince-nez or cufflinks, breathings and even blowings. Blackburn maintained that Myers and Gurney were especially anxious to get corroboration of their pet theories, that they were extremely gullible and very lax in their testing procedures. Time and again they gave the benefit of the doubt to experiments that were complete failures. They allowed Blackburn and Smith to impose their own conditions, and they accepted without demur their explanations for failures. For example, reported Blackburn, in describing one of the experiments, Myers and Gurney maintained emphatically, "In no case did B. touch S., even in the slightest manner." Yet, said Blackburn, "I touched him eight times, that being the only way in which our code worked." Blackburn was amazed that two young men, with only a week's practice and mischief in their hearts could so inveigle Gurney and Myers (and later Podmore and Sidgwick) to accept them as genuine psychics. What, he asked, could a more experienced cheat be able to do with the same investigators?

When Smith read Blackburn's 1911 confession, he emphatically denied that he had cheated. Blackburn said in response that he was totally unaware that Smith was still alive and that he regretted embarrassing

27. *Daily News,* London, Sept. 1, 1911.

him, but he insisted that they *had* cheated and he went on to describe in great detail how they were able to hoodwink the S.P.R. The S.P.R. Committee, he said, often tried to get them to transmit "irregular things," for these could not be easily signaled by a code. Blackburn and Smith failed so often in communicating irregular things that the committee abandoned the use of these tests. One test was devised, however, in which Blackburn and Smith acquitted themselves successfully; it involved the transmitting of "irregular patterns." The test greatly impressed the S.P.R. as a genuine confirmation of telepathy. Yet, said Blackburn, it was pulled off in an ingenious way.

In this test, Smith would sit at a large table, his eyes padded with wool and bandaged with a thick dark cloth. His ears were filled with cotton wool and covered with putty. Seated on a chair, he was entirely covered by two heavy blankets. Underneath his feet and the chair was a thick soft rug, which was intended to deaden sound and prevent any sound signals from being transmitted by his feet. Blackburn was standing at the opposite side of the room. He was shown a drawing by Myers, which he was supposed to transmit to Smith. It was a triangle, intertwined with lines and curves, which could not be described in words or easily conveyed by a signal. Blackburn reported that he gazed on the irregular figure for several minutes, pacing back and forth, but always keeping out of touching distance of Smith. During this time, Blackburn drew and redrew the figure several times, openly in front of the observers, in order, he said, to better fix his mind on it. He then described how he was able to convey the message directly to Smith by sleight of hand:

> I also drew it, secretly, on a cigarette paper. By this time I was fairly expert at palming, and had no difficulty while pacing the room collecting "rapport," in transferring the cigarette paper to the tube of the brass projector on the pencil I was using. I conveyed to Smith the agreed signal that I was ready by stumbling against the edge of the thick rug near his chair.
>
> Next instant he exclaimed: "I have it." His right hand came from beneath the blanket, and he fumbled about the table, saying, according to arrangement: "Where's my pencil?"
>
> Immediately I placed mine on the table. He took it and a long and anxious pause ensued.
>
> This is what was going on under the blanket. Smith had concealed up in his waistcoat one of those luminous painted slates which in the dense darkness gave sufficient light to show the figure when the almost transparent cigarette paper was laid flat on the slate. He pushed up the bandage from one eye, and copied the figure with extraordinary accuracy.
>
> It occupied over five minutes. During that time I was sitting exhausted with the mental effort quite ten feet away.
>
> Presently Smith threw back the blanket and excitedly pushing back the

eye bandage produced the drawing, which was done on a piece of notepaper, and very nearly on the same scale as the original. It was a splendid copy.[28]

This test is one of those that was cited as incontrovertible evidence for telepathy. It indicated also that gullible scientists are sometimes easy targets for frauds intent on deceiving them.

The scientists and philosophers associated with the S.P.R. in time, however, became aware of the widespread fraud—if not in the Blackburn-Smith case, then in others. And they did their part in exposing some of it. Thus Richard Hodgson was sent to India to investigate the reported occult and psychic powers of Madame Blavatsky, founder of Theosophy. Helena Petrovna Blavatsky was the daughter of a German nobleman named Hahn, who had settled in Russia, and a granddaughter of a Russian princess. She was married at the age of 17, and left Russia to visit India, the United States, and Canada. In 1873 she visited New York, and two years later in collaboration with various well-known individuals interested in spiritualism she founded the Theosophical Society. She left for India in 1878 and established her headquarters in Madras. She was allegedly capable of many supernormal powers, which her followers took as miracles. Upon her death in 1891, it is said that there were approximately 100,000 Theosophists.

Hodgson went to India with an open mind. He returned to issue a blistering report in 1885, accusing Blavatsky of brazen fraud. Sidgwick, president of the S.P.R., along with others had been impressed by Blavatsky and considered her a "remarkable" woman. Yet they were persuaded by Hodgson that her alleged marvels were based on fraud from beginning to end.[29] Basic to the teaching of Theosophy were some letters supposedly written by a Tibetan religious man, but these, Hodgson reported, were penned by Madame Blavatsky herself. The shrine in which she performed her miraculous feats was situated next to her bedroom, and Hodgson discovered a fake panel between them. This exposure by the S.P.R., however, had little effect on those true believers who accepted Madame Blavatsky and the weird philosophical-theological underpinnings of her system.

There were other mediums whom the S.P.R. investigated, notably Eusapia Palladino in Great Britain and Leonora Piper in America. Mrs. Piper went into a trance state and claimed, by means of her control, Mr. Phinuit, to be able to convey messages from the "other side." Phinuit, a

28. Ibid.
29. Ruth Brandon, *The Spiritualists*, p. 23.

Frenchman, curiously never spoke in French, and although he was supposed to have lived in Marseilles, there is no record of such a person having lived there. Yet William James was impressed by her alleged powers. No doubt the most thoroughly tested medium of her time was Eusapia Palladino. We shall concentrate on her in the rest of this chapter in order to illustrate the curious tendency for people—even those in science—to be charmed by the appearance of the supernormal.

Eusapia Palladino

Eusapio Palladino was born in the province of Bari, Italy. The daughter of peasants, she was unable to read or write. She had been married at a young age to a traveling magician, who no doubt taught her the arts of legerdemain. After the death of her father, she was employed as a domestic in the household of a family that practiced spiritualism. Taking part in a séance, she pronounced herself to be a medium and was, according to her account, able to stimulate spontaneous manifestations. She claimed to be fearful of these, yet she was persuaded to develop her spiritualistic talents further. She went through an extended apprenticeship in mediumistic circles in Naples and was tested by the famous Italian criminologist and psychiatrist, Professor Cesare Lombroso, who, although a skeptic, was taken in by her powers, which he thought were genuine.

Lombroso invited a group of scientists, including Professor Charles Richet of Paris (who later became president of the S.P.R.) and others to participate with him in various sittings in 1892. Richet at first expressed his uncertainty about Eusapia. Nevertheless he found many of the phenomena produced in her presence to be impressive, and he found it difficult to write all of them off as deception. Her impressive demonstrations included the levitation of tables, the movement of adjacent objects, the alteration of her weight on a balance, and the appearance of "spirit hands." In these séances Eusapia was in constant contact with her "spirit guide and control," John King. Richet admitted that he could provide no assurance that fraud was not being perpetrated by Eusapia or that simple illusion was not present.

In 1893-94, forty sittings were held in Warsaw by Dr. Ochorowicz, in which twenty-three experimenters participated. The results were mixed: some attested to Eusapia's paranormal powers, while others expressed doubt and said that they saw her using her hands and feet to evoke manifestations.

In 1894, under the auspices of Professor Richet, various British researchers were invited to take part in four sittings. These included Sir

Oliver Lodge, a noted physical scientist, and F. W. H. Myers. Professor and Mrs. Sidgwick later took part in the sittings. A report of these séances was published by the S.P.R.[30] Lodge and Myers, strong believers in psychic phenomena, were convinced that they had observed authentic supernormal manifestations. The Sidgwicks, who did not attend the same number of séances as Lodge, where truly extraordinary things allegedly occurred, nevertheless maintained that if the medium's hands were well controlled then the phenomena could only be accounted for by nonnatural means.

The general conditions at the sittings were dictated by Eusapia. She insisted that her rules be followed, and if they were not, she would fly into a rage and refuse to continue. The séances were generally held in darkness or semidarkness, usually late at night, when many of the sitters were tired. Eusapia was usually a late riser and came well prepared. Attired in a long dress or skirt, she would sit at a small table at one end of the room. Behind her, in a corner of the room black curtains were stretched. This small corner section (about three feet deep) was called a "cabinet." Inside was a table and various musical instruments. Eusapia sat only one or two feet from the curtains. Two controllers would sit on either side of her. One sitter would hold her hand, and she would hold the other sitter's hand. Eusapia would place each foot on the feet of the witnesses; or they would put their shoes against hers or encircle her feet. She would begin by going into a trance-state, with her body twisting and writhing. In the course of the séance, rappings would be heard, the table would levitate, and objects such as a little table would emerge from the cabinet and rise. Musical instruments would play, and strange hands would appear; at times sitters would be pinched on their arms and legs.

Serious doubts were expressed, particularly by Richard Hodgson, about the adequacy of the controls. Were Eusapia's hands and feet held securely, or did she manage to produce phenomena by sleight of hand or foot? A sufficient number of questions were raised so that the S.P.R. invited Eusapia to Cambridge University to give twenty-one additional sittings. Eusapia again imposed serious restraints on the proceedings. She determined the lighting conditions; she was insistent that her hands be held only in a certain way; she would not permit the witnesses to feel about in the darkness, and they were expressly forbidden to grab a materialized hand or any of the objects that levitated or moved about. Those who participated concurred that Eusapia was sometimes seen to be using her hands and feet. She would, for example, release one hand so that one or both sitters might hold a different portion of her other hand or even

30. *Journal of the S.P.R.,* vol. 6, 1894.

their own, unaware that Eusapia's hand was now free to perform tricks. They concluded that she was engaged in blatant fraud, systematically developed over the years by arduous practice, and that this accounted for some of the phenomena; and they inferred that it applied to the others as well. The light in the séance room was usually so poor that Eusapia's hands were rarely completely visible, and they could not be held very well because of her contortions. For experimental purposes the committee had given Eusapia ample latitude to cheat so that she could be observed; and she took advantage of the opportunities. Spiritualist critics of the Cambridge sittings admitted she evidently resorted to fraud on occasion. But they insisted that not all of her manifestations could be so explained and that some of them were supernormal. The S.P.R., however, was convinced of her roguery and decided not to have anything more to do with her.

Eusapia so intrigued the world of science that she continued to be tested by distinguished scholars and men and women of science, such as Henri Bergson and Pierre and Marie Curie. Eusapia received handsome fees for her sittings. Though highly controversial, she became a celebrity. Some scientists, such as Professor Enrico Morelli of the University of Genoa, declared unreservedly that Eusapia Palladino possessed paranormal powers, but others continued to insist that she was a fraud. Thus, the scientific world was divided in their verdict.

After many years, the S.P.R. decided in 1908 to investigate her powers again. A committee of three was appointed, comprising Everard Feilding, Hereward Carrington, and W. W. Baggally. Albert Meeson, an employee of the American Express Company, was engaged to take notes of the sittings. Eleanor Sidgwick, president of the S.P.R., explained why the society's council decided to conduct new tests. She said that the S.P.R. had a policy of not concerning itself with mediums who had been detected in deliberate fraud. To do so, would encourage "a mischievous trade" and stand in the way of scientific inquiry. Although the S.P.R. wished to discourage the flowing of fees with "unabated abundance" into the hands of tricksters, still it was important that an exception be made in the case of Eusapia; for it had been repeatedly alleged that she manifested genuine powers. In spite of the fact that she had been caught cheating by the S.P.R., enough scientists with distinguished reputations had testified to her powers, Mrs. Sidgwick concluded, that it was important that these claims be reexamined by the S.P.R., particularly since they affirmed that not all of the phenomena could be explained by trickery.[31]

The three sitters, although dedicated psychical researchers, were ama-

31. *Proceedings of the S.P.R.*, vol. 23, 1909.

teur conjurers; they had been involved in exposing the trickery of others and had published accounts of how the tricks were done. Feilding came from a distinguished family. A committed Roman Catholic, intent on discovering whether immortality existed, he nonetheless wished to use careful methods of investigation. Carrington, an American, had written a book showing in detail how tricks were perpetrated by fraudulent mediums.[32] He was able to reproduce by conjuring techniques many of the miracles of the mediums; indeed, the three investigators declared that they had never met a medium who could produce supernormal phenomena that could not be explained by trickery. Feilding had encountered so many blind alleys in his investigations that he had reached a state of complete skepticism "of the possibility" of ever finding a genuine spiritualist force. Indeed, he had investigated one hilarious case, in which a well-known lawyer had materialized small objects seemingly out of nowhere. He and his accomplice were carefully searched, and it was finally discovered that small objects, such as screws, pens, and sealing wax, were adroitly concealed in a suppository tube, seven inches in length![33]

The Palladino sittings took place from November 21 to December 19, 1908, in the Victoria Hotel in Naples. The set-up of the séance room was similar to that used before. Detailed notes of eleven séances were taken and published by the S.P.R.[34] Feilding was involved in all eleven séances, Carrington in all but the last, and Baggally in seven. In addition, other people, including Italians known to Eusapia, took part. Some evidence of trickery was discovered. Baggally reported that at least three times she used an arm or leg by substitution, and that a kiss allegedly made through the curtains seemed to be a thumb and finger pinching him. Nonetheless, all three declared that in many instances there was no discernible evidence of trickery.

The degree of control Eusapia permitted them to exercise varied greatly, depending on her mood. If she was in a good temper, Eusapia allowed her feet and hands to be well controlled. If she was in a bad temper, she was uncontrollable. The committee found that even when the controls were stringent, Eusapia was able to manifest spiritualistic phenomena. During the séance, she would slip into a deep trance and be in communion with her spirit control, John King. Sometimes she would talk in deep tones, as if transmitting his voice. At other times she became passive. Yet time and again they reported that physical phenomena ap-

32. Hereward Carrington, *The Physical Phenomena of Spiritualism* (London: Werner Laurie, 1907).

33. Everard Feilding, "The Haunted Solicitor," reprinted in *Sittings with Eusapia Palladino and Other Studies*, by E. Feilding, with an Introduction by E. J. Dingwall (New Hyde Park, N.Y.: University Books, 1963).

34. *Proceedings of the S.P.R.*, vol. 23, 1909. Reprinted by Feilding, ibid.

peared. In addition to raps, levitations, and table tippings, hands materialized, her dress bulged forth, a bulge appeared in the cabinet curtains, and cold air was emitted from a scar on her forehead.

The Feilding committee reached the conclusion that "some force was in play which was beyond the reach of ordinary control, and beyond the skill of the most skillful conjurer."[35] The conditions of the séance were such, they added, that it was impossible for Eusapia to have accomplices. The question had been raised as to whether the sitters were collectively hallucinating; this hypothesis was also emphatically rejected. Alice Johnson, a skeptical member of the S.P.R., wondered whether Eusapia was able to influence their perception in some way so that they imagined they saw the phenomena. No, they replied. The only conclusion they could draw, and with "great intellectual reluctance," was "that there does actually exist some hitherto unascertained force liberated in her presence."[36]

Hereward Carrington added a note in which he summed up his own attitude:

> I have to record my absolute conviction of the reality of at least some of the phenomena; and the conviction, amounting in my own mind to certainty, that the results witnessed by us were not due to fraud or trickery on the part of Eusapia.[37]

Carrington was a key sitter, who participated in ten of the séances and was responsible for holding Eusapia's hand and foot on one side of the table. Was Carrington a bad observer? Or was there some other explanation for Eusapia's powers? Carrington, on occasion, even lay on the floor, claiming that he held her ankles and that they did not move. Curiously, the next year, after publication of the Feilding report, Carrington served as Eusapia's agent and took her on a tour of the United States. He became her strongest defender and chief apologist. Did he profit from heralding her powers to the world? Carrington's entire life was devoted to spiritualist and psychical research. He wrote dozens of books and evidently made his living in this field.

Some of Eusapia's critics in the S.P.R. were not fully satisfied with the Feilding report. Frank Podmore analyzed the report with great care and speculated that virtually all of the major phenomena reported could have been produced normally, if Eusapia had been able to get a hand or leg free. He wryly observed that the conviction of Baggally, Carrington, and Feilding "that their senses were not deceived" is irrelevant to the

35. Ibid.
36. Ibid.
37. Ibid. "Final Note" by Hereward Carrington.

scientific issue.[38] If a man knows that he is deceived, the deception is incomplete. If he believes that he is not deceived, then it is effective. In any case, Podmore observes, the matter is not to be decided by argument but by experiment.

Science, deception, and the predisposition to believe

In the history of religion, we have seen that gurus, shamans, prophets, and saints performed their wonders with few, if any, controls. Most of what we know about the founding of the great religions is so veiled in obscurity that post hoc scientific or historical analysis is well-nigh impossible. It can only be done by analogical reconstruction and interpolation from what we have since learned of human behavioral processes. Most of the supernormal feats proclaimed in the sacred books are immune to skeptical scrutiny, and millions of devout believers have gone to their graves expressing a steadfast faith in the occurrence of such miraculous events and convinced that the only interpretation that could be placed on these events was a supernatural one.

Today, however, we are able to submit mediums and psychics to careful experimental investigation. This is certainly an advance over accepting the uncritical "eyewitness testimony" of credulous believers. Yet surprisingly, a new obstacle has been encountered. For one should not assume that religious superstition has been cultivated only by untrustworthy and uneducated people and that sophisticated intellectuals are immune to infection by gullibility or willful belief. On the contrary, it is intellectuals or quasi-intellectuals who have spread the word of new religions far and wide. Often they were the only ones who knew how to read and write, and they served as propagandists for the new faiths. Moreover, it is they who have prostituted philosophy, made it the handmaiden of theology, and sought a "higher" intellectual apologetic for the absurd faith. It is wrong to assume that if a person calls himself a scientist, he or she will be open-minded and impartial. It is often the case that we simply do not know what the cause of an event might be. To discover it may involve a slow and laborious process. The term *miracle* thus is a mask for our ignorance of natural causes. Unable to discover a natural explanation, some investigators are prone to invoke some supernormal, occult, or paranormal one. All too often this willingness is the product of a passionate predisposition to invest existence with hidden meaning, based upon a glimmer of hope that there may be an afterlife. Some scientists are more easily deceived than the proverbial man-in-the-street. The ordinary

38. Podmore, *The Newer Spiritualism,* p. 142.

person, who is busily engaged in a day-to-day struggle for a livelihood or survival, may be much less likely to fall for a con man than might an aristocrat or a distinguished professor or scientist, who lives in a cloistered tower. Some of these are unable to imagine ignoble motives, such as fame or fortune.

A perplexing factor confronting the scientific investigation of the supernormal is the persistence of deliberate fraud. This is not the judgment of skeptics alone. Even devoted disciples of psychical research have come to the same realization, often with considerable dismay. Indeed, it is rare to encounter mediums and psychics, so pure in motive and sincere in interest, that they are willing to cooperate fully with scientific inquiries. If the supernormalists' claims were demonstrated to be true, what profound and far-reaching implications this would have for our understanding of the universe and our moral life. Yet in virtually every case either (1) the medium is found to engage in fraud, or (2) the experimental conditions are unreliable. In spite of the endemic presence of fraud, a curious rationalization is still offered to discount it. Many believers expostulate that though a medium may be caught cheating on some occasions, he or she does not cheat all of the time. And until the skeptic can prove in every case how cheating occurs, believers will continue to insist that there are genuine psychic phenomena mixed in with the fraudulent. There are still glimmering psychic gems, we are told, that stand out in the shoddy rough.

How are we to cope with this specious argument, which is still being offered today? It is, I submit, the last vestige of a deep-seated faith, which remains as a kind of intransigent will-to-believe in someone or something in spite of evidence to the contrary. It illustrates the persistence of magical thinking in human psychology.

One may ask: Are the moral characters of the subject and the experimenter relevant in evaluating the evidence for the supernormal? What, for example, would be our response, if it could be shown clearly that Jesus, Moses, or Mohammed at times resorted to chicanery? Would we allow that they *still* evinced divine qualities? Or would the case for them be weakened, particularly since the three great religions founded in their names brought forth great moral systems devoted to their memory? Surely any defects in character, as evidenced by chicanery, would make chinks in their armor—or at least it should, though believers would, no doubt, invent new rationalizations for their continued belief.

Now Eusapia Palladino is no saint, and we do not hope to derive a moral system from her performances. (Though whether she had the ability to break through the known laws of the natural universe and to unlock a transcendent, spiritualist one is of momentous import for those who pant for it.) The real question is this: If a person cheats some of the time, a few times, half the time, or most of the time, is it still possible to believe that on some occasions he or she did not cheat and hence that a truly supernatural event was manifested? If we cannot figure out how the medium cheated, should we assume that no cheating was going on? We have two options: (1) to adopt the stance of the agnostic and admit that the medium was so clever that we cannot say how or whether he or she cheated, or (2) to adopt the position of the believer and maintain that, since we cannot tell how she cheated *in every instance,* we have a right to infer that she did not and that at least some of the phenomena were genuine. Believers insist that the second option is justifiable, especially when distinguished investigators were present at a sitting or conducted an experiment and could not fathom how the medium might have tricked them. Ergo, the only solution to the quandary is a spiritualist one. Skeptics who refuse to accept this are accused of being closed-minded.

But does the burden of proof rest with the skeptic? Must the skeptic show in complete detail how the medium cheated? Is it the case that, unless a skeptic can reconstruct the historical situation post hoc and indicate how he or she operated, the believer is entitled to believe that a genuine effect was present? I think not. The only sensible position to take is that of the agnostic or *a*-spiritualist.

Two examples will illustrate my argument. First, let us say that we go to a magic show, that the performer does not tell us he is a magician, and that we cannot figure out how he did his act. Let us say that the magician controls the conditions of his performance, that he is given free rein, and that the light under which he carries out his performance is not the best. Because we cannot dope out his clever act, are we entitled therefore, to believe that magic is "real"? What may be at stake here is the status of magical thinking. Do things suddenly go bump in the night without cause? Or are there natural causes at work, even if we do not know what they are? Magic fascinates us because the magician is able to make surprising things happen, contrary to our normal expectations. Surely we are not entitled to infer on the basis of our inability to explain how he performed magic that the cause is supernormal.

Second, let us imagine for the moment that money is missing from a bank vault, and we discover that a bank teller has concealed money in his briefcase. Suppose the teller pleads innocence, saying that it was an absent-minded mistake, and he expresses profound remorse. Let us say that since he has worked for the bank for many years, the bank president decides to give him the benefit of the doubt. A month later a considerable sum of money is again missing from the vault. This time we discover that the teller was the only one to have entered the vault. He is thoroughly searched and this time the money is found hidden in his lunch box. When confronted, the teller says that he has special psychokinetic powers, that money sometimes mysteriously moves from place to place when he is around, and that he cannot always control this power. Should we assume that the teller is honest, and take him at his word? Let us stretch the case and say that he is given still another chance, and that a third time money is missing, but when he is searched, it cannot be found. Are we entitled to conclude that the cause was "supernormal"? Or has the teller outsmarted us the third time? To maintain that the cause for the money's disappearance is supernormal would be an outrageous violation of all principles of common sense and logic. Yet this kind of reasoning seems to be acceptable in the world of psychical research.

Many psychic investigators were apparently impressed by the Feilding report on Eusapia Palladino. Here at last, they said, hard-nosed skeptics had taken precautions against cheating, and they had witnessed the phenomena that could not be explained by normal means. Were some of the medium's manifestations genuine? Do we at last have a "white crow"? Or may we surmise that Eusapia was more clever than Feilding and his associates? Did Eusapia have accomplices—perhaps Italians, scientists, and friends who had attended several séances, or even Carrington? Did she use every trick in the book, changing them to suit her purposes? Since she was a voluptuous woman, were her male sitters taken in by her erotic charms and did they fail to take the proper precautions? Eusapia was obviously a master illusionist, well-versed in her craft; and those who sat with her, though skilled in their specialties, may perhaps have been outsmarted by her. The Feilding report denies the possibility of accomplices or prearrangements in the hotel room. But should we accept this denial?

The decisive exposure of Eusapia occurred on her visit to America. Here the results were totally negative. Moreover she was caught red-handed in one of the most blatant acts of chicanery. The advice of professional magicians, however, became essential in ascertaining *how* she cheated.

As we said earlier, Carrington arranged Eusapia's trip to the United States, acting as her impresario. He declared that he was bringing her to America so that other scientists could test her. There was great fanfare and intense public interest. Numerous stories appeared in the newspapers. Two series of tests were held. In November and December of 1909 a number of scientists participated in séances, but with no apparent results. The most revealing séance was held on December 18, 1909, for which Hugo Munsterberg, Harvard professor of philosophy and psychology, had carefully laid a trap. Eusapia was thoroughly searched before the séance began, including every article of her clothing. During the séance, as usual convened late at night, Munsterberg sat on Eusapia's left side; on her right side sat another scientist. Eusapia was held under strict control:

her right hand was clasped by the scientist and her left hand by Munsterberg. Her right foot was pressed by the foot of the scientist and her left foot rested on Munsterberg's. For an hour, Eusapia carried on her performance, while Carrington implored John King, the spirit control, to touch Professor Munsterberg's arm and then lift the table in the cabinet behind her. And sure enough John King complied. Munsterberg reports that first "he touched me distinctly on my hip and then on my arm and at last he pulled my sleeve at the elbow; I plainly felt the thumb and finger. It was most uncanny."[39]

They waited in anticipation as John King was expected to lift the table in the cabinet behind Eusapia, and indeed the table began to scratch the floor, as if to move. Suddenly, there was "a wild, yelling scream." "It was such a scream," reports Munsterberg, "as I have never heard before in my life, not even in Sarah Bernhardt's most thrilling scenes. It was a scream as if a dagger had stabbed Eusapia right through the heart."[40] What had happened, unknown to Carrington or Eusapia, was that by prearrangement a man had been lying on the floor and had managed to slip very quietly below the curtain into the cabinet. Once behind the curtain he was shocked when he witnessed the medium move her right foot from her shoe and with a backward thrust of her leg begin fishing with her toe for the little table in the cabinet. At that point, the man on the floor seized her foot and held onto her heel. With her loud and piercing scream, Eusapia knew that she had been detected at long last and that her method was now uncovered. In an article published later, Munsterberg reconstructed what had happened in the séance. Eusapia apparently lifted her free foot to his arm and had pinched him with her toes. Just before penetrating the cabinet she had leaned forward heavily over the séance table, and in so doing, she was able to stretch her foot backward and reach into the cabinet. Munsterberg thought that he had felt her foot on his all during the sequence, but it was only her empty shoe. It was then that her heel was captured. Eusapia had clearly indulged in fraud and been caught, but what about the other times? Were any of the various phenomena induced in her presence authentically supernormal?

In a second series of ten séances, conducted at Columbia University between January and April of 1910, a second committee of ten professors especially appointed were unable to detect any trickery and could not reach a decision. At that juncture they decided to turn to experts in conjuration. Joseph F. Rinn, a noted magician, was invited along with others well-trained in the craft: W. S. Davis, James L. Kellog, and John W.

39. *Metropolitan* magazine, Feb. 1910. Quoted in C. E. M. Hansel, *ESP and Parapsychology* (Buffalo: Prometheus Books, 1980), p. 62.
40. Ibid.

Sargent. Rinn provides a detailed account of what ensued.[41] A controlled test was carefully arranged, with a dress rehearsal even held beforehand. Rinn and a Columbia student, Warner C. Pyne, were to dress in clothing the same color as the carpet and secretly position themselves on the floor beneath the chairs of the sitters.

On the evening of the séance, Eusapia arrived dressed in black. Since she spoke no English an interpreter was brought in. The sitters, Davis, Kellog and Sargent, were introduced to her as professors. Standing about two feet away was a group of other professors and their assistants. Eusapia sat down at the séance table, her back to the curtains of the cabinet. The séance table, noted Rinn, weighed only eleven pounds and could be easily lifted by one finger. The room was put into semidarkness, and Eusapia began to moan, as if beginning her trance. At that point Rinn and Pyne crawled furtively into place. The plan was to allow Eusapia to go through her act in the first part of the séance so that they could see exactly what she did. Kellog, seated on the right of Eusapia, held her right wrist and she placed her left hand in Davis's hand. As the séance continued she began to writhe and to slap her feet up and down on the feet of Kellog and Davis. She also began slapping her hands above the table. Under the table Rinn saw Eusapia give a slap with her foot that was resting on Davis, and at the same time she twisted her right foot sideways, and as it came down, she allowed her toe to rest on Davis's toe and her heel on Kellog's toe. Her left foot was now free. Rinn and Pyne had devised a code to inform Davis and Kellog what was happening. As they lay there they saw her make the curtain bulge out, as if by a spirit breeze, and the musical instruments on the table behind the curtain began to play—all accomplished with her foot and toes. She was able to release one hand on the table in a similar way and was able to make the table wobble to and fro with her left hand. She tilted the table, her toe under one of the legs to give it a boost, and with the use of her hand she caused the table to rise, as if by levitation. She again put her foot into the cabinet and played musical instruments.

Thirty minutes transpired, while Eusapia was given free rein to perform. At an agreed-upon signal, Davis and Kellog tightened the controls. They shifted and moved. Eusapia's free hand and foot were now completely controlled, and they would not allow her to shift or free either hands or feet. She cursed and shrieked. As a result, however, nothing occurred during this period of tight control. Again at an agreed-upon signal the controls were loosened, and she was given thirty minutes of freedom to do as she pleased. Manifestations were again produced by the skillful use of a free hand or foot. During this period, Kellog and Davis

41. Joseph F. Rinn, *Searchlight on Psychical Research*, chap. 21.

reported that they felt touches and their hair was pulled, not by a spirit, but by Eusapia's fingers and toes. Again they tightened controls and again nothing transpired. Thus a kind of conditional relationship was established. Whenever the controls were loosened and Eusapia could use a free hand or foot, phenomena occurred. When they were securely tightened, there were no manifestations. Rinn observed that it was truly remarkable to see how many stunts Eusapia was able to do with a free hand or foot. Like a trapeze artist or juggler, her skills were highly developed and she used them with abandon.

In the Feilding report, it had been claimed that a cool breeze came from a scar on her forehead. Her face was covered by her fingers and hands, as she gulped in air and exhaled through them. Supposedly, she was coming out of her trance state. Some of the professors present noticed that a breeze was being emitted above her head. Evidently, as she sucked in air and breathed out, an air current was blown over her forehead. This readily explained how she was able to produce the "mystical breeze" that so stunned her sitters in Naples. Rinn said that it was incredible that Eusapia was able to fool so many astute scientists with this trick for so many years.

The séance was concluded without either Eusapia or Carrington aware that detailed observations of her performance had been made by Rinn, Kellog, and Davis. Even some of the professors at the test were unaware, and they believed that genuine paranormal phenomena had occurred. Only Professors Jastrow and Miller knew beforehand what the plan was. After Eusapia and Carrington left, Davis and Kellog took the same positions they had earlier, and Rinn took the place of Palladino. Rinn was able to replicate everything that Palladino did. The professors present were satisfied that the phenomena, though produced by trickery, were similar to those produced by Eusapia, and that she was a fraud.

The following day the full exposé was published in the press. Professor Dickenson S. Miller of the Columbia philosophy department also published a letter in *Science* magazine, in which he described what had happened at the ten sittings that had been conducted with Eusapia Palladino. At these sittings, he said various phenomena had been observed, including levitations, rappings, touches, breezes, lights, materializations, and movements in and about the cabinet. Miller states, however, that "conclusive and detailed evidence was gained as to the method by which typical specimens of them were repeatedly produced." He says, "when the medium was securely held, they were not produced at all." He concluded that "no substantial evidence remains that her feats were spiritualistic."[42]

42. *Science*, May 20, 1910.

Carrington responded in a letter to the *New York Times,* insisting that he remained "quite unaltered" in his belief that Eusapia Palladino produces "genuine phenomena." "Why do I continue to believe in her?" he asks. And he replies: "because I have seen levitation when both the medium's ankles were beneath the table" (only by Carrington and no other). And he went on to say that "Eusapia herself says that she will cheat if allowed," but that her powers are still real.[43]

Carrington proposed that further tests be held and that stringent conditions be set forth by those who were to do the testing. These were arranged. Eusapia balked at these conditions, finally agreeing to them, but never turned up for the tests. She eventually returned to Italy, her American trip having ended in disaster.

One final series of five sittings, however, were conducted in Naples in December 1910, this time by Everard Feilding, whom she had deceived before. This time, Feilding was aware of the findings of the American tests and Eusapia's methods of cheating. Carrington was not present at these tests. Feilding, W. Marriott, and Count and Countess Petovsky-Petrovo-Solovovo got only negative results. In their report to the S.P.R., they concluded that "all the sitters agreed that the . . . phenomena were entirely fraudulent." At the termination of the séances they informed Eusapia of their negative results and the spurious nature of the phenomena. The medium did not dispute the justice of these conclusions, but she stated that she could not remember anything that occurred while she was in a trance state. She pleaded "ill health" as an excuse for her failure to give satisfaction. She accepted her full fee, however.[44] Eusapia's career was now virtually over. Although she did occasional séances thereafter, she was able to arouse little scientific interest in testing her again. She died in 1918 in Italy, with a number of scientists convinced that she was a fraud but other believers still willing to testify to her supernormal powers.

Hereward Carrington fits into the latter category. In a book written many years later, he summarized his attitude about Palladino. If Eusapia could produce spiritualistic phenomena, he asked, then why did she need to resort to fraud? And he replied that in a really *good* séance practically no trickery was used but that at a *poor* séance some 50 to 90 percent of the phenomena might be spurious. But why was any trickery necessary if she was genuine? He responded that Eusapia depended for the production of the phenomena on an "inner energy" over which she had very little control. Sometimes this energy was very strong. When it was weak, however, she "would endeavor to produce them artificially rather than disappoint

43. *New York Times,* May 13, 1910.
44. *Proceedings of the S.P.R.,* vol. 25 1911, "Report on a Further Series of Sittings with Eusapia Palladino at Naples," by Everard Feilding and W. Marriott.

her sitters." If she found that her trickery was undetected, he then rationalized, she would continue the fraud throughout the rest of the séance. Moreover, he adds that Eusapia "took a mischievous delight in seeing how far she could hoodwink her sitters."[45] If she couldn't get away with trickery, she would then endeavor to produce the phenomena psychically. Eusapia had a good deal of vanity and felt duty-bound to produce the phenomena. But genuine phenomena "exhausted" her; hence she resorted at times to trickery.

What a lame excuse for her outrageous behavior. Yet many believers accepted this pretext at face value. And it appears over and over again in the literature. Eusapia was a highly skilled conjurer and was able to dupe team after team of scientists, without being detected—particularly if there was poor lighting. Carrington explained that the bright lights demanded by skeptics inhibited the phenomena!

Carrington also stated in the conclusion to his book that his own conviction in Eusapia's supernormal powers "has remained unshaken" over the years, following the first tests in Naples. But he added that his conviction also rests upon "certain unofficial sessions," with the medium. According to Carrington, Eusapia often asked some of her sitters to remain behind after the séances for informal sessions. And, said Carrington, it was at these meetings that "the most startling phenomena developed."[46] He reported that she could, for example, develop a "psychic waterspout," a vortex of forces which would hover over the center of the séance table, and everyone present could feel it. However, more importantly, Carrington reported that on several occasions Eusapia "transferred her telekinetic powers" to him, and that he then had the power to move objects himself without any physical contact. He reported that Eusapia might cause a small stool in bright light to move simply by placing his hand over it. Then when she touched Carrington's shoulder, he could also get the stool to move as if some magic or occult force were being transmitted through him. He also reported that on at least one occasion he held a "materialized hand" in his and "felt it dissolve" within his own grasp!

Carrington's testimony was uncorroborated by other witnesses. We only have his word for it. Was Eusapia's chief disciple in America—her St. Paul to the heathen—a true but naïve believer? Or was he, like her, a fraudulent hoaxer? Each explanation has some rationale, though the latter seems especially compelling.

45. Hereward Carrington, *The American Seances with Eusapia Palladino* (New York: Garrett, 1954), pp. 7–8
46. Ibid., p. 271.

XIII: The paranormal: Psychics, ESP, and parapsychology

The role of science and its view of claims of transcendence has taken a surprising turn in the postmodern world, for science has been enlisted anew in the quest for a paranormal universe. Growth in the influence of science—and particularly its technological impact on society—is impressive; as a result, its reputation has been greatly enhanced. Instead of opposing science as being limited in dealing with transcendent reality, paranormalists now attempt to demonstrate in a scientific manner the validity of their claims.

It is always a surprise to witness the emergence of bizarre new religious cults and sects, which compete with ancient orthodoxies. In the nineteenth century, Mormonism, Adventism, spiritualism, and Theosophy were spawned by the religious imagination; in the twentieth century, other religious sects have sprouted—Scientology, the Church of Satan, Rev. Moon's Unification Church, a westernized Hare Krishna—and we have also seen the growth of Protestant fundamentalist sects and Muslim fundamentalism. A rather new development is the growth of paranormal belief systems. Science fiction has helped to create novel mythological beliefs by weaving exotic tales and embroidering them with fanciful scientific symbols. There is also a new cast of miracle workers and superminds. We now have psychic superstars, who are able to generate wondrous mental acts, and semidivine creatures from other solar systems and galaxies, who hover over us in space vehicles. Pseudosciences proclaiming new truths have emerged. The devotees of the paranormal assert that a shift has occurred and that an entirely new picture of the universe has emerged. The skeptics, they assert, are dogmatic men of little faith, limited by the conceptual framework of an earlier materialistic and naturalistic science.

The growth in belief in the paranormal is truly astonishing; it involves belief in extrasensory perception (ESP), precognition, clairvoyance, psychokinesis, and other psychic powers, but it also entails a new space-age religion, which postulates other forms of life in the universe. Intelligent beings, we are assured, have visited us in the past—ancient astronauts in space chariots—and are viewing us today in UFOs. They even have landing bases, such as the Bermuda Triangle, where numerous ships and planes are said to have mysteriously disappeared. In the paranormal universe there are monsters of the deep (the Loch Ness monster and other sea serpents), and land-based creatures (Sasquatch, Bigfoot, or the Yeti).

We must keep an open mind about anomalous phenomena and be willing to examine the claims of new discoveries. Unfortunately, many of the paranormalists are tempted to promote claims that outstrip the evidence, which is not dissimilar to the quest for the transcendental or the occult of earlier ages. In this chapter I will concentrate especially on parapsychology and its elusive quest for ESP and on the appearance of superpsychics, capable of seemingly miraculous feats.

Parapsychology and paranormal religion: J. B. Rhine

The term *paranormal* was invented by parapsychologists; its literal meaning is that which exists besides, beyond, or alongside the normal. The paranormal has replaced the supernormal, supernatural, or transcendental as a subject for study. The paranormal refers to that which supplements the existing conceptual framework of science.

The key person in the development of parapsychology is Joseph Banks Rhine. Rhine played an important role because of his desire to provide an experimental and behavioral basis for investigation. We have seen how the members of the Society for Psychical Research attempted to study psychical phenomena empirically. They dealt with clairvoyance and telepathy, collected anecdotal information, and devoted a good deal of their efforts to the survival question. We have also seen the spiritualist motivation in much of their inquiries. By the early part of the twentieth century efforts to communicate with discarnate persons in darkened séance rooms had been discredited. The field was found to be full of fraudulent opportunists and one medium after the other was exposed. Rhine gave up on any attempt to verify scientifically the existence of discarnate spirits, and turned his attention instead to extrasensory perception, including clairvoyance, telepathy, precognition, psychokinesis. His goal was laudatory: to try to establish in the laboratory the reality of these phenomena; he planned to use only rigorous experimental tests and procedures.

The investigation of ESP undoubtedly grows out of puzzles that arise in ordinary life, where there is a persistent tendency to believe that strange psychic phenomena are present. Many individuals are convinced that they can read the minds of others or predict events. Just before the telephone rings, Henry suddenly gets the idea that Martha, whom he has not heard from for months or years, will call him; sure enough she does, thus confirming his belief in precognition. Whether this is due to chance or some mysterious process is what is at issue. A mother has a strong premonition that her daughter, Jane, living many miles away, is in great distress. The sudden news that she has had a serious accident stuns her. She forgets the fact that she often worries about her daughter and has had premonitions that did not come true. A man has a dream about a plane crash; excited and distraught he calls the airport to notify an official. Eventually he claims that a crash like the one he saw in his dream occurred, as they sometimes do. Lovers and friends claim to be able to read the thoughts and feelings of their loved ones, as they learn to read body language and other cues.

These are common human experiences, and they raise the fascinating question of whether it is possible to know things by other than normal sensory means. Can the mind see things that happen, though there is no one present to observe it? Can two minds separated by distance communicate? Can someone peer into the future, or have a genuine premonition of what will happen? Intriguing questions such as these are found throughout human culture. They are often basic, as we saw, to religious faith; for the great prophets and seers claimed to have such psychic powers. These are independent of or in addition to the claims of revelation, where explicit messages are conveyed; they point to possible psychic gifts of a supernormal character. Some think that these gifts are granted not only to extraordinary psychics but also to the average person. How shall science deal with the claims for paranormal phenomena? Can the phenomena be tested experimentally in the laboratory? Can naturalistic explanations be given for them? Are they due in some cases to coincidence and in others to self-fulfilling prophesies? Often what seems unusual or strange is simply due to coincidence. People are surprised by these and say they involve some hidden or mysterious cause, or with Carl Jung they postulate some underlying or preexisting synchronization of all events.

Surely interesting coincidences need to be studied. The establishment of a parapsychology laboratory at Duke University marked a critical turning point for psychic research, for there was an attempt to test these claims systematically. The laboratory was initiated when William MacDougall, professor of psychology at Harvard, moved to Duke in 1927, and was joined shortly afterward by J. B. Rhine and his wife, Louisa. The task of parapsychology, as they conceived it, was to provide a quantitative and statistical approach to psi-phenomena, using the best techniques of experimental and behavioral inquiry. Experimental psychology then prevalent in American universities was largely based on behavioristic theory. According to Rhine and others, it had banished the "mind," "self," or "soul" from consideration, and only dealt with observable behavioral processes, which were considered to be a function of the physical body or organism as it interacted in the natural environment. Rhine believed that this physicalist interpretation of the world was limited. He wished to show the existence of nonphysicalist, extrasensory processes in human experience and to demonstrate these by using verifiable testing procedures. The evidence would not be anecdotal or qualitative; nor would it be based on untrustworthy, secondhand testimony; the only data to be considered were those that were amenable to strict statistical control and interpretation.

The term *paranormal* refers to that which could not be readily explained by natural, material, or physical causes. Nineteenth- and twentieth-century science, said Rhine, being predominately materialistic, left no room for the paranormal, which could not be explained in terms of the existing categories of science but only by introducing new kinds of explanation. Thus *extrasensory perception* was postulated to account for certain apparent abilities of the human mind that could not be accounted for by the usual means. *Clairvoyance* refers to the alleged ability of the mind to discern events that are far distant, without the presence of any sensory intermediaries. *Telepathy* refers to the alleged ability of two or more minds to communicate by other than normal sensory means, including language. *Precognition* refers to the alleged ability of some minds to know or predict events before they occur.

The very notion of extrasensory perception seems to fly in the face of our standard modes of perceiving and understanding the world. Our knowledge of external objects and events depends on sensory input. We perceive qualities and properties, interpret sense-data, and make inferences from them. The colors, sounds, tastes, smells, and tactile qualities we perceive come through our sensory receptors. Our perception is also related to motor behavior, as we intervene in the world of objects and seek to use or modify them to suit our purposes. Rhine said that some forms of paranormal perception do not follow the known channels of communication, and that this is what occurs, for example, in clairvoyance or clairaudience, in which we are sensitive to reality in an unusual way and in which there is a paranormal mode of apprehension. To say that a person is "psychic" means that he or she can see, hear, or discover some information, but that it is not received by or is dependent on the material or spatial configurations in the environment or the sense organs.

Moreover, ESP allegedly can operate over long distances and thus transcend spatial barriers. Clairvoyance can occur whether the psychic is a few feet away, several hundred, or even thousands of miles removed; the phenomena does not weaken with distance. Similarly, human beings normally communicate experiences or ideas by being in the same place and co-observing similar phenomena, or by transmitting images, ideas, or thoughts in language. Telepathic transmissions do not depend on either, said Rhine; they enable us to leap-frog beyond visual signs or verbal symbols, transcend the known channels of communication, and overcome distance.

Precognition violates both common sense and scientific concepts of time because it implies that one can know about a future event before it occurs. Our knowledge of the world is based upon the memory of past events and present perceptions. On the basis of this we build up habits of expectation and infer future events. We normally make predictions based on generalizations. Here, we presuppose that the future in some way will resemble the past and that similar effects will have similar causes. This is the principle of induction, upon which causal predictions are based in science and everyday life. But the paranormalist goes beyond this by claiming that it is possible to have a prevision of future events, without reference to causal explanation, prediction, or inference. The future literally can be known before it happens. Paradoxically this involves *several* possible futures, for if a person can know that an event is going to occur, he or she can try to prevent it from happening or take appropriate measures to avoid its harmful aspects. This seems to suggest a form of backward causation: The real or possible future causes the present cognition and forewarns us to make an adjustment in our behavior. If this is the cause, precognition transcends the limits of time. Moreover, we can not only precognize but also "retrocognize" events of the past, so that the mind can actually go backward or forward on the time scale in some mysterious way. Einsteinian relativity physics made some people think this could be possible. But does it occur? Are minds capable of such psychic power?

There are other strange aspects to the paranormal universe of parapsychology that do equal violence to our scientific and common-sense view of the world. Thus psychokinesis or telekinesis points to the ability of the mind to cause physical objects to move without any form of physical intervention. Rhine sought to investigate this directly. Could "mental energy" change the state of physical entities without the interposition of physical forces? If so, this would mean a major exception to the laws of physics. Rhine investigated the capacity of the mind to influence the throw of dice. Some gamblers have always believed they had this knack. It could now be tested in precise probability terms.

Did the anomalistic phenomena that Rhine investigated have any relationship to the supernatural? Was there a religious motive underlying his inquiry, and was this similar to the kinds of religiosity we have already discussed? Rhine explicitly stated that one of his primary motives was to reject what he called the "metaphysical doctrine" of the "mechanistic philosophy of man." Rhine made it clear that "it was to refute . . mechanistic philosophy that parapsychology arose."[1] Among Rhine's tasks was to show the limitation of the "physicalist theory" of the universe and of man. According to Rhine, parapsychology attempts to deal with "non-physical causation in nature." Psi-phenomena, he stated, involve the behavior of living organisms "that fail to show regular relationships with time, space, mass, and other criteria of physicality."[2] Rhine argued that psi was "mental" not physical:

> It is now fairly clear that psi-phenomena are identified by the fact that they defy physical explanations and require a psychological one. They also happen to people (or animals) or involve some associated or at least suspected agency or experience; but at the same time they do not follow conventional physical principles.[3]

Rhine thus presupposed a dualistic view of human reality in which mind was independent in some sense of its material basis. He admitted that this gave some support for religious experience and the religious impulse. He denied, however, that the data of parapsychology were occult or transcendent, for although they could not be understood as natural in the same sense as other phenomena, they were capable of experimental description and interpretation in the laboratory. Unlike many other parapsychologists, Rhine clearly sought to dissociate his point of view from a purely transcendental or supernaturalistic one, for he thought that man was part of nature and could be investigated as carefully as any other natural phenomenon. Still, there appeared to be an immaterial aspect to human nature. He claimed that he had demonstrated the existence of ESP and psychokinesis (PK) experimentally. He did not bring in vague mystical or intuitive notions; nevertheless, there was room for the spiritual dimensions of experience. If there were a dualism between mind and body and if the mind displayed psychic powers, then our theories of human nature would have to be altered and the materialistic view of the universe would have to be abandoned. Religious belief would not simply be a matter of faith but could have some basis in empirical evidence.

According to James A. Hall, Rhine had a deep concern with the relationship of parapsychology to religion. In bringing forward evidence against the physicalist theory of man, he believed that he was indirectly

---

1. J. B. Rhine, "Comments on 'Science and the Supernatural,' " *Science* 123, no. 3184 (1956), pp. 11–14.
2. J. B. Rhine, "The Science of Non-Physical Nature," *Journal of Philosophy* 51 (1954), pp. 801–10.
3. J. B. Rhine, *New World of Mind* (New York: William Sloane, 1953), p. 150.

supporting the religious view. Moreover, according to Hall, he stressed "the similarity between religious forms and types of psi."[4] PK was possibly related to "omnipotence," ESP to "omniscience," clairvoyance to the "all-seeing eyes" of a divine being, precognition to "prophecy" or knowledge of things to come, and other forms of psi to religious "miracles." Rhine himself states that "on the whole, the types of psi that have been quite independently outlined by laboratory research closely resemble the kinds of exchange that religious men have assumed in the theologies that arose out of human experience long before the laboratories of parapsychology began their work."[5] Rhine then goes on to say that "no matter what one thinks about the theological claims of these religions, he can now at least see that their founders must have built those great cultural systems on a rather good acquaintance with the same powers that have now been independently established as parapsychological."[6]

Rhine thus believed that parapsychology was relevant to religion and that it offered a way to reconcile the claims of religion with the principles of science. Among the religious questions it might help to resolve were whether persons were free moral agents, whether we have a postmortem destiny, and whether the universe is a "personal universe, with a type of intelligent purposive agency within it to which man can with rational confidence turn for helpful communication."[7] In this regard, Rhine's interests were similar to the earlier founders of the Society for Psychical Research, who had hoped to provide a scientific basis for the "spiritual principle in man."

Does ESP exist?

There are two major issues that need to be raised: (1) Does parapsychology violate our basic conceptual scheme? Is it logically impossible? (2) What is the evidence for ESP? Can the so-called experimental data amassed survive scrutiny?

Many skeptics have raised logical objections to ESP on philosophical grounds, saying that the postulates of parapsychology contradict the general categories by which we interpret and understand the world. C. D.

    4. James A. Hall, "The Work of J. B. Rhine: Implications for Religion," in K. Ramakrishna Rao, ed., *J. B. Rhine: On the Frontiers of Science* (Jefferson N.C.: McFarland, 1982).
    5. Hall, ibid., pp. 137–38. See also J. B. Rhine, "The Parapsychology of Religion: A New Branch of Inquiry," *Journal of the Texas Society for Psychical Research and the Oklahoma Society for Psychical Research,* 1977–1978, p. 6.
    6. Ibid., pp. 7–8.
    7. Ibid., p. 9.

Broad called them "basic limiting principles."[8] They generally take the form outlined above, namely that (1) our knowledge of the world depends upon sensory observation and inferences from them and presupposes physical causes in the material world, (2) the human mind presupposes a brain and nervous system, and (3) events in the future cannot cause changes in the present and past to occur. The critics have charged that if parapsychology were true, then both physics and common sense would have to be radically altered or abandoned. Thus, they say, we can reject a priori its conceptual framework as nonsensical.

These logical objections are very attractive for some skeptics. It would be an elegant argument if we could prove the invalidity of parapsychological ideas and simply reject the entire bag and baggage of parapsychology antecedent to inquiry. Although I can appreciate this line of criticism, I think that we have to take it with considerable caution, for there is always the danger that in so arguing we would limit or restrict novel or radical departures in knowledge. Many new theories, when introduced, wreak havoc with older and more comfortable concepts. The history of science provides ample testimony to the fact that long dead systems in the past blocked new systems of thought. We cannot simply define the fixed structure of reality a priori in terms of the existing level of scientific development. New data and discordant, anomalous, or bizarre experiences or facts can destroy the best explanations. Thus we cannot say with absolute confidence that the data and theories of parapsychology must be false because they contradict the existing body of physical theory. Indeed, the saga of the contemporary physical sciences vividly demonstrate that new concepts are offered that may undermine our commonsense view of the universe: relativity theory and quantum mechanics have readjusted our views of space, time, and causality. Conversely, many parapsychologists argue, mistakenly I think, that the new physics leaves room for psi. I don't deny that some theories introduced in science may so violate our canons of logic that they can be ruled out, particularly if they are self-contradictory. But this is not the case when new empirical claims are being made based on the uncovering of new data, which may require new conceptual schemata.

The central issue of the truth or falsity of parapsychology can only be decided by the evidence. If we discover a body of discordant facts and if these cannot be explained by the prevailing theories, then so much the worse for the dominant theories, which must be either modified or abandoned. This was the unfortunate case in the history of science, where familiar theories were converted into received dogma. But this is alien to

8. C. D. Broad, *Philosophy* 24 (1949), pp. 291–309.

the entire spirit of science, which must be open, fallible, tentative, and willing to revise even its most basic principles.

There is another kind of logical objection, however, that warrants careful consideration, and that is Hume's argument from miracles, which we examined in Chapter XI. Where a body of hypotheses, concepts, and theories are well established on the basis of abundant evidence and have been tested repeatedly and found to apply to broad bodies of data, and where these hypotheses and theories have an internal coherence, a wide range of applicability, and explanatory power, then we may radically alter or abandon these theories only if the evidence for the anomalist data is so strong that we are faced with no other alternative. If we do so, however, then we require a considerable amount of evidence based on trustworthy testimony and experimental data, which is inexplicable by known concepts. At some point new schematic explanations may eventually be introduced. But extraordinary claims that violate our basic limiting principles demand extraordinary amounts of evidence. These claims are not to be dismissed out of hand, but if they are accepted the evidence cannot be fragmentary or questionable.

Unfortunately, at the present juncture in parapsychological research this does not appear to be the case: (1) The amount of data establishing the presence of paranormal phenomena is at best weak and inconclusive. Moreover, experimental bias and bad experimental protocol, sensory leakage, and even fraud provide alternative explanations for many or most above- or below-chance runs. (2) A basic criterion in the experimental sciences is that there must be a replicable experiment that can be repeated in any laboratory by independent or neutral observers. This condition has not been satisfied in the parapsychological laboratory so that the very existence of psi-phenomena is still in question. (3) Even those parapsychologists who think that ESP has been demonstrated have yet to offer a general causal theory to satisfactorily explain the data they claim to have uncovered. Although this is not a fatal objection, it shows the elementary character of parapsychology as a science.

In my judgment, it is the empirical question that is crucial at the present time. We can speculate item for item about the implications of psi to our conceptual framework, but if psi does not exist or if a perfectly natural explanation can be given for it, then there is little need to probe its ontological meaning. In my view, though parapsychology has, since Rhine, the goal of attempting to be experimentally grounded, it has not achieved any kind of breakthrough, and its findings are as inconclusive as they were at the inception of the Society for Psychical Research more than a century ago.

Rhine considered that his experimental work at Duke University in the early thirties was remarkable, and that it had decisively established ESP. His findings were first summarized in his book *Extrasensory Perception,* to which there was an immediate positive public reaction and a strong scientific counterreaction.[9] Rhine's colleague in the psychology department at Duke, K. E. Zener, had devised a series of cards, which were to be used in testing ESP. These consisted of five symbols: circle, rectangle, star, cross, and three wavy lines. Since they came in packs of twenty-five, the chance probabilities could be easily calculated. Various types of experimental runs were carried out over the years. First there was a test of *clairvoyance,* in which two techniques were used: (1) the *basic technique,* in which the cards are shuffled and laid face down and the subject guesses each card as it is removed from the deck; (2) *down through,* in which the cards are shuffled and all placed face down and the subject guesses each of the cards, without moving any of them. Second, there were various tests of *telepathy,* in which a sender removes the cards one at a time and attempts to transmit each one to a receiver, whose task it is to divine the correct sequence. Third, still another series of runs were devised to test *precognition.* Here the receiver is supposed to guess the correct sequence of cards before they are removed from the deck or before they are even shuffled.

The whole point of the experimental battery is that precise probability statistics could be used. No longer would we have to depend upon anecdotal accounts of ESP or bizarre coincidences. The parapsychologists could determine quantitatively whether certain individuals could consistently beat the odds and hence demonstrate psychic ability. Thus, for the first time psi-phenomena could be given an operational interpretation.

Rhine announced early that he had located this uncanny ability in at least eight of his students, whose performances were extraordinary, far exceeding chance expectation. Indeed, with the possible exception later of Samuel G. Soal's work in Britain, it has been rare to find such a concentration of psychic talent as appeared in the early work of Rhine. A. J. Linzmayer, the first of these high-scoring subjects, received some extraordinarily high early test scores. Rhine reported that Linzmayer, when seated alone with Rhine in his car with the engine running, was able to guess twenty-one out of twenty-five trials (the odds being 30 billion to one). In repeated trials of cards, he continued to score high. However, as C. E. M. Hansel notes in his critical analysis of Rhine's early work, when Linzmayer was tested with cards at a distance from him he was unsuccessful.[10] Eventually Linzmayer was beset by the "decline effect"; that is,

9. J. B. Rhine, *Extrasensory Perception* (Boston: Bruce Humphries, 1934).
10. C. E. M. Hansel, *ESP and Parapsychology: A Critical Re-evaluation* (Buffalo: Prometheus Books, 1980), p. 91.

in time his results began to decline and to approach the chance level. The same effect was noted in all of Rhine's subjects: there was an early show of extraordinary "ability," but in an extended sequence of trials this ability declined and eventually disappeared. The decline effect, according to John Palmer, a noted parapsychologist, was to haunt Rhine throughout his career. And it applied even to his most noted subjects, such as Hubert Pearce and Charles Stuart.[11] Indeed, according to Palmer, "the consistent success of Rhine of the early 1930's was not to be duplicated in subsequent decades. There never again appeared a group of star subjects that rivaled Pearce, Linzmayer, et al., although occasionally a few stars appeared on the horizon only to quickly fade away."[12]

The reasons for this need some analysis. Was it due to the fact that the protocol of the early tests was too loose? Were normal sensory cues being read? Were there poor grading techniques? Did the experimenters' and subjects' will to believe in the reality of ESP cloud the testers' judgment? Was fraud being perpetrated on naïve experimenters? Or was the seeming success due to the enthusiasm of early experimenters?

Let us examine the tests of Hubert E. Pearce, Jr., a Methodist divinity student. Rhine embarked on extensive testing with him. In the early tests the observer and subject sat opposite each other at a table on which about a dozen ESP card packs lay. One of the packs was handed to Pearce. He shuffled them, and it was cut by the observer. Pearce removed the top card, laid it on the table face down, and continued calling the cards without looking at them. Rhine said that "there is no legerdemain by which an alert observer can be repeatedly deceived at this simple task in his own laboratory."[13] How naïve Rhine was, for it would have been possible for a skilled conjurer to use trickery, possibly reading the backs of the cards or marking them in some way.

Over a period of extensive testing with Rhine, Pearce scored very high—unless visitors were present, at which time there was a drop in Pearce's scores. Rhine reports that they noticed "that when someone dropped in to watch Pearce work the score at once dropped down."[14] On one occasion when Rhine had a magician present, Pearce's scores were low, but he attributed this to the fact that Pearce was somewhat ill with tonsillitis. Was Pearce deceiving Rhine and getting consistent results by some trick? Rhine discounts the fraud hypothesis in explaining the high test scores of his student subject, but fraud should not be so easily dis-

11. John Palmer, "Review of J. B. Rhine's ESP Research," in Rao, ed., *J. B. Rhine,* pp. 38-39.

12. Ibid., p. 44.

13. Rhine, *Extrasensory Perception,* p. 98.

14. Ibid., p. 101.

missed. Rhine reports in *Extrasensory Perception* that "all of Pearce's work has been carefully witnessed" and that he had the "fullest confidence in his honesty."[15]

But this confidence in Pearce's honesty cannot be so easily accepted by others. Had Pearce devised some way of reading the cards by normal means? The early Rhine tests were done with extremely weak controls. Since Rhine permitted Pearce to shuffle the cards, he could have peeked at the top or bottom cards, thus improving his scores. Moreover, critics have suggested that it was possible to read the cards from the back. The results of the tests were thus questionable.

In a telepathy test George Zirkle, a graduate psychology student, received symbols sent by Sara Ownbey. Rhine reports that these scores were also astronomically high. However, the test conditions were extremely lax. Miss Ownbey did not use cards but thought of a symbol. Zirkle called out the symbol from the next room, and Miss Ownbey then wrote down Zirkle's call and the original image in her mind. There was not even a third party to record Zirkle's call. The scores all depended upon the accuracy and honesty of Miss Ownbey—hardly a rigorous test procedure. There were great weaknesses in all of Rhine's early experiments. There was the possibility of the subjects learning to identify the cards, from irregularities or specks on their backs, especially where the subject could lift the cards or witness the grading. It was found that some of the commercially prepared cards were warped and that this enabled a person under the right lighting conditions to read their design through the backs. Some cards apparently showed irregularities in the design at the edges. When psychologists pointed this out, Rhine gave instructions that the cards must be screened from the subject. Other critics pointed out that grading errors might also intervene to some extent, but this would hardly seem to account for all the deviation from chance.

The most important series of tests were devised to see if Pearce could read the cards at a distance. This is one of the most famous tests in the annals of parapsychology: the Pearce-Pratt tests conducted in 1933-34 under the supervision of Joseph G. Pratt, a psychology graduate student. Since this experiment is often cited in parapsychological literature as a classic proof of ESP, it is instructive to focus on it. The test, called the "Campus Distance Series," was supposed to demonstrate whether ESP operated at a distance. A test of clairvoyance, it did not depend upon someone's being in close proximity to the deck of cards.

Pratt was in a room in the social-sciences building, and Pearce sat in a cubicle in the library one hundred yards (in one part of the experi-

15. Ibid., p. 97.

ment) or two hundred and fifty yards away. The procedure was as follows. Both students synchronized their watches. Pratt observed Pearce leave the building and cross a quadrangle to enter the library building. At an agreed-upon time Pratt would draw the top card from the deck without looking at it and place it face down on a book on the table. He would do this every minute. Pearce, in his cubicle, was supposed to write down his guess every minute. At the end of a run of twenty-five cards there would be a five-minute break, and Pratt would then begin another run of twenty-five cards with a fresh deck. At the end of a sequence, Pratt would record the runs. Both Pearce and Pratt would seal their answer sheets, after they had made a duplicate copy for themselves, and they later transmitted them to Rhine. There were thirty-seven sittings. Only in the last three was Rhine present with Pratt.

The scores obtained at these sittings were extraordinary. According to chance, we would expect five calls out of each of the twenty-five runs, but in four series of sittings the calls were 9.9 (100 yards), 6.7 (250 yards), 7.3 (100 yards) and 9.3 (100 yards). The odds against these were astronomically high: 10^{22} to 1. Was this definite proof of ESP, as some parapsychologists have maintained?[16]

Various skeptics have pointed out that the experiment was defective in many respects and could have lent itself to cheating. Thus C. E. M. Hansel deems it unfortunate that Pearce had no supervision during the experiment. If Pearce had wished to, he could have cheated. He knew exactly where Pratt was at any moment in the experiment. Thus Pearce could very well have sneaked out of the library and possibly have observed Pratt when he turned the cards over at the end of the sitting and read them as he was recording their scores. Hansel maintains that he could have peeked through a window in the corridor. There was also a trap door in the ceiling overlooking Pratt's table. It could have been possible for an intruder to have seen the cards from a hole in the trap door. Thus the experiment was not airtight, and if Pearce had wanted to, he might have (with or without an accomplice) contrived a method of trickery. He would not have been the first to do so. It is difficult at this point to say whether he did, but some interesting facts about Pearce emerged, which aroused great suspicion.

Pratt gives some background and information on the circumstances that led to the Pearce-Pratt experiment and on what happened subsequently. He reports that Rhine told him that one day he had given a talk at the School of Religion and that Hubert Pearce remained after the

16. See, for example, Charles Tart, *Psi, Scientific Studies of the Psychic Realm* (New York: E. P. Dutton, 1977).

lecture to ask questions of Rhine. Pearce told Rhine that he thought that he might have ESP, and he also related remarkable psychic things that his mother was capable of doing.

> Hubert said he had seen his mother place her hands lightly on one end of a heavy, solid oak table while some man tried to hold the other end to keep it rising from the floor. His mother had warned them that if they did not release the table it would break. Hubert said that he had seen it buckle and split across the middle.[17]

Did Hubert believe this or was he leading J. B. Rhine on?

Pearce agreed to be tested by Rhine, and he was paid for his services. Pratt, who until then reported that he invariably received no positive results on ESP tests, now found that while testing Hubert in his dormitory room he began to achieve test scores at the very high double-chance rate. From then on Pearce was tested in Rhine's laboratory. Pratt reports that on one occasion "Hubert got all twenty-five cards right in one run."[18] This testimony should have been enough to raise anyone's suspicion about chicanery, for it was an incredible feat against the laws of chance. The tests continued for eighteen months. Louisa Rhine confirmed that at one stage Pearce was being paid a hundred dollars a month by Rhine. Continued financial support during the Depression years depended on his being successful—here was a possible motive for legerdemain.[19] The most important stage of the tests were the "Campus Distance Series," which we have already described. Did Pearce use a method suggested by Hansel: tiptoe and peek through the window or trap door? Pratt's description of the tests suggests still another possibility. He reports as follows:

> When the day's work was finished, Hubert and I, before we moved from our places of working, made copies of our lists of calls and cards, respectively. (These were for our own personal checkup in case we met later during the day.) Then each of us immediately sealed up the original copy of his record in an envelope, signed his name, put the date on the outside, and kept this sealed record of his part of the day's results in his own possession until he could hand it over to Dr. Rhine directly.[20]

Apparently there was an interval (for thirty-four of the sittings) between the time that they completed the day's sittings and the time that they each delivered their scores to Rhine. Since they each kept a copy of the calls and cards, it might have been possible for Pearce to have changed

17. J. Gaither Pratt, *Parapsychology: An Insider's View of ESP* (Metuchen, N.J.: Scarecrow Press, 1977), p. 55.

18. Ibid., p. 57.

19. Louisa E. Rhine, *Something Hidden* (Jefferson, N.C.: McFarland, 1983), p. 197.

20. Pratt, *Parapsychology,* p. 58.

some of his scores after seeing Pratt and comparing their notes, and only then to have sealed the scores in the envelope that was to be delivered later to Rhine. We do not know where Pratt kept his copy, and if Pearce might have surreptitiously obtained Pratt's copy and used it before filling out his copy for Rhine. Since there were several sittings at which Pearce scored at below chance, it was not necessary for him to have encountered Pratt each time.

The sittings with Pratt went on over a period of several months. Professor MacDougall had followed the earlier stages of the experiment with great interest, and he warned Rhine that skeptical scientists would prefer to believe that there was some form of collusion between Pratt and Pearce rather than that ESP was present. This prompted Rhine to become involved at the final subseries of three sittings, at which time he was present with Pratt in the room. This time we are told that Pearce appeared immediately after the day's work with the results. But, incredibly, Rhine did not have Pearce watched in the library, and he could have sneaked out of the library, as Hansel suggested, or used some other form of trickery. In any case, at the conclusion of the tests Rhine pronounced the entire series bona fide, and he considered, as a whole, the chance odds were 1 in 10,000,000,000,000,000,000,000,000!

What happened to Pearce after these experiments was similar to what happened in virtually all other cases where psychic stars have emerged—the decline effect set in; Pearce lost his psi-ability and was never again able to demonstrate it. Pratt reports tersely:

> Shortly after this experiment was finished, Hubert lost his ability to score above chance in the card tests. This happened rather suddenly, when he came into the laboratory one day and said he had received some very disturbing news from home.[21]

Rhine informs us that after the Pearce-Pratt test the next step involved a test where Pearce was taken a distance of two miles. Rhine reports what happened in these tests: "things went wrong from the start . . . After things were finally straightened out, there was no appreciable success."[22] Further tests are also cited by Rhine in which Pearce was taken in a car to different places, but again the results were negative. One can speculate why he suddenly lost his powers—if they were ever genuine. If the hypothesis that Pearce had used fraud was true, it would explain his sudden lack of results, since he was removed far from the target and would not be able to use subterfuge to achieve success. Or did his conscience tell him he should give up hoodwinking a naïve professor,

21. Ibid., p. 61.
22. J. B. Rhine, *New Frontiers of the Mind* (London: Faber & Faber, 1938), p. 165.

who so desperately wanted to believe in the reality of ESP that he could be taken in easily by his students?

Pratt reports that Hubert Pearce went on to take up his profession as a minister. Pratt visited Reverend Pearce many years later, in 1959, in his home state of Alabama. He tried some ESP tests, but there was no success. The formerly gifted psychic displayed no ESP at all. The same fate befell all of Rhine's other gifted students.

John Beloff, the noted parapsychologist, states that "never again were we to witness such an auspicious concentration of psychic talent."[23] He says that cynics have attributed Rhine's success to lax conditions in the experimental lab, and notes that once conditions are tightened up in deference to critics, the powers disappear. Rhine, however, believed that the reason he was so successful was because of the intense early excitement in the laboratory. Was the famous "sheep/goat effect" at work? Gertrude Schmeidler reports in a famous paper that those who believe in ESP are more likely to get results (sheep) than those who do not (goats), and that the goats (skeptics) tend to inhibit results.

In this regard, a key question that can be raised is this: Have other researchers—neutral, skeptical, objective inquirers—reached similar results? Were they able to replicate Rhine's experiments? The need for replication is essential in science. Before we can accept the existence of an effect, we need corroboration by objective and impartial observers. Do such exist? Rhine and other parapsychologists have indicated that ESP has been demonstrated in the laboratory and that there have been numerous decisive replications—though high-scoring subjects have proved very rare indeed. Skeptics have pointed out that this is all too infrequently the case and that experimenter after experimenter—sheep as well as goats—have tried to get positive results, only to have them end in abysmal failure; further, where there is any success, some fatal flaw is usually found in the experiment, or outright fraud is discovered. Several other tests have been cited in the parapsychological literature as evidence for ESP, but in each instance skeptics have raised some objections to the tests and have insisted that before they can accept the reality of ESP as proven fact, there must be some replicable experiment to which they can refer.

The Soal Scandal

Most likely, the most important single development in contemporary parapsychology since Rhine was the effort by Samuel G. Soal to replicate

23. John Beloff, "Historical Overview," in *Handbook of Parapsychology*, ed. Benjamin B. Wolman (New York: Van Nostrand, 1977), p. 18.

the work of Rhine in Great Britain. Soal had a strange career. He began as a believer in psychical phenomena, and at one time he was an "automatist" and even a sensitive known anonymously as "Mister V." As he became more absorbed in the subject, he became one of the best-known investigators in Britain. In the 1920s and the 1930s he participated in radio tests of telepathy and clairvoyance, though without any statistical success. He also conducted tests on Fred Marion (Josef Kraus), a well-known vaudeville telepathist, and he demonstrated that Marion was using normal sensory cues to effect his remarkable act. Marion could read the backs of cards by touching them, and he could locate objects especially hidden in boxes by carefully observing the reactions of people in the audience who knew in which box they were located. Soal resolved to see whether he could replicate Rhine's test, using the Zener cards. He never allowed the subjects to see the backs of the cards and never used the same deck twice during a test sitting. Beginning in 1934 and for several years thereafter until 1939, he tested a great number of people for ESP. In the test pool were many individuals who considered themselves to be spiritualists, mediums, or psychics, including Mrs. Eileen Garrett, a well-known American psychic. All together he tested a hundred and sixty subjects for clairvoyance and telepathy in a total of 128,350 trials. The results were largely negative. Indeed, Soal confessed that it was virtually impossible, by using the Zener cards, to find anyone in England who could demonstrate that they possessed ESP.[24]

In 1939, Whately Carington, a psychical researcher, suggested that Soal reexamine his data to determine whether or not any of his subjects manifested precognitive abilities. Soal was to examine the card immediately before and after the card was called, to see whether a displacement effect was present. One must be very cautious about reanalyzing the data after a test originally designed for another purpose has been completed. Nevertheless, although Soal was apparently himself opposed to this process, he decided to follow Carington's advice; he found that two of his subjects, Basil Shackleton, a London photographer, and Mrs. Gloria Stewart, displayed above-chance scores when recalculated in this way. On the strength of this, he resolved to test them at greater length. What followed was perhaps the most perplexing and controversial drama in parapsychological research to date. Soal achieved a kind of breakthrough with extraordinary results, so much so that the tests were widely cited as the mainstay of evidence for ESP. The implications of these tests (if allowed to stand) were to shake the very foundation of the sciences.

---

24. S. G. Soal and F. Bateman, *Modern Experiments in Telepathy* (London: Faber & Faber, 1954).

The first series of tests with Basil Shackleton were supervised by Soal in cooperation with Mrs. K. M. Goldney, a member of the Council of the Society for Psychical Research. They were conducted during difficult conditions at the height of World War II, from 1941 to 1943 in London. The second series of tests were with Mrs. Gloria Stewart and lasted from August 1945 until January 1950. They were done in collaboration with F. Bateman and are known as the Soal-Bateman tests.

We will devote most of our attention to the tests with Basil Shackleton.[25] Most of these were held in Shackleton's studios, as Soal believed that this would provide a more propitious setting. The setup of the tests was ingeniously contrived to guard against loose protocol and the possibility of fraud; and they were supposed to test telepathy, precognition, and clairvoyance. Soal abandoned the Zener cards and produced a new set of cards, each of which depicted the face of an animal in color: elephant, giraffe, lion, pelican, and zebra. In general four persons were involved in each sitting: (1) Basil Shackleton, the telepathic percipient (P); (2) a sender or agent (A); (3) an experimenter controlling the agent (EA); (4) an experimenter controlling the percipient (EP); (5) in addition on most occasions a fifth person was present, who assumed the role of observer (O).

The agent (A) was seated at a card table in one room. The table was lit by a strong photographic lamp. Placed on the table was a large plywood screen with a hole (three inches square) in the center. Resting on the table facing the agent was an open rectangular box. Lying in the box in a row were the five animal cards. On the opposite side of the screen sat the experimenter (EA). He had five cards, though these were numbered one through five.

Seated in an adjoining anteroom was Shackleton, observed by another experimenter (EP). There was no physical contact between Shackleton and the agent. Sometimes the door between the rooms was shut, though it was usually slightly ajar. The purpose of the experiment was to see whether the agent, looking at a card, could transmit the image telepathically to Shackleton. Could Shackleton pick up the image at the same moment that the agent was looking at the cards or beforehand (a precognitive test) or afterwards (a postcognitive test)?

The experimental design was supposed to be foolproof and to guard against chicanery. There were two ways in which the cards to be viewed by the agent were selected. In the first set of sittings the cards were selected by means of a list of random numbers (1 through 5) prepared

25. These are described in S. G. Soal and K. M. Goldney, "Experiments in Precognitive Telepathy," *Proceedings of the Society for Psychical Research* 47 (1943), pp. 21-150.

beforehand by Soal (or on occasion by someone else) and brought to the experiment under lock and key. In this series of sittings, Soal selected numbers following the random lists and presented them in sequence at the aperture. Before each run of fifty calls, the agent would shuffle the animal cards on the other side of the screen and lay five in a row face down. The experimenter (EA) would determine which card the agent would look at. Shackleton, in the next room, would write down his guesses of the animals (using letters) on a score sheet. The calls were generally of two or three seconds duration. After each run of fifty, the code designating the order of animal cards was written down on the score sheet. The cards would be reshuffled and a new list of fifty random numbers would again be produced. This procedure was generally followed for eight times (400 guesses) for each sitting. At the conclusion of each sitting, the two sets of sheets (the agent and the percipient) were compared by the experimenter and the observer present. The letters designating animals were translated into numbers. The guesses were graded in terms of straight hits, or those displaced by one or two cards before or after. This was designed to test for precognition and to see whether Shackleton could guess a card before it was drawn and seen by the agent. After the grading was completed, witnessed, and signed, a duplicate set was made and in the presence of three persons was stamped and mailed to C. D. Broad, professor of philosophy at Cambridge University.

A second procedure for selecting the cards to be called was as follows: instead of using the lists of random numbers provided by Soal, the experimenter (EA) drew a color counter (1 through 5) from a bag or a bowl and showed it at the aperture. This was to designate the card that the agent was to turn over and look at. At many sittings Shackleton did not always record his guesses in writing, but this was done by the experimenter observing him (EP) as he called them off.

All together forty sittings were conducted. Many distinguished psychical researchers and professors observed them as they were being carried on, including philosophers H. H. Price and C. E. M. Joad, Professor C. A. Mace, a psychologist, and Sir Ernest Bennett, a member of Parliament. Different agents were used, and the roles of EA and EP and of Soal and Mrs. Goldney were shifted.

The results of the telepathy and precognitive experiments were declared no less than incredible, though Shackleton scored only at the chance rate for clairvoyance. However, the probabilities for precognitive telepathy ranged anywhere from 10^{11} to 10^{35} to 1. C. D. Broad, who monitored the score sheets, gave a full account of the results. He stated emphatically:

There can be no doubt that the events described happened and were correctly reported; that the odds against chance-coincidence piled up by billions to one.[26]

Broad also said that these findings conflict with the "basic limiting principles" by which we normally interpret the world and that as a result these principles would have to be radically modified. The Soal-Goldney tests thus had profound implications for both science and philosophy. If true, we would have to modify our concepts of mind, matter, causality, time, and ways of knowing. Those basic limiting principles are based upon our past observations and the generalizations we make from them. We come to infer that like causes would have like effects. This is the common-sense view of reality. Yet if the Soal-Goldney experiments were to be taken as veridical, then these governing principles must be questioned, for an event could be known before it occurs, and information could be telepathically conveyed without direct sense perception, verbal communication, or inferences from them. It was clear that our most basic conceptual schemata would have to be altered, for there now seemed to be sufficient, indeed overwhelming, evidence to require it. In this regard, the evidence seemed to meet the demand that the witnesses be trustworthy and that the experimental conditions be rigorously carried out. They also appeared to be fraud-proof.

Broad reaffirmed a widely held view of those involved in the Soal tests when he declared: "The precautions taken to prevent deliberate fraud or the unwitting conveyance of information by normal means are . . . seen to be absolutely water-tight."[27] Many scientists agreed with this interpretation and considered the Soal experiments to have been the most carefully conducted investigations ever carried out in this area. Moreover, the results obtained with Basil Shackleton, according to Soal, were replicated by the tests of Gloria Stewart and thus corroborated the existence of psi-phenomena. During the Stewart tests a hundred and thirty sittings were conducted, most of them at Mrs. Stewart's house. There were a total of 37,100 trials. The experimental conditions and protocol were similar to the Shackleton tests, though according to Hansel, they were much less stringent. The results obtained with Mrs. Stewart on telepathy tests were equally impressive; they reached as high as 10^{70} to 1 for some runs!

Had ESP finally been definitely confirmed in the laboratory, as many psychical researchers held, or were there still problems with the results?

---

26. C. D. Broad, "The Relevance of Psychical Research to Philosophy," *Philosophy* 24 (1949), pp. 291-309.
27. C. D. Broad, *Philosophy* (1944), p. 261.

Soal was hailed for his work and elected president of the S.P.R. in 1950-52 in recognition of his contributions. Some skeptics were still uncertain about the results, however, notably George R. Price and Hansel, both of whom expressed their doubts about the rigorous character of the tests. In a famous article published in 1955 in *Science* magazine, Price outlined his misgivings.[28]

Price opened by saying that believers in telepathy, clairvoyance, precognition, psychokinesis, and other psychic phenomena appear to have achieved a major victory and silenced their opposition. This victory, he said, was due to an impressive amount of experimentation and argumentation. Price cited the list of the card-guessing experiments of Rhine and Soal, which showed enormous odds against chance. The statistical methods used by parapsychologists are valid; thus it is difficult to reject these data. Skeptics who do, usually do so out of ignorance of the testing that has been done. Moreover, if parapsychology is true, Price said, then there are not only important philosophical implications but also practical applications of great benefit.

Price, however, said that although at one time he believed in ESP, he now rejected it, the reason being that he had been persuaded by Hume's argument against miracles. Price also quoted Lucian, the Greek author, who noted long ago that even though one might not be able to detect the precise trick by which an illusion is produced, we should nevertheless retain our conviction that the entire thing is a lie and an impossibility.

Price maintained that there are alternative explanations for ESP; for example, the findings of parapsychologists may be dependent on clerical and statistical errors, the unintentional use of sensory cues, mildly abnormal mental conditions, or even elaborate fraud. The last explanation, he thought, is especially compatible with the basic limiting principles of modern physics and common sense as outlined by Broad. If this is the case, he asked, why should we not reject parapsychology? The alternative hypothesis that seems most applicable to explain above-chance results is that some men lie or deceive themselves. This explanation fits within the existing framework of science and enables us to reject parapsychology rather than the body of modern scientific knowledge.

Defenders of parapsychology claim that they accept both science and the supernatural. This means that they accept the existence of magical or animistic phenomena and of things happening in nature without a known cause. The parapsychologist must admit that there are no physical mechanisms by which messages are transmitted and objects can be changed.

28. George R. Price, "Science and the Supernatural," *Science* 122, no. 3165, (1955), pp. 359-67.

When he postulates that a "mind" accomplishes these "events," this does violence to everything we know about the world. Why should we accept this kind of explanation, Price asked?

We need to come back to an alternative hypothesis, said Price, namely that there may have been fraud in the experiments. He stated emphatically that he knew of no evidence showing whether or not Soal cheated. He simply wished to demonstrate that "Soal *could* have cheated if he wanted to" and that Soal's experiments were not watertight but were full of loopholes. He went on to suggest several methods by which fraud could have been perpetrated. Soal could have cheated by himself or he could have had one or more collaborators, whether the agent, the percipient, or the other experimenter. Price reviewed various tricks that could have been used to achieve Soal's results. He said that it was clear that Soal's work was *not* conducted "with every precaution that it was possible to devise." He added that the work would have been much more fraud-proof if Soal "had simply had many different Agents 'send' directly from lists prepared by outsiders and given directly to the Agent at the start of each run." According to Price, we cannot accept the findings of parapsychology until one test is devised where the opportunity for fraud or error is entirely absent. Until such an experiment is carried out, we should not accept any evidence for ESP that has thus far been adduced These were strong words and they raised quite a furor, for they implied the strong possibility of fraud without directly accusing Soal.

Both Soal and Rhine replied to Price in the pages of *Science*.[29] Soal expressed his "amazement" at Price's accusation, claiming that Price was unable "to produce the least fragment of factual evidence that any such fraudulent malpractice ever took place." He says that no English scientific journal would have supported "such a diatribe of unsupported conjecture." Soal said that it was implausible to suppose that he could have persuaded so many distinguished men and women "to enter into a stupid and pointless collusion to fake the experiments over a period of years." They would have had everything to lose by the besmirching of their academic reputations. He insisted that Price had "not taken sufficiently into account the high quality of the personnel connected with these experiments." He stated that in certain of the Shackleton experiments the lists of random numbers were prepared "two times" by Wasserman, a mathematical physicist, and that he himself did not see them until the experiment was concluded. He adds that belief in ESP will be strengthened after more experiments by competent experimenters achieve more high

29. S. G. Soal, "On 'Science and the Supernatural,'" *Science* 123, no. 3184 (1956), pp. 9-11; J. B. Rhine, "Comments on 'Science and the Supernatural,'" ibid., pp. 11-14.

scores. In retrospect, the reply by Soal is rather pathetic, given the revelations that were to ensue.

J. B. Rhine replied to Price by pointing out that his paper illustrated the head-on collision that is inevitable when the findings of parapsychology come face to face with the "physicalist," "mechanist," or "materialistic theory of man." According to this theory, ESP is impossible; hence the parapsychologists who claim to have results must be fakers. But, said Rhine, the metaphysical doctrine of mechanistic philosophy defended by Price has been contradicted by the experimental results of parapsychology. Hence, it is the dominant physicalist model that must be abandoned, not parapsychology. Rhine called for a more tolerant attitude by American science toward parapsychology and asked that it be given an open hearing.

The crucial issue is whether the experimental findings of parapsychology hold up under careful scrutiny. Were Soal's experiments properly conducted? Was the implication of fraud simply a red herring? Can we infer, on the basis of Soal's experiments, that ESP has been demonstrated and that our materialistic and/or physicalist views of the universe are thus in need of modification?

The responses of Soal and Rhine did not satisfy their skeptical critics. Hansel continued to believe (1) that ESP was impossible because it violated the laws of physics, and (2) that therefore there must be some explanation, including bad experimental design or fraud, that could account for Soal's extraordinary results. In his book *ESP: A Scientific Evaluation* he continued to speculate about how Soal and others might have cheated.[30]

Of special interest was the fact that one of the three agents, Gretl Albert, with whom Shackleton had received above-chance results, had complained after one sitting that she had seen Soal, while serving as EA, altering the figures several times while checking Basil Shackleton's guesses. Apparently, Mrs. Albert had remarked, "Oh, you are altering the figures?" But she said that no one at the sittings took notice of her remark. Mrs. Albert had taken part in the fifteenth and sixteenth sittings, when the normal sitter, Ruth Elliot, was absent. Mrs. Albert was not a professional or academic observer, but took part at the behest of Mrs. Goldney, who needed a substitute for the normal agent. Immediately after the sixteenth sitting Mrs. Albert informed Mrs. Goldney what she had seen. Mrs. Goldney checked the original score sheet a few days later and had a colleague do the same, but she could find no evidence of alterations in the

---

30. C. E. M. Hansel, *ESP: A Scientific Evaluation* (New York: Scribner's, 1966), and *ESP and Parapsychology: A Critical Re-evaluation*, rev. ed. (Buffalo, N.Y.: Prometheus Books, 1980).

test scores. She reported this to Soal, who became indignant at Mrs. Albert and refused to have her participate in the experiments again. Mrs. Albert's complaint was not included in the original Soal-Goldney report published in 1943 in the *Proceedings of the Society for Psychical Research*. Soal did not agree to include mention of it. It only came to light in 1959, when Christopher Scott heard about the incident and interviewed Mrs. Albert. Mrs. Albert claimed to have seen Soal changing 1's into 4's and 5's several times. An exchange was published in 1960 in the *Journal of the Society for Psychical Research* with Goldney and Soal responding. But one could not easily prove whether Mrs. Albert's allegation had merit; perhaps Soal was simply tidying up the sheets and not deliberately altering them.

For the first sixteen sittings, the practice had been for Soal to bring to the experiment lists of random numbers from which the targets were selected by him. After Mrs. Albert's allegations, the protocol was changed. After the sixteenth sitting Soal never again acted as EA. There was a switch to the counter method of selecting the target and Soal assumed the role of EP, supervising Shackleton.

The Albert charges remained unresolved for many years. In 1971 R. G. Medhurst attempted to throw some light on the affair, claiming that Mrs. Albert had seen Soal altering the figures not once but four or five times. Moreover, he showed that there was an excess of hits on targets 4 and 5 in the sixteenth sitting. He thought that the alterations were made not in the Shackleton guesses but in the random lists of targets.[31] However, he was unable to identify the method Soal said that he used in preparing his random lists of targets. In 1974 Christopher Scott and Philip Haskell published a new study, "Fresh Light on the Shackleton Experiments?", in which they confirmed the fact that for the sittings in which Mrs. Albert was present, there was strong evidence that the number sequences had been tampered with by Soal.[32] Unfortunately, Soal claimed in 1956 that he lost the original data sheets, having left them on a train in 1945, so that the only records that could be examined were handwritten or typed copies of the original or photocopies. Scott and Haskell, using computer analysis, were able to confirm that Mrs. Albert was correct in claiming that 1's were changed to 4's and 5's in the check-up period. And they found that there was a deficiency of 1's in the targets where the guesses of 4's and 5's were made, a factor that was below-chance expec-

31. R. G. Medhurst, "The Origin of the 'Prepared Random Numbers' Used in the Shackleton Experiments," *Journal of the Society for Psychical Research* 46, no. 39 (1971), pp. 44-55.
32. *Proceedings of the Society for Psychical Research*, vol. 56, part 209 (Oct. 1974), pp. 43-72.

tation. Moreover, they suggested that Soal had probably prepared lists of random numbers in which there were an excess of 1's and a deficiency of 4's and 5's, which would be changed later. For the sittings prior to Mrs. Albert's appearance, the witness for Soal was Miss Elliott, who was later to become Soal's wife; she either did not monitor him carefully or was in on the trick. She had no official status professionally outside of these tests. Scott and Haskell found, however, that this analysis applied to only three of the sittings. Perhaps, they said, Soal used other methods of counterfeiting data at the other sittings. In any case, they concluded that there was statistical evidence for manipulation of the target lists by Soal and hence independent corroboration of Mrs. Albert's allegation.

Following this article, various parapsychologists replied; all of them rejected this interpretation, some of them indignantly. Mrs. Goldney reviewed the events as she recalled them and stated, "At no time did I have cause to doubt Soal's entire honesty"; she insisted that she had never seen Soal altering any of the figures."[33] "I . . . find it almost impossible to believe that the Soal whose work I knew so well had cheated."[34] For someone who intended to cheat, Soal was surely asking for trouble, she said, by inviting herself and twenty-one well-known people to become involved. If he wanted to cheat, she said, it would have been far easier if he had conducted the tests alone. Similarly, Soal invited Bateman and many other qualified observers to work with him in the spectacular 130 sittings with Mrs. Stewart over four and one-half years. Moreover, following the Stewart tests, for several years thereafter (1954-58) he received negative results in other tests he conducted at Birkbeck College. (Though, I should add, Soal later claimed to have achieved excellent results when he tested the Jones brothers, two lads from Wales, for telepathy. We will not review these tests here, but it has been shown that the experimental conditions were not very rigorous, and Hansel has raised the question as to whether the Welsh boys used codes and signals.)[35]

Following Scott and Haskell's paper, other parapsychologists came to Soal's defense. C. W. K. Mundle, a philosopher, questioned the reliability of Mrs. Albert's testimony and found inconsistencies in it. He pointed out that the critics claimed to find anomalies in three sittings, but if there was a trick they had not showed how it was done in the other sittings.[36] Robert H. Thouless, a noted parapsychologist, said that the

33. "The Soal-Goldney Experiments with Basil Shackleton (BS): A Personal Account," ibid., p. 78.

34. Ibid., p. 80.

35. Hansel, *ESP: A Scientific Evaluation*, pp. 169-87; see also Soal and Bowden, *The Mind Readers*.

36. C. W. K. Mundle, "The Soal-Goldney Experiments," *Proceedings of the Society for Psychical Research*, vol. 56, part 209 (Oct. 1974).

case against Soal was based solely on the unsupported testimony of a single witness on a single occasion. He also did not find it easy to imagine that Soal would have been interested in carrying out a fraudulent ESP experiment. He said that, as late as 1939, Soal was strongly committed to the view that there was no experimental evidence for ESP and that an individual guessing cards cannot better the laws of chance. He added that even if it were proved that there had been some manipulation of the scores "this would not justify a conclusion that none of the results in the Shackleton experiments were obtained by ESP."[37] Here we have the same curious argument repeated—that even if it is revealed that *some* cheating is going on, this does not invalidate other parts of an experiment where ESP is real! John Beloff's reply is entitled "Why I Believe That Soal Is Innocent."[38] He admits that Soal made two tactical blunders: (1) He never gave anyone else an opportunity to test Shackleton and Stewart. (2) He did not provide sufficiently explicit accounts as to how the target sequences were derived. However, he suggested that the odd sequences of numbers that Scott and Haskell reported could have been due to Shackleton's ESP and not to trickery.

J. G. Pratt also defended Soal against the charge of dishonesty. He asserted that if Soal had perpetrated a hoax, "it was one of much greater magnitude and one supported with more cleverness and more years of tedious work than anything heretofore known from the history of science," "If Dr. Samuel George Soal is to go down in scientific history," he said, "as the greatest scientific hoaxster of all time, let us have the decency to render our verdict in the light of all the evidence."[39] Pratt also reported that he visited Shackleton in October 1973 in Johannesburg, South Africa, where the latter resided, and that Mr. Shackleton "was horrified" at hearing the charges that Dr. Soal had cheated. Shackleton said, "I am sure that Dr. Soal did not cheat and that he is absolutely innocent of the charge."[40] Moreover, Shackleton told Pratt that during the tests he had been interested in the scores he had achieved and that he himself was present as an observer during the check-ups. And so we may ask: Were the skeptics unnecessarily petty and picky and had they constructed a flimsy case against Soal based upon the uncorroborated testimony of one woman? Shall we accept the argument that, even if there was some manipulation of the data in some sittings, this should not invalidate the remainder?

---

37. R. H. Thouless, "Some Comments on 'Fresh Light on the Shackleton Experiments,'" ibid., p. 91.
38. John Beloff, "Why I Believe That Soal Is Innocent," ibid., pp. 93 ff.
39. J. G. Pratt, "Fresh Light on the Scott and Haskell Case against Soal," ibid., p. 108.
40. Ibid., p. 111.

The next event that unfolds in this dramatic debate is the publication of a paper in 1978 by Betty Markwick, which presents clear evidence of data manipulation in at least some of the sittings.[41] Markwick relates that her own enthusiasm for parapsychology had been aroused by reading Soal and Bateman's book *Modern Experiments in Telepathy*. Since the validity of the research had been seriously called into question by critics, she had a special interest in seeing the issues resolved. Although she found Scott and Haskell's paper impressive, it was hardly conclusive.

Like Medhurst, Markwick attempted to determine how Soal obtained his original random-number sequences. Soal had said, for example, that he used Chambers logarithm tables and Tippets random-number tables (which Soal said had been used three times), but even though she used extensive computer analysis she was unable to locate the chart of random numbers. By chance she discovered that there were some duplications in the long sequences in the target numbers but these came from different sittings. Using a computer search, she reports that "a very suspicious feature came to light." Some of the repetitions that she uncovered in target sequences were not exact, and on occasion they exhibited extra digits in the sequences "as though digits had been inserted into one of the pairs of sequences (or omitted from the other)." The remarkable thing about this was that "these extra digits showed a marked tendency to correspond to hits." She noted that in one sequence, for example, after every five digits an extra digit was inserted, and that "every one of the five extra digits corresponded to a hit." After further investigation she found that "the suspicious hits virtually accounted for the 'ESP' effect." She also found that about three or four of the extra single digits corresponded to hits and that the odds against such a high rate occurring by chance were thousands of millions to one. When the suspicious extra digits were omitted, the scoring rate fell to the chance level. Thus, she found two sequences of nineteen numbers and another two of twenty-four numbers from different sittings that match, except for the inserted digits which were hits.

Markwick also found that at one of the sittings in which Dr. Wasserman was supposed to prepare the random-target list, the same phenomena occurred, "thus raising the question of possible substitution of sheets by Soal." The real question at issue was: Where did Soal derive his random-number sheets from? She found that he often repeated the sequences in different sittings. She speculated that to save time Soal had at times prepared lists by copying from earlier record sheets, leaving out the

41. Betty Markwick, "The Soal-Goldney Experiments with Basil Shackleton: New Evidence of Data Manipulation," *Proceedings of the Society for Psychical Research*, vol. 56, part 211 (May 1978), pp. 250-77.

digits that corresponded to hits. Markwick was thus able to compare only those sittings where she found sequences that were similar in overlap, the differences noted in the extra digits being anticipated. This was the case in about 20 percent of the total lists of numbers. In all of these cases she found decisive evidence of digital insertion and manipulation.

How should we interpret these puzzling anomalies? The evidence seems consistent with the hypothesis advanced by Scott and Haskell that target lists were prestacked with ones written in pencil and later subsequently altered in such a way as to achieve hits. In other cases perhaps, said Markwick, Soal was in collusion with Shackleton and had prearranged blank spaces for the hit. "I shall not dwell on *how* the supposed manipulation might have been effected, or whether it could have been done single-handedly."[42] A number of manipulation methods could have been devised and their uses varied to suit the particular conditions encountered in the sittings.

Interestingly, in sittings 1 through 17, when Soal brings in his random sheets and serves as EA, there is an unbroken bond of significant scores. After sitting 17 Mrs. Goldney informs Soal of Mrs. Albert's allegations. The next sitting, number 18 at which H. H. Price was an observer, shows no significant result at all. Soal changes his role at sitting 19 and thereafter to EP, whereby he observes Shackleton. Here Shackleton indicates or calls off his guesses and Soal records them; Soal is not observed or only intermittedly so. Only in sitting 24, in which Soal is observed throughout, does Shackleton again have only chance scores. The question which Markwick asks is whether Soal has "misrecorded" Shackleton's guesses. Or, as Hansel suggests, did the score sheets given to Shackleton by Soal as the sitting began contain slight pencil marks identifying the targets to call? This would need to have been done in only three or four positions in each column. Added to this, D. J. West points out that Soal took charge of every session of the experiment and would not let others have a try at it. The only exception was when Mrs. Goldney unexpectedly, and by her own decision, visited Shackleton and conducted some tests without Soal. The results of this sitting were negative.[43] Markwick dismisses the protestations of Soal's defenders that he was a respected scientist and that the Shackleton series precluded fraud. "We can rarely fathom how conjurers achieve *their* feats," she observes, "and perhaps Soal was as clever."

Did Soal consciously cheat? It is apparent to those who knew him that he had a strange personality; he was obsessive, secretive, and subject

42. Ibid., p. 272.
43. *Journal of the Society for Psychical Research* 49, no. 777 (Sept. 1978), p. 898.

to spells of dissociation. One might argue that perhaps Soal unwittingly had been behaving precognitively when he prepared his target lists, anticipating them in advance. This interpretation is unlikely. A more likely explanation is that Soal engaged in "data massage." All the evidence now points in that direction. Falsification of results have been perpetrated by other scientists (such as Cyril Burt, who falsified his data on I.Q., and Walter Levy, who was caught cheating in Rhine's laboratory in the 1970s). Many researchers do not set out with a deliberate deception plan, but when things go wrong with negative results and pressures mount there seems to be some temptation to improve on the results.

Markwick speculates that after Soal embarked upon the Shackleton tests, perhaps the scoring rate began to fade and so he succumbed to the temptation to rectify a temporary deficiency. Believing passionately in parapsychology and wishing to gain scientific recognition, he then went all the way in falsifying his data. She concludes that since data manipulation existed in certain sectors of the target lists and in the absence of a convincing innocent explanation, "the sad and inescapable conclusion remains that all of the experimental series in card-guessing carried out by Dr. Soal must, as the evidence stands, be discredited." Thus the doubts of Price, Hansel, Scott, and Haskell now seem justified.

The Markwick paper was so powerful in its use of analytic techniques of disproof that most of the principal parapsychologists who had formerly defended Soal's honesty were forced to recant, including Mrs. Goldney, Thouless, Pratt, Beloff, and Ian Stevenson. Still, Pratt tried to salvage something of Soal's reputation. He speculated that perhaps Soal's precognition allowed him to select the target sheets! But he conceded that he could not prove this and that "we must set aside, at least for the time being, all of the Soal experimental findings as lacking scientific validity."[44] Robert H. Thouless was mystified as to the motives of Soal's fraud and why he went about it the way he did, but he concludes that "the fact of fraud seems too clearly demonstrated to be doubted."[45] John L. Randall admits that there is "indisputable evidence of experimenter fraud." He tries to save Soal's reputation by speculating that what may have occurred was "psi-evasiveness." Magical thinking again intervenes to rescue the a priori belief in psi. Randall reasons as follows: It has been reported by a number of experimenters that Uri Geller is able to cause scientific equipment to malfunction; in the presence of a psi source (PK) such as Geller's, electronic equipment is likely to malfunction so as to frustrate the experimenter, and in particular the brains of experimenters and observers

44. Ibid., p. 279.
45. Letter, *Journal of the Society for Psychical Research* 49, no. 778 (Dec. 1978), p. 966.

may malfunction. Thus it is possible, he says, that Dr. Soal developed patterns of behavior in the presence of Shackleton that caused him to deviate from his normal integrity as a scientist. The psi-phenomena tends to be "camera shy" and evasive; there is a tendency for it to cover its tracks.[46] How can a scientific rationalist respond to such "reasoning"?

Christopher Scott points out the devastating implications of the entire Soal affair.[47] First, he says, it is clear that even the most respectable academics *may* cheat in their own experiments. Second, the defenders of ESP and other paranormal phenomena have usually argued that the conditions of the experiment were so tight that fraud was ruled out and that therefore we are forced to accept ESP as the only explanation. Betty Markwick did not attempt to respond to this argument. She did not show that the conditions were not tight, nor did she seek to demonstrate that they left room for fraud, nor did she even attempt to explain how the fraud was achieved. She simply provided the evidence that fraud *did* in fact occur. The Soal-Goldney-Shackleton experiment looked fraud-proof, yet fraud occurred. The lesson that Scott draws is that human beings are very poor at deciding whether or not deception is possible in a given situation. Professional conjurers are more skilled at detecting fraud than most people, but even they are not infallible.

The entire matter is even more depressing when we note that a number of respected and distinguished scientists and scholars were intimately involved as observers in the experiments over an extended period of time and had attested to their "water-tight" legitimacy.[48] We do not wish to argue that all human beings are prone to fraud and deception, nor that all work in the area of parapsychology is illustrative of human deception. There are honest, sincere, and well-intentioned men and women of high moral integrity in this and other fields. Many are committed to the standards of ethical behavior requisite in the quest for scientific truth. No one is perfect. Other fields of science have not been immune to fraud. However, there is something basic and endemic about parapsychological fraud. It should be abundantly clear by now that there is a powerful hunger in human beings for the supernatural, occult, magical, spiritualistic, or paranormal—all part of the quest for the transcen-

46. Letter, ibid., pp. 968-69.
47. Letter, ibid.
48. I should add that I debated J. B. Rhine at the Smithsonian Institution in 1978. At that time I raised the question of fraud in parapsychology and cited the case of Soal, having learned about Betty Markwick's remarkable paper. Rhine's reply was, "Why pick on poor Dr. Soal? He is a sick man." Rhine himself had faced a scandal of gigantic proportions when his chief research officer and heir apparent was caught cheating in 1973. After considerable pressure Rhine accepted Walter Levy's resignation and published a report on the entire affair in the *Journal of Parapsychology*.

dental. No doubt Rhine and Soal, even though they were committed to the goals of science, shared the transcendental temptation with many other parapsychologists. The underlying tendency is the will to believe. Perhaps this explains the passionate willingness to accept paranormal realities without sufficient evidence (most likely the case of Rhine) or even to fudge or fabricate evidence (the apparent motive of Soal).

The similarities between historical religious phenomena and spiritualistic and paranormal developments in recent years should be obvious by now. The difference between the religious and the scientific investigation of psychical or paranormal phenomena is that the latter is amenable to careful empirical and analytic controls. It confirms our hypothesis that the transcendental temptation is so strong that it not only beguiles the ordinary man to accept untested religious claims, but it can also cloud the vision and overwhelm the judgment of even the most dedicated researchers otherwise committed to objective and dispassionate inquiry.

Uri Geller: Superpsychic

In this chapter thus far I have focused on two parapsychologists, Rhine and Soal. I have not analyzed the motives or personality of their psychic subjects, Pearce and Shackleton, nor have I demonstrated that either of the latter two were responsible for the extraordinary results or engaged in cheating—though skeptics have inferred or implied that some of Rhine's and Soal's subjects probably did.

It is time that we redirect the focus of this chapter by examining the subject rather than the observer. Let us look at the modern-day "psychics" or "super-minds," who claim to have extraordinary mental powers. The appearance of a great number of psychics in the twentieth century is a remarkable development, given our high level of scientific knowledge and technology. Yet we still encounter a number of individuals who claim to have special psychic powers ranging from ESP and psychokinesis to astroprojection and faith healing; often they are not unlike the prophets, mediums, miracle workers, or spiritual gurus of earlier ages. The difference is that we now have a full, contemporary record and an opportunity to study them carefully under rigorous scientific conditions. No doubt the most famous and remarkable of these individuals in recent years has been Uri Geller, an Israeli psychic who has traveled throughout the world and been seen or read about by millions of people; he also has been tested extensively by numerous scientists.

The Uri Geller story has a good deal of romance and legend associated with it. A strikingly handsome and charming individual, exhibiting great flamboyance, he has managed to attract numerous devotees and followers by performing incredible feats. He appears to be able to bend and break keys, spoons, and rings; to start watches that have stopped; to jam computers; to divine the contents of sealed envelopes and boxes; to materialize objects; and even to teleport himself. A consummate showman, he displays childlike wonder at his own feats.

Geller was born in Tel Aviv, on December 20, 1946. In his book *My Story,* he claims that he was "born with these powers" and that he "was actually given them from some source."[49] "I don't want you to think that I'm a Moses or a Jesus, but according to the Israeli account, Jesus was born on the 20th of December. Maybe it is a coincidence," he says. "I believe that Jesus had powers and so did Moses, and so did all those others in history."[50]

According to Uri's uncorroborated testimony, his psychic mission manifested itself when he was only three or four years old, when he was visited by a flying saucer while playing alone in a garden near his house:

> It was late afternoon but still light . . . Suddenly there was a very loud, high-pitched ringing in my ears. All other sounds stopped. And it was strange, as if time had suddenly stood still. The trees didn't move in the wind. Something made me look up at the sky. I remember it well. There was a silvery mass of light . . . This was not the sun, and I knew it. The light was too close to me. Then it came down lower, I remember, very close to me. The color was brilliant. I felt as if I had been knocked over backwards. There was a sharp pain in my forehead. Then I was knocked out. I lost consciousness completely. I don't know how long I laid [sic] there . . . Deep down, I knew something important had happened.[51]

As a six- or seven-year-old, Uri discovered that he could move the hands of his wrist watch by ten or fifteen minutes solely by the power of his mind. His mother knew that he had strange gifts because he was able to tell her after she came home from a card game exactly how much money she had won or lost. When he was nine silverware began to break in his hands. He said he kept quiet about these strange occurrences because of fear of ridicule.

In 1967, Uri was wounded during the Six-Day War. While the young soldier was recovering, he met Shipi Shtrang, a lad seven years younger, and his sister, Hannah. The three became intimate friends and were together during Uri's meteoric rise to fame. Shipi and Uri apparently found a book that dealt with magic, and they began working together to

49. Quoted in John L. Wilhelm, *The Search for Superman* (New York: Pocket Books, 1976), p. 10.
50. Ibid.
51. Uri Geller, *My Story,* (New York: Praeger Publishers, 1975), p. 95-96.

develop a night-club act. Uri was booked into theaters throughout Israel and became an instant star, performing his one-man act innumerable times. The routine began with Uri asserting that the mental energy of the audience was essential to his success and that he could not perform with negative vibes. Uri's show, according to his critics, was standard magic, with psychic claims thrown in to heighten the drama. He followed the mentalist routine that others, such as Kreskin and Dunninger, have employed—though with clever new twists. He would usually begin by asking someone from the audience to write a color word on the blackboard for the audience to see; he named the color, even though he presumably could not see the word. Similarly, he guessed the names of foreign capitals. It is easy to do these tricks with an accomplice in the audience, provided prearranged signals are given. Next Uri would bend or break a metal object—razor, key, nail, or spoon—or he would start a watch that had stopped. All these tricks can be done by magicians by sleight of hand without any pretence that they are supernormal. Geller was eventually brought to court in Israel for using the words *psychokinesis, ESP,* and *parapsychology* in his promotion and prohibited from doing so in the future, since he was doing nothing more than conjurer's tricks.

Uri's career took a new turn when he and Shipi met Dr. Andrija Puharich, an American psychic researcher and physician, who went to Israel to investigate Uri's gifts. Puharich was duly impressed. Hypnotized by Puharich, Geller identified himself as "Spectra," a computer aboard a spaceship from a distant galaxy. Under the control of "Hoova," who came from a remote planetary civilization, Geller was the intermediary, and Puharich was sent to assist Geller in this great task.

How much of this was due to Puharich's and Geller's fantasies and how much was pure fabrication is difficult to say. But the myths are perhaps no worse than the fairy tales concocted by or about Moses, Jesus, Mohammed, Joseph Smith, and other prophets who claimed to have been visited or sent by a divine being. The "intelligences" that Uri drew upon were from outer space. For many, UFOlogy has become a new religion—replete with science-fiction imagery—of the postmodern world. Like countless others Uri has embellished his act with fanciful space-age symbols. The language of transcendent revelation in previous ages was clothed in the imagery and metaphors of the times in which they appeared. The paranormal revelations of the present are charged with scientific technology and space-age gadgetry. Uri's fanciful accoutrements appear brazen to skeptics today, for they can be easily investigated and refuted. The point is that although there have been thousands of reports of UFO sitings and many cases of alleged abductions or direct visitations, after decades of intensive scrutiny not one case stands up to close scrutiny.

Still Geller managed to play the extraterrestrial fantasy for all it was worth early in his career.

Puharich represents the growing number of psychic "investigators" on the fringe of science, confusing the "phenomena" of pseudoscience with genuinely tested data. Puharich accepted practically everything being touted by paranormalists who were claiming that a tremendous paradigm shift was occurring, that a whole new paranormal universe was opening up, and that an impressive breakthrough of momentous proportions was in the offing. He had explored the use of hallucinogens found in mushrooms, brought the Dutch psychic entertainer, Peter Hurkos, to America, and even accepted as genuine the psychic surgery of Arigo, the Philippine charlatan. Puharich believed it was essential that Uri be tested by scientists, and he convinced Uri to come to the United States for that purpose.

John Wilhelm, a science writer, interviewed Geller and Puharich after they arrived in America. It is a hilarious account. Who was putting on whom? Puharich claimed that his own tests with Uri convinced him that Uri was genuine. He told Wilhelm that both he and Uri were in contact with extraterrestrial beings who used various channels to communicate, and he reiterated how he had taped Geller's transvoice and how it sounded as if it was synthesized by a computer. Moreover, the tapes that he often used had vanished (was Geller hiding them from Puharich?)—due to the intervening intelligences from Hoova. Puharich said many inexplicable things happened in Uri's presence: radar jammed, computers reprogrammed, motors shut off in cars, and gears shifted—all without any physical intervention. Puharich even claimed that The Voice would at times speak to Uri on the phone.[52]

At times Geller seemed to wish to distance himself from some of Puharich's wilder ideas. Yet he went along with the game. Thus Geller related that he and Puharich witnessed UFOs in several instances, but no one could corroborate this or any of their tales. On one occasion Geller rushed from his Tel Aviv apartment at night, taking Puharich and a female friend with him. Driving into the suburbs, Geller stopped the car; he saw flashing blue lights in the distance. Cautioning the others not to accompany him, he approached the lights just behind the dip in the terrain. After a few minutes Geller returned carrying Puharich's pen, which, according to Puharich, had mysteriously "dematerialized" a few days earlier. Were either or both having hallucinations, or was this a hoax perpetrated by Uri on Puharich or by both on their listeners? Geller also relates that one day he teleported himself virtually instantaneously from the streets of Manhattan to Puharich's home in Ossining, New

52. Wilhelm, *Search for Superman*, p. 32.

York, some thirty miles north, coming right through the upper portion of the screened porch.[53] Wonders never cease with Uri around, and like Jesus walking on water he can transport himself in some mysterious way. Uri has claimed that he didn't think the power he evinced was coming from him but that it was being channeled *through* him. "What I am able to do," he confides, "is maybe part of a much greater plan which concerns more than the earth and mankind."

No doubt the most dismaying development was the testing of Geller by numerous scientists and Geller's ability to convince them that his powers were genuine. Geller had by then become an international media personality transfixing audiences worldwide. Famous television and radio personalities witnessed him and were flabbergasted at his ability to bend keys or read the contents of sealed envelopes. They became convinced of his reliability. But skeptics asked, were his powers real or fraudulent? Had a modern-day superpsychic emerged, one who had been tested in objective, controlled laboratory situations?

Inasmuch as science has been heralded in the modern world, it was clear that for Geller to make his way and prosper in his stardom, he needed scientific approval. He soon learned that many scientists were willing to submit him to tests and to certify his psychic credentials: the "Geller effect" was so named by scientists, who pointed to strange contracausal phenomena that had appeared in Geller's presence. Not simply had the word taken on flesh, but also the mind had achieved new forms of psychokinetic energies. The fact is that Geller was able to hoodwink scores of scientists. He appeared to have the powers of Dunglas Home, Eusapia Palladino, and even Joan of Arc. But this only reconfirms the thesis that the scientific mind is often as gullible as that of the ordinary person.

Geller agreed to be tested at the Stanford Research Institute (SRI) in Menlo Park, California, by Dr. Harold Puthoff and Russell Targ, two physical scientists. Puthoff was identified with Scientology, a new space-age religion, which presupposes astroprojection, reincarnation, space travel, and other psychic powers. Targ and Puthoff have tested other Scientology psychics, such as Pat Price and Ingo Swann, who they claimed also manifested psychic powers, especially clairvoyance, telepathy, and the ability to identify an object outside the laboratory and transmit it to a psychic within. They called this "remote viewing," as if a television camera of the mind could pick up a distant signal by some strange means. The experiments with Geller were carried out at various times between November 1972 and August 1973. They were supposed to test Geller for clairvoyance, telepathy, and psychokinesis. The results of the clairvoyance

53. Geller, *My Story*, pp. 266 ff.

and telepathy experiments were published in *Nature,* Britain's most prestigious scientific magazine, though the editor expressed numerous doubts
about the experimental design and the lack of rigor.[54] The paper was
published, we are told, in order to allow parapsychologists and other
scientists to assess the quality of the SRI research and to see how much it
contributed to the field of parapsychology.

One of the tests reported dealt with Uri's ability to telepathically read
target pictures chosen at random by the experimenter or others. Of thirteen targets, Targ and Puthoff maintained that Uri had scored good on
six and fair on two. According to them, the probability of this happening
by chance was one in three million. Skeptics have criticized the test for
lacking stringent controls. They have pointed out that the pictures drawn
by Geller did not match what they were supposed to correspond to but
appeared to be responses to verbal cues. What constitutes a "hit" is open
to dispute. The conditions under which the experiments were conducted
were extremely loose, even circuslike and chaotic at times. The sealed
room in which Uri was placed had an aperture from which he could have
peeked, and his confederate Shipi was in and about the laboratory and
could have signaled him.

There was another test of clairvoyance in which Geller passed twice
but guessed eight out of ten times the top face of a die that was placed in
a closed metal box. The probability of this happening, we are told, is one
in a million. Critics maintained that the protocol of this experiment was
poorly designed, that Geller could have peeked into the box, and that
there were dozens of other unreported tests in which there were no astonishing results. Even Targ has admitted that "Geller will cheat if he can."

Geller failed a third test in which he was supposed to draw target
pictures, which were inside envelopes. Targ and Puthoff also reported in
Nature that although metal bending by Geller had been observed in their
laboratory, they had not been able to combine such observation with
adequately controlled experiments to obtain data sufficient to support the
paranormal hypothesis.

The Targ and Puthoff tests enhanced Geller's career and his popularity spread even further. Wherever Geller went he repeated his act,
bending objects galore. The "Geller effect" took on psychological dimensions; for if people believed that he possessed psychic powers, they tended
to believe that what they saw was miraculous, a self-confirming prophecy.
The least little suggestion that anything unusual was going on was taken
as psychic. In his many telecasts performed throughout the world, Uri
would ask his viewers to concentrate on bending metal. After the pro-

54. Russell Targ and Harold Puthoff, *Nature* 252, no. 5476 (Oct. 18, 1974).

gram, hundreds of calls would pour into the station reporting miraculous psychokinetic effects: spoons, keys, clocks, and watches all displayed mysterious effects.

Geller describes the psychokinetic effect he was able to produce on metal:

> What happens is very simple but also very startling. The key begins to bend slowly as I either rub it lightly with my fingers or hold my hand over it. Then it continues bending after I take my hand away. Sometimes it bends only slightly and stops. Other times, it continues up to a 45 degree angle, or even to a right angle. Sometimes it will seem to melt, without heat, and half the key will drop off.[55]

All of Geller's tricks can be performed by magicians. James Randi, a professional conjurer, has bent keys by physical manipulation or pressure, unseen by the observer; he holds them in his fingers and turns them gradually so that they appear to bend. Similarly, Randi is able to exploit metal fatigue on a specially prepared spoon or key, which after gentle stroking seems to melt and then break. He is able to advance or reverse the minute or hour hands on watches by deft flicking of the winder. Similarly he has been able to divine the contents of sealed envelopes either by holding them up to a light or by surreptitiously peeking into them when no one is looking.

On one occasion before Geller was tested (as on the "Johnny Carson Program"), Randi had been consulted to ensure that the tests were foolproof. When he did so, Geller failed to produce results. It is when Geller controls the conditions of the performance or experiment—as was the case with Eusapia Palladino—that he is successful. If he is unable to succeed, he blames the negative vibes of skeptics.

Two psychologists, David Marks and Richard Kammann, carefully monitored Geller's visit to New Zealand.[56] Marks was permitted to test Geller in his hotel room. He placed a previously prepared drawing of a sailboat in a sealed envelope and handed it to Uri. Marks was mystified when Geller was able to reproduce the figure. Later Marks retrieved the discarded envelope and found that Geller had evidently peeled it open when Marks was in the bathroom and another observer was distracted by a phone call; Geller had apparently peeked into the envelope to find the correct answer. Kammann also tested Uri's ability to read drawings and found that Uri could succeed if he could observe the top of the pencil or if he etched the drawing on a pad; but if the pencil top was concealed he could not. In testing their students, Marks and Kammann also demonstrated that an astute person could perceive the outlines of a folded drawing through an envelope, if allowed to put it up to the forehead and finger it as Geller had done. As to the thousands of reports of bent cutlery that stations had received, Marks observed that virtually anybody can find a bent fork or spoon in his house if he makes a thorough search. Marks and Kammann also performed an interesting test on watches that needed repairs. If a watch is held in the palm for a few minutes, the body heat is able to warm the watch and it will often begin to tick. They examined watches in seven jewelry stores awaiting repair, and found that sixty out of 106 could be started again by this perfectly normal method. The success rate was 57 percent.[57] There is no evidence that Uri could repair watches by the power of his mind, not where the mainspring was broken or some other structural mechanism was in need of repair.

It is the uncanny ability of Geller to convince scientific observers of his powers that deserves our special attention. There are any number who have fallen into his trap; this is especially the case where there is a predisposition to believe in psychical powers, a naïve ignorance of how trickery or conjuration could occur, and an implicit trust that the person under observation is honest, reliable, and would not cheat.

A remarkable collection of papers was assembled by Charles Panati (at one time a science editor of *Newsweek* magazine) and published under the prestigious title, *The Geller Papers: Scientific Observations on the Paranormal Powers of Uri Geller.*[58] A reading of these papers written by fourteen scientists and three magicians might convince a naïve reader that the "Geller effect" had been confirmed in the experimental laboratory. The book contained the paper by Targ and Puthoff already discussed, as well as three other papers by Wilbur Franklin, Eldon Byrd, and John Taylor, which seemed to indicate some kind of experimental confirmation of Geller's psychokinetic powers.

John Taylor, physicist and professor of applied mathematics at King's College, University of London, wrote a brief report on Geller's visit to his laboratory. He details how Geller apparently caused objects to "fly through the air" or disappear, how a "compass needle had been caused to rotate without the intervention of a visible mechanism," and other strange phenomena.[59] He deemed these events impossible to comprehend, and said they left him in a state of "mystification." Taylor had observed Geller perform on a television program and was so baffled that he became convinced of Geller's paranormal gifts. He was unaware of conjuring

57. Ibid., p. 107.
58. Charles Panati, ed., *The Geller Papers* (Boston: Houghton-Mifflin, 1976).
59. Ibid., p. 217.

techniques at that time. In pursuing his research, Taylor also tested a great number of young children, who he claimed were able to bend metal spoons and forks. He devised an experiment whereby a piece of aluminum rod six inches in length is placed in a transparent plastic tube sealed at both ends by red rubber stoppers and held in place by a screw covered in wax. Much of this was written up in Taylor's book, *Super Minds.*[60] Taylor noted that there was a "shyness effect," i.e., children could not perform their feats when skeptics were watching or they were under direct observation. But Taylor always saw evidence of disformation after the fact. He attributed these phenomena to unknown causes.

James Randi once visited Taylor incognito at his office, and he was able to perform psychokinetic feats whose cause Taylor was unaware of. He found that he could easily break the seal, bend the aluminum rod, and return it undetected. Obviously Taylor had been taken in by simple conjuring tricks, for he was an honest man who could not believe that young teenagers could possibly cheat by bending metal objects behind his back. Two scientists at Bath University tested the "shyness effect." They allowed six metal-bending children an opportunity to perform. An observer who was present in the room did not note anything unusual, but a hidden television camera showed that the children cheated when the observer was not looking. Geller evidently did the same thing. To his credit, Taylor has since withdrawn his views on psychokinesis, admitting to his former errors and now claiming that the existence of psychic phenomena is doubtful.

The paper by Eldon Byrd in the Panati volume has been cited as strong evidence for psychokinesis. When I first read it I was impressed by what seemed to be independent physical corroboration that Geller was capable of bending metal by nonnormal means. Byrd used a sample of an unusual alloy, nitinol. This metal wire has a physical "memory" for the shape in which it was formed at the time of manufacture. Only by heating the wire to a very hot temperature can it be reshaped. Geller rubbed the nitinol wire in Byrd's presence. When he removed his fingers the wire had a definite bump or kink in it. When Byrd placed it in boiling water, instead of snapping back to its original straight shape, it began to form a right angle. The bends that Geller had produced, reports Byrd with apparent awe, had "permanent deformations."[61] How had Geller achieved such results? Byrd has been quoted as saying "Geller altered the lattice structure of a metal alloy in a way that cannot be duplicated. There is no

60. John Taylor, *Super Minds: A Scientist Looks at the Paranormal* (New York, Viking, 1975); see also his later book, *Science and the Supernatural* (New York: E. P. Dutton, 1980).

61. Eldon Byrd, "Uri Geller's Influence on the Metal Alloy Nitinol," *The Geller Papers*, p. 72.

present scientific evidence as to how he did this."[62] Byrd's paper has been cited by parapsychologists as a decisive verification of the "Geller effect."

Martin Gardner, however, has refuted these claims. He attempted to replicate Byrd's experiments, and much to his surprise he got the same effect but by normal means. He bent the wire using pliers. Then bending it back into shape, he caused a bump in it by pressing it with his thumbnail. He then put the wire into a bowl and poured boiling water on it; lo and behold, the wire took the form of an angle, similar to that described by Byrd. It was entirely possible, said Gardner, that Uri, who would arrive with Shipi, had done the same thing as Gardner when Byrd was not watching or had even come with a prepared wire—whose properties, incidentally, are well known to magicians.[63]

Another paper by Wilbur Franklin, chairman of the department of physics at Kent State University, entitled "Fracture Surface Physics Indicating Telenural Interaction"[64] also seemed to provide physical proof that Geller was able to cause changes in physical objects by unknown means. In this case, Franklin reported that a platinum ring had spontaneously developed a fissure in its surface in Geller's presence, "without his having touched the ring." Franklin submitted the fractured ring to metallurgical analysis. He found that the surface of the fracture on the ring and also a needle broken by Geller were "distinctively different" from known types of nontelenural fracture surfaces. He concluded that "it would have been extremely difficult to fabricate these surfaces by known laboratory techniques."[65] He also said that such "telenural interaction with matter" points to the necessity of developing new theoretical constructs to account for the "Geller effect." In the Panati book there are photographs of the various fractured metals—all presented as impressive verification of the preceding.

Franklin has since withdrawn his conclusion and admitted his error. The circumstances were as follows. As chairman of the Committee for the Scientific Investigation of Claims of the Paranormal (CSICOP), I invited Professor Franklin to Buffalo, New York, to appear on a television program on the paranormal and parapsychology that I was moderating. Also participating were three skeptics, James Randi, Ray Hyman, and Ethel Romm. Mrs. Romm expressed her dismay that Franklin did not include a picture of a normal fractured ring along with the Geller ring so that we could compare the similarities and differences. He later went back to his laboratory and fractured a platinum ring by physical means and compared

62. Quoted in Martin Gardner, *Science: Good, Bad and Bogus* (Buffalo: Prometheus Books, 1981), p. 160.

63. Martin Gardner, "Geller, Goals and Nitinol," *The Humanist,* May-June 1977.

64. Panati, ed., *The Geller Papers,* pp. 75 ff.

65. Ibid., p. 80.

it with the photo of the Geller fracture and found, much to his embarrassment, that they were virtually identical![66] The last major piece of
physical corroboration was now withdrawn.

No doubt one factor in Franklin's decision to reexamine the evidence
was the bag of "paranormal tricks" that Randi performed in his presence.
He bent Franklin's key, broke a spoon, correctly guessed the contents of
a sealed envelope, moved a watch hand ahead. He even asked Franklin to
pull a book from his own briefcase and open it; he then told him a word
on the top line. All of this flabbergasted Franklin. He had witnessed
Geller do similar things and had been convinced they were "telenural"
and "paranormal." But now a professional conjurer had duplicated the
"Geller effect."

It should be abundantly clear by now that there is strong evidence
that Geller is a magician, not a psychic. Geller has been decisively refuted,
and belief in his powers have been thoroughly discredited. Yet Geller's
reputation lives on, and countless numbers of people continue to accept
him as genuine. This even applies to those who have read or heard of the
refutations by skeptics. Like other religious or quasi-religious myths, once
given root, no matter how many times they are plucked out they may
continue to grow and perhaps even to flourish in the soil of gullible belief.

Conclusion

The point to my discussion of Uri Geller and others involved in the
psychic field should now be evident: it is a fairly simple matter to deceive
people, especially (1) if someone is intent on perpetuating a fraud, (2) if
no known or apparent cause is readily observable, and (3) if there is a
predisposition to belief. This tendency may even apply to otherwise skeptical people who are not ordinarily given to gullibility or self-deception.

One last illustration of these points concerns Tina Resch and the
alleged poltergeist manifestation in her house in Columbus, Ohio. This
case has already become a classic, for it illustrates the central conflict between skeptics and believers concerning evidence and the will to believe.

The details are as follows. Strange events suddenly began to happen
in the home of John and Joan Resch when their adopted daughter, Tina,
was present. Lights and faucets went off and on without any apparent
cause, and objects flew through the air as if possessed by some psychic
force. A reporter for the *Columbus Dispatch,* Michael Hardin, who had

---

66. See the *Humanist,* September 1973, p. 55. He states: "This analysis indicates that
the fracture surface, for the most part, appears similar to well-known mechanical modes
of fracture "

written a story about Tina before, was called in, and he brought a photographer, Fred Shannon. They reported that there were indeed eerie events taking place: eggs, lamps, telephones would suddenly fly through the air and fall to the floor. Shannon photographed a telephone in mid-air, and this photo appeared in newspapers worldwide. An avalanche of press people descended on the house and many journalists and cameramen swore that they too had witnessed paranormal events.

Called upon by a skeptical journalist in Boston to provide another point of view, CSICOP sent a three-man scientific team—composed of Professors Nicholas Sanduleak and Steven Shore, and James Randi—to Columbus for an on-site investigation. They were denied admittance to the house, though they were able to question many of the key figures and to piece the "evidence" together.

Two parapsychologists, William Roll and his associate, were allowed into the house, however. Roll promptly announced that psychokinetic activity had occurred. Roll is one of the leading experts in the field of poltergeist research. Reports of poltergeists (literally, "rumbling spirit") have abounded throughout history; strange rappings and moving objects are attributed to a spiritualistic agency. In an article in the *Handbook of Parapsychology* Roll attributes the phenomenon to psychokinetic energy released by an adolescent who is in the vicinity.[67] He surveyed 116 cases, and estimated that approximately one-third of the cases were genuine, one-third were fraudulent, and one-third were uncertain in interpretation. In the Resch house he maintained that he observed a picture fall off its hook to the floor. When he went to get a pair of pliers lying nearby to nail the hook back in, a tape recorder on the other side of the room flew through the air. As he went to pick that up, the pliers on the table also took off. Tina was the only one present in the room with him.

Our investigators came up with the following explanation: Tina Resch was a disturbed fourteen-year-old who had dropped out of school and was being tutored at home. The Resches had provided a foster home for more than 250 children over the years. Tina was intent on finding her natural parents. She had also seen a movie on poltergeists and had learned how to hurl objects into the air unobserved. These feats brought her great attention. Here was another patent case of fraud being perpetrated on the gullible public.

Could this skeptical solution be corroborated? Yes it was, but nonetheless believers persisted in their conviction that a semisupernatural event had occurred. First, a television crew ensconced itself in the house in

---

67. William G. Roll, "Poltergeists," in *Handbook of Parapsychology,* ed. Benjamin B. Wolman (New York: Van Nostrand, 1977).

order to photograph flying objects. Though they waited hours, nothing occurred when they had the camera on Tina and were observing her directly. On one occasion, when they left the room, they left the camera on, unbeknownst to Tina. Suddenly a lamp came crashing to the floor, doing so twice. People in the room with Tina believed that it was a poltergeist. When the camera crew returned to the studio to develop the film, they were astonished to discover that Tina had furtively yanked the lamp to the floor. No one in the room saw her do it. But the film, if played in slow motion, gave incontrovertible evidence of fraud. The film was shown on the evening news for everyone in Columbus to observe. Tina's mother accepted Tina's pretext that she was tired by all the commotion and wanted everyone to leave so that she could meet with her friend. She told everyone that Tina had knocked the lamp down so that people would be satisfied. Her supporters accepted this rationalization, and even Hardin and Shannon accepted the story at face value, thus confirming the principle that if a person is strongly predisposed to believe in a supernatural event, often little can be done to dissuade him.

Was the lamp incident an isolated exception? There is a principle of caution that seems eminently sensible: if a person is once caught cheating, then other displays of his or her powers should be highly suspect. Alas, believers resist this warning in respect to their favorite psychic, prophet, or miracle worker.

Was there additional evidence of fraud in Tina's case? If we go back to the original photographs taken of the flying phones, we find that the photogrpaher, Fred Shannon, said that if he looked at Tina directly, the "force" would not manifest itself. Only if he turned his head and squinted out of the corner of his eye would the phone take off; then he would suddenly turn and photograph it directly while in the air. It turned out that under careful questioning no one could testify to having seen an object first standing at rest and then taking off; they had seen the objects only after they were airborne. Anyone can try the following experiment and carefully observe people's reactions: as soon as they are not looking, quickly throw an object into the air. If you tell them that a poltergeist is at work and they did not see you throw the object, they may accept the claim as genuine. This would explain Roll's misperception and that of other reporters in the house. If there is a predisposition to believe and the situation is charged with drama and emotion, an event claimed to be supernatural is more likely to arouse an affirmative response.

Is this skeptical critique unfair to Tina? Perhaps she did have marvelous powers and perhaps it is the skeptics who refuse to accept the testimony of believers because of their closed minds.

In the case of Tina, we have at least four other incidences of chicanery. Shannon had taken a whole series of photographs of other kinds of poltergeist or kinetic manifestations besides the flying telephones. A careful analysis of these photographs reveal the following facts. In one case a chair seems to be rising in the air. People on the scene swore that it was a paranormal manifestation. A close inspection of the photograph, however, clearly shows Tina's foot under the chair and apparently responsible for lifting it—shades of Eusapia Palladino! Another photograph of a moving chandelier shows a hand striking it—though it is not clear if the hand is that of an accomplice, perhaps her brother, who was also in the house.[68] The last two instances are the testimony of Joel Achenbach, an investigative reporter sent to interview Tina later in North Carolina.

Tina was taken by William Roll to his parapsychological laboratory for further testing. While there she broke her leg and was admitted to the hospital to have it set. Achenbach visited her in the hospital. While he was in the room, her plastic wrist band suddenly flipped through the air; later a bottle of nail polish did the same. In both cases the reporter saw her hurl them across the room. When confronted with this she simply giggled and apologized. "Sorry," Tina said, smiling.[69] Since that time the Columbus poltergeist case has become part of the literature. The skeptics have published their critiques. Numerous publications—including the *Reader's Digest*—have presented the account of what transpired in Tina's house, with a clear implication that genuine psychic events had occurred. Thus another paranormal manifestation has become part of folklore and legend.

Tina Resch and Uri Geller are like Eusapia Palladino, Dunglas Home, Joseph Smith, and perhaps even Jesus and Moses of earlier generation. The Tina Resch case is not unique; it is strongly reminiscent of the Fox sisters in the nineteenth century. The same psychological processes are present in our age and in earlier ages. There is willful deceit, fraud, and an exacerbated and dissociated psychological sense of one's own special mission on the part of the psychic or prophet, and a hunger to believe in a spiritualistic, transcendental universe and superpsychics by a gullible public. Why does the transcendental temptation have such powerful roots in the human breast? Why does it appear and reappear in virtually every age? Can an explanation be given for its persistence? It is to these questions that we must now turn.

---

68. For a complete account of the Resch case, see James Randi, "The Columbus Poltergeist Case, Parts I and II," *Skeptical Inquirer,* 9, no. 3 (Spring 1985) and 10, no. 5 (Fall 1985).

69. Joel Achenbach, "The Teen-Ager and the Poltergeist," *Buffalo Magazine,* October 28, 1984.

XIV: Is there life after life?

Is there life after life? This is a perennial question that spiritualists and parapsychologists have investigated, and it is perhaps the most momentous one that anyone will have to face. The existential dilemma is that death happens to everyone; though for a good part of one's life one may choose to ignore it, one cannot in the end escape it. The moment a person is born, he or she is old enough to die. Thus the survival question has been a source of intense human curiosity, philosophical speculation, and scientific interest; and it has been the wellspring of the religious impulse. Indeed, the quest for God has its strongest impetus in our fear of death and the unknown, and our desire to transcend it. Belief in God is rooted in the longing that God will save us from extinction and provide us with everlasting life. It has thus been difficult for most humans to treat the question dispassionately, when so much is at stake.

Those who deny that there is life after death have often been treated as pariahs or heretics, outcasts of society. Those who promise immortality, on the contrary, are deified; religions are founded in their names—Jesus is the best example of this. The more ambitious the promise and the more dramatic the myth, the more likely are people to believe it; when it concerns the doctrine of the afterlife, it can attract countless millions of believers who will go to death secure in the belief in an eternal life. Although the hope for life beyond the grave has been a profound source of religious and aesthetic inspiration, unfortunately the wish for it far outstrips the evidence.

A number of arguments have been offered historically in support of the survival hypothesis:

1. *Inferential argument related to religious beliefs.* If God exists, it is said, immortality follows as the fulfillment of some divine plan. This argument is open to criticism. If belief in God ultimately is only an article of faith, then immortality of the soul has not been demonstrated, and those who do not share the religious conviction can hardly find the case for it convincing. It is, moreover, possible to believe in a deity and not in personal immortality.

2. *Empirical claims.* Can we establish the existence of immortality independently of the existence of God? As we have seen, there have been

many efforts to confirm the survival hypothesis scientifically, particularly after the founding of the Society for Psychical Research in 1882. There has been a marked increase in interest in this question in recent decades, especially in parapsychology. We shall review some of the paranormal evidence below.

3. *Value considerations.* Whether or not God can be demonstrated to exist or the soul proven to have a discarnate existence, the *doctrine* of immortality, it is alleged, is a necessary postulate of the ethical life. Without it, life would lack all meaning, and ethical obligation would lack any secure foundation. The secularist maintains that ethics can be autonomous and that responsibility does not require a doctrine of divine salvation. He maintains further that if there were an afterlife and everything that happened in this were preparation for the next, then this life might be deflated in meaning and all our plans and projects would be senseless. Can human beings lead significant lives without assuming that the universe has an overall purpose or that there is an afterlife? Perhaps life has no meaning per se. But does it not present us with innumerable opportunities, which we may choose to invest with meaning? Nonetheless, believers insist that life would be meaningless without the hope for immortality.

Skeptics generally have raised at least three main objections to the survival hypothesis: (1) logical objections, (2) evidential questions, (3) the argument from value. We shall examine each of these in turn.

Logical objections

The first objection to immortality, one which is very popular today, is logical—that is, there is no clear meaning of the term *soul*. Before one can prove whether immortality of the soul exists, one must know what one is talking about. Classical philosophers, from Aristotle to Hume and Kant, have raised serious doubts about the idea of a substantial soul separable from the body. Many contemporary philosophers have raised more fundamental issues. John Dewey and the behaviorists, for example, rejected "mentalism" and the existence of a mind-body dualism as empirically unwarranted. Gilbert Ryle, Antony Flew, Peter Geach, and other linguistic philosophers have found the concept of "soul" logically unintelligible. Ryle thinks that we do not make any sense when we talk about the "ghost in the machine," and he indicts Descartes and all who hold a dualistic or Platonic view for committing a category mistake. The human being is an integrated unity; what we call the "soul" or the "mind" is simply a functional aspect of the physical organism interacting in an environment. If this is the case, it is difficult to know what it means to say that something called the soul survives the dissolution of the body. The notion of a "disembodied self" or "nonmaterial soul" is based upon abstraction and reification and a puzzling misuse of language.

Hume pointed out the problem of personal identity. One may ask: Is there an independent substantial soul or self underlying my particular experiences? If my body no longer exists after my death, and if I survive it, will I remember my past self? Will I continue to have feelings in my limbs, genitalia, and stomach, etc. after they are gone? How can I have memory if I have no brain in which to store experiences? Will it be I who survive or only a pale shadow? Antony Flew asks: Can I witness my own funeral? As my corpse lies in its coffin, can I, lurking in limbo, view it? Does it make any sense to say that I can see if I do not have any eyes to receive impressions or a nervous system to record them?

Thus many contemporary philosophers point to many linguistic confusions about what it means to say that the "soul" survives the body. The basic issue for them does not involve the factual claim as to whether the soul exists so much as it involves the logical puzzle about the definition of the concept. For some believing Christians the only sense that can be given to immortality is to say with St. Paul[1] that at some point in the future the body, including the soul, will be physically resurrected. This stand at least avoids the issue of whether the soul is separate from the body, and the believer merely needs to say that some divine being will in the future ensure the survival of the whole human being. But this is entirely a matter of faith, not proof.

Thus the meaning of the question, "Is there life after death?" as it is traditionally formulated, is confused, even contradictory. *Life* is defined as that which is characterized by the processes of "metabolism, growth, reproduction," and is capable of manifesting "internal powers of adaptation to the environment" (Webster's Dictionary). Death is the absence of organic functions. We have no evidence that life in the sense we have defined it can survive the death of the organism. All the evidence seems to point to the disappearance of these functions with the dissolution of the body. Thus it is questionable whether consciousness, the "self" or the "soul," can survive in some discarnate form and whether we can reify psychological functions. Would there be memory and a sense of personal continuity without a body, brain, or nervous system? Those who are committed to some form of the "ghost in the organism" doctrine respond in the affirmative. Skeptics are puzzled as to how this could be possible. Of course, something remains after an individual's death: a person's descendants will carry his genetic endowment, and his or her accomplish-

1. Rom. 6:5; 1 Cor. 15:42-58.

ments, whether good or bad, may have an influence upon the society or culture for a period of time.

A number of alternative, nondualistic theories of the mind have been proposed: materialism, phenomenalism, epiphenomenalism, the identity theory. Perhaps none of these is satisfactory at this stage of inquiry. Perhaps some form of existence survives physical death. We must not be imprisoned by our existing linguistic categories, and we must be prepared to redefine them in light of new discoveries. Still, there are serious conceptual problems.

Evidential questions

Aside from the meaning muddle, there is still a more basic question: What is the *evidence* for the claim that I can survive my death in some form, whether as a separated soul or as a being resurrected in one piece? Do people who die reappear on earth at some time immediately after death in some form or another, or in some remote future will they be resurrected in some new guise?

Many cultures and individuals have believed that something survives the physical death of the body; some have thought it to be part of the world soul, others to involve personal identity. Is there sufficient evidence for these assertions—objectively confirmable evidence, not wish-fulfillment or supposed sightings of the dead by credulous people? Can the dead communicate with the living?

We should not blithely assume that that which survives death is better than what we experience during life. Bertrand Russell is said to have observed that he would find the afterlife boring, if it meant singing hymns throughout eternity! If there is a future existence, it could be full of suffering and torment. I might spend all of my time fixated at the scene of my death or aggravated that my wife has remarried and that her new spouse is squandering my estate or in sorrow because my children are not living up to expectations. We may be totally impotent, incapable of enjoying fine wine, good food, embracing our loved ones, engaging in sexual activity or a debate, running in the gym or playing pinochle—all of the things we found exciting or satisfying in this life.

Aside from these considerations, the basic issue concerns the evidence for the claim that something survives the death of the body. As you examine the "evidence" carefully, it can often be reduced to wishful thinking, self-deception, even fraud. Let me review some of the chief areas of research. The early psychical researchers investigated the hypothesis that discarnate entities might communicate with us. I have treated this subject at length in Chapter XII. There have been extensive investigations of alleged mediums, who, in specially prepared séances, attempted physical or mental interaction with spirits of the dead, allegedly communicating with the departed by means of a control, who spoke through the medium. Unfortunately, so much trickery has been uncovered that the whole field has been thrown into disrepute. Any number of mediums were exposed, such as the Fox sisters and Eusapia Palladino. Houdini later unmasked Marjery, a noted Boston medium. After the death of his beloved mother, Marjery claimed to put Houdini in touch with her, but he was dismayed when his "mother" did not speak Yiddish, a language Marjery obviously did not know.

Arthur Ford, in a famous séance, claimed to put Bishop James Pike in touch with his deceased son. According to Pike, Ford related facts about the family that was privileged information and which "he could not have known otherwise." Yet after Ford's death, extensive files were uncovered that indicated he did preliminary research on his subjects, including Pike and his son, before a séance and that "private" facts were contained in these files. William James was horrified by the amount of skulduggery in the field, although he thought that Mrs. Piper, a noted psychic, might have been his white crow. Even if conscious fraud does not account for every case, alternative hypotheses may explain the information conveyed in the trance state; subconscious processes may be at work, and some have suggested telepathy, etc.

The discarnate hypothesis is not the only, nor indeed the most parsimonious, explanation available. I myself carefully investigated a much-heralded claim by Patricia St. John, allegedly derived from a séance (or spirit control), that Terrapin Point at Niagara Falls would collapse on July 22, 1979, at 4:56 P.M. and that a *Maid of the Mist* boat carrying a group of deaf schoolchildren would capsize. On that fated day at the appointed hour I took the boat; nothing happened. It became clear that Mrs. St. John was a confused individual, whose prediction, allegedly derived from a discarnate spirit, proved to be false.

Reports of apparitions and poltergeist phenomena have also been investigated. Ghost stories have been common down through history and must be sorted though carefully. Unfortunately, from the standpoint of science the evidence is largely anecdotal, highly subjective, and not available to controlled inquiry. It is difficult to produce replication of the phenomena, which is essential to scientific inquiry. In some cases, the stories are fiction, as the Amityville Horror hoax. In other cases, however, reputable witnesses seem to be involved. In no sightings that I am aware of, however, do we have a decisive case, either because the observations cannot be corroborated by authoritative testimony or else alternative explanations are available. Apparitions may be due to a rich imagination, hallucination, faulty perception, etc. Poltergeists often involve youngsters engaging in pranks, as in the Tina Resch case cited earlier. Some parapsychologists have attributed poltergeists to psychokinesis, but I am skeptical of this explanation. Curiously, in very few, if any, of the classical cases of apparitions do the entities appear unclad; though it is difficult to believe that if one *were* capable of survival, his clothing would also survive rot and decay. A more likely explanation is that the apparitions are in the eyes of the beholders.

Still another kind of evidence that has been adduced in support of the survival hypothesis is the use of photography. For over a century we have seen innumerable photographs of alleged spirits. But again no photograph exists that clearly requires a spirit hypothesis; other alternatives are more plausible: imperfections of the film, double exposure, or doctoring of the film. The use of photographs to prove anything is notoriously unreliable. Today they are used for UFOs, the Loch Ness Monster, and Bigfoot. In the 1920s it was fairies. Sir Arthur Conan Doyle collected actual photos of the wee folk, which he said proved they existed.

Such investigations of discarnate survival were widely discredited by the 1920s. As we saw, J. B. Rhine thought that parapsychology could best make strides by leaving the question aside, since he did not think that it was amenable to empirical research. Instead, he concentrated on ESP and PK.

Today the survival issue has returned strongly. Reports of séances, apparitions, and spirit photography abound. Hope springs eternal in the human heart, particularly when the surrounding religious culture constantly affirms the reality of human survival. Several new approaches have opened up, however, which according to their advocates enable us for the first time to provide strong scientific evidence for survival. These researchers claim to be using the latest techniques and tools of science. It is interesting that astrologers and biorhythm theorists use computers to frame horoscopes and predictive charts, as if this would make their field more "scientific." Many scientific endeavors may have the trappings of science, but not the substance.

Illustrative of this is the startling "voices from the dead" phenomena first revealed in 1959. This research is summarized in the book *Breakthrough,* by Konstantin Raudive.[2] In it, Raudive claims he has evidence that deceased souls can send us messages by mysterious means, such as tape recorders, microphones, and radios. D. Scott Rogo, a prolific parapsychologist and writer, recently reinforced the "significance" of that re-

2. Konstantin Raudive, *Breakthrough* (New York: Kensington Books, 1971).

search by asserting that "this phenomena is a breakthrough of the first magnitude."[3] Careful analyses of the noises have brought forth a variety of interpretations of their meanings. They are less likely to be messages from afar than they are random signals picked up from background noises by the electronic devices, which serve as transmitters. Julius Weinberger reports on another experiment on communicating with discarnate persons, in which he uses a Venus' flytrap wired to electrical devices. He tells us that he knows initially that discarnate entities are present because of the familiar scalp-tingling sensations that he experiences. The experiment begins with a recital of the Lord's Prayer. Then he poses questions to the plant and observes its reply. There have also been reports of telephone calls from the dead!

Recently, the most widely hailed "evidence" for postmortem survival is the study of deathbed visions and near-death experiences of those who have been resuscitated, and the use of hypnotic and other means to regress to past lives. (The latter is used, of course, as evidence for reincarnation.) Karlis Osis and E. Haraldsson, in *The Hour of Death,* report their extensive studies of people on their deathbeds who have had visions.[4] Osis, head of research for the American Society for Psychical Research, sent several thousand questionnaires to doctors and nurses in the United States and India. Only 20 percent of those receiving the questionnaire responded; of those some reported that in a number of cases dying persons claim to "see" into the beyond, meeting deceased relatives, friends, or religious figures who come "to take them away." They also reported an uplift of mood, even a kind of blissful peace and radiant calm following the experience. Many of them reported seeing a bright light.

Osis admits that most dying patients drift into unconsciousness and oblivion, and do not report such an experience. He also reports that the religious figures who appear in a limited number of cases to Indian patients are different from their American counterparts (given the different religious traditions). In many cases, the vision appears immediately prior to the moment of death; in other cases, as much as one or two days earlier. In some cases, it occurred to people who did not die. Osis believes that this is nevertheless evidence for another life. Admitting that many forms of hallucinations are often present at the moment of death, he says that many patients with visions were fully conscious and rational. Perhaps the patient is experiencing a schizoid reaction to alleviate the stress of dying. No doubt many have acquired an expectation of immortality from

3. John White and Stanley Krippner, *Future Science: Life Energies and the Physics of Paranormal Phenomena* (New York: Anchor Books, 1977).
    4. Karlis Osis and E. Haraldsson, *The Hour of Death* (New York: Avon, 1977).

the cultural environment, and this offers the patient sustenance to face the moment of death.

The basic question here is whether such subjective experiences are evidence for the thesis that some aspect of the human psyche separates from the body after death, or whether there are natural psychical or psychological explanations. Osis rejects the influence of drugs, elevated fever, brain damage, wish-fulfillment, or hallucinations as explanations.

A highly popularized version of near-death experience is by Elisabeth Kübler-Ross and Raymond Moody.[5] Moody questioned 150 patients. He claims that many individuals who have been pronounced dead—their hearts or lungs have stopped—and were resuscitated, or who in accidents came close to death, reported similar experiences. There was an out-of-body sensation of floating above the bed, operating table, or accident scene. In the experience, a person may feel himself going through a long tunnel. There is a rapid panoramic recall of his life, the meeting of deceased relatives and friends, a shining light, a sense of peace and then a return, often reluctantly, to the body. (Incidentally, neither the tunnel experience nor panoramic flashback seems to be present in Osis' study.)

Serious methodological criticism of Moody's study has been raised. It is a very imprecise collage. We do not know the conditions under which the subjects were questioned, the procedures used, nor do we have data on the patients' physical or psychological states. Not everyone who had a near-death experience reported the same phenomena; most who died in similar conditions experienced nothing that we know about. What are we to make of this experiential phenomenon? It apparently has been reaffirmed by other researchers in other cases, such as Dr. Michael B. Sabom, a cardiologist. Dr. Sabom conducted his own investigation of patients who had been resuscitated, and he reported the out-of-body experiences they underwent.[6]

What can we make of such reports? First, we have no hard evidence that the subjects had in fact died. Such proof is not impossible to obtain: rigor mortis is one sign, brain death is another. Rather, Moody, Sabom, and others seem to be describing either the dying process or a near-death experience, not death itself. Noyes and Kletti obtained accounts of near-death experiences from individuals who *thought* themselves close to death, at least psychologically: people rescued at the last moment from drowning, falls, or accidents.[7] Many of them reported similar out-of-body ex-

5. Raymond Moody, *Life After Life* (Covington, Ga.: Mockingbird Books, 1975).

6. Michael Sabom, *Recollections of Death: A Medical Investigation* (New York: Harper & Row, 1982).

7. Russell Noyes, Jr., and R. Kletti, "Depersonalization in the Face of Life Threatening Danger," *Psychiatry* 39:19-27.

periences, altered states of consciousness, and panoramic reviews of their lives. Noyes and Kletti postulate that at the time of impending death tremendous emotional stress can trigger such an experience.

John Palmer has presented an alternative hypothesis: that an out-of-body experience can be triggered by a change in the person's body concept.[8] Impending death threatens the self-concept, which in turn activates deep unconscious processes, attempting to reestablish the person's sense of individual identity as soon as possible. Hence, the belief emerges that one's self will continue in spite of the frightening conditions in which the body finds itself. The out-of-body experience is only one option that the unconscious mind can employ to restore some sense of identity.

Sabom and others are surprised that unconscious patients, after coming out of a coma, can report events that transpired around them; they can even recount conversations between doctors and nurses present. Moreover, they feel that they are "hovering above" the bed or operating table and looking down. However, it seems apparent that even though a person is in a comatose state, some auditory and perhaps visual sensations (where the eyes are open) can occur. The "hovering" experience is similar to other out-of-body experiences that people undergo when death is not threatening. Some have likened these deathbed experiences to other kinds of out-of-body experiences, in which some people claim that they were able to leave their bodies. James Alcock believes that these states are similar to dream states encountered in hypnagogic sleep, that is, the state experienced by practically everyone at one time or another during the period between wakefulness and sleep, or hypnopompic sleep, which occurs between sleep and wakefulness. It involves dreamlike fantasies, sometimes mixed with elements of reality.[9]

Interestingly, Sabom reports that three of his subjects experienced similar out-of-body experiences many years after the resuscitation. These after-experiences suggest that a psychological process is at work and not that the person has really left his or her body.

Moody, Sabom, and others also maintain that near-death experiences are not colored by the cultural environment (even though both admit to their own strong Christian faith). This claim is not entirely true, however, for although out-of-body experiences contain common psychological dimensions, many report meeting the particular religious figure of their faith. A Christian sees Jesus, a Hindu, Krishna, etc. I have carefully questioned several individuals who claimed to have had such an experience and have found that the phenomenological content varied consider-

---

8. John Palmer, *Parapsychological Review* 9, no. 5 (Sept./Oct. 1978).
9. James Alcock, "Psychology and Near Death Experiences," *Skeptical Inquirer* 3, no. 3 (Spring 1979).

ably. One woman reported entering shining gates and a splendid palace, where rubies, diamonds, and other gems were strewn on the floor; she encountered a man attired in a white robe and sandals, the usual depiction of Jesus. A man who said he was drowning maintains that he floated peacefully in the clouds. On the other hand, a woman who had a heart attack and apparently lay unconscious for days before being discovered by her son said that her experience was full of occult and demonic forces.[10] Indeed, Dr. Maurice Rawlings, in a book that runs counter to the usual claims of beatific experiences, reports that a great number of patients he resuscitated insisted that they had descended into Hell and that their near-death experiences were extremely frightening.[11] But he concludes: "I feel assured that there is life after death, and not all of it is good."

Reincarnation: Past lives

Of considerable relevance to the survival issue is the question of reincarnation. Hinduism and Buddhism postulate belief in a cycle of rebirth, the transmigration of souls, and the law of *karma,* by which souls can be eventually liberated from previous errors. Most of this is based upon pure faith, though philosophers as far back as Plato defended a doctrine of the preexistence of the soul grounded in memory recall and the theory of ideas. Many devout believers in immortality reject reincarnation as contrary to their Judaic-Christian faith. Others use it as added support for the survivalist hypothesis. They share in the common hunger for transcendence.

Recently there has been an attempt to give some scientific support to the preexistence hypothesis. A parapsychologist, Ian Stevenson, has studied the empirical evidence for reincarnation, focusing on the memories that some young children, especially in India, supposedly have of their prior lives.[12] The conscious and unconscious influences of the surrounding environment on the young child seem a more likely explanation of the tales he relates. Further evidence is the alleged ability of some adults to trigger recall of past lives. Hypnosis has been used by psychologists to elicit strange reports, but whether hypnosis really occurs has been debated by the specialists in the field. In any case, the material generated by the hypnotized person in a past-life regression seems, on analysis, to contain

10. These interviews were in connection with a television program that I did with Barry Farber in 1982.

11. Maurice Rawlings, *Beyond Death's Door* (Nashville: Thomas Nelson, 1978).

12. Ian Stevenson, *Twenty Cases Suggestive of Reincarnation* (Charlottesville: University of Virginia Press, 1974).

a blend of fictional tales and information that had once been learned and apparently forgotten. C. S. Zolick reports on a study that shows that subjects reconstructed a past life (or lives) by combining events from their own lives with stories taken from books and plays they had read or seen. It is difficult to use these so-called reports as evidence for past-life existence, for hypnosis is an unreliable basis for such an inference.[13]

Some writers claim that regression can occur even without hypnosis. Frederick Lenz reports cases of people who believed that they were Civil War soldiers or medieval knights.[14] Some people went even farther back to more primitive reincarnations in which had they lived as birds, owls, or turtles. Like Moody and Osis, Lenz finds a common theme running through such experiences. Basing his material on 127 cases, Lenz says that a person at first hears a high-pitched ringing sound, he feels his body begin to float in the air, the room begins to vibrate, and he subsequently loses the sensation that he is in his body. He beings to feel ecstatic and has a profound sense of well-being. Again, there is a rapid panoramic flashback to his former life or lives. He becomes totally absorbed in the experience. Eventually the vision begins to fade, and he returns to an awareness of his physical body.

In these cases, there is no near-death experience, nor is there a radiant light; yet the person is taken out of himself and supposedly experiences "another world." Lenz is especially careless in providing any data to corroborate his findings. We only have his word to vouch for the authenticity of the cases, which are based on subjective accounts. We may ask: Are these experiences evidence of reincarnation? Or are they stimulated by a fertile imagination, hallucinations, or an abnormal state? They appear to be forms of out-of-the-body experience, but there is no clear evidence that the souls actually leave the body, and alternative, naturalistic interpretations can be given for them.

There is no doubt that some people have had out-of-body experiences, whether in ordinary life, under hypnosis, facing death, or at other times. Does something leave the body on such occasions? The parapsychologist Susan J. Blackmore has sifted through the literature on out-of-body experiences in general, and has examined the claim that people can project astrally, whether in the near-death experience or otherwise. The parapsychological literature is full of reports of people who have insisted that it is possible for them to leave their bodies and travel to another place, such as Rio, Hoboken, or the planet Jupiter, and to report what is occurring.

13. C. S. Zolick, "An Experimental Investigation of the Psychodynamic Implication of the Hypnotic 'Previous Existence' Fantasy," *Journal of Clinical Psychology* 14:177–82.

14. Frederick Lenz, *Life Times: True Accounts of Reincarnation* (Indianapolis: Bobbs-Merrill, 1979).

Should we take their reports as veridical? Or should there not be efforts made at independent confirmation? Simple tests can be performed to confirm whether or not such a claim is true. Allow the subject to go to sleep and have him or her travel only to the next room and name a six-digit number placed on the top shelf of a bookcase. It is essential that there be at least two witnesses at all times in both rooms to see that the subject does not take a peek. I know of no positive tests—singly or in replication—performed under such rigorous conditions. Blackmore believes such experiences can be accounted for by a vivid imagination. She then concludes that since "nothing leaves the body in an OBE . . . there is nothing to survive."[15]

What are we to infer from all this research on the survival question? Has there been a scientific breakthrough? Do we have new evidence for some discarnate existence? Unfortunately, the evidence is still inconclusive, even scanty. It has not been rigorously corroborated. Moreover, alternative explanations can be given for the reports. No doubt if there is a wish to believe in survival or reincarnation, the evidence may appear plausible.

One might perhaps claim that we cannot decisively confirm what happens after death until we die and find out, but we ought at least to suspend judgment until then. But this seems to me an act of desperate folly, for the preponderance of evidence from all that we know about human and animal death as a biochemical phenomenon points to the high improbability of any sort of significant survival. Of course, some things do survive death in one form or another. There is the physical body, which decomposes, and the skeleton, which, under proper conditions, can be preserved for thousands of years. It may even be that just as urine is expelled from the bladder at death, some discharged energy remains—hovers, haunts, or whatever else—for a period of time until the body is fully decomposed; but maintaining that this discharged energy has personal consciousness or identifiable experiences is another matter. We would have to submit this proposition to careful measurement, and nothing like it has even been confirmed.

A further problem for survival concerns the time scale. Perhaps energy patterns survive for a brief time, say ten minutes—or a few years or at most a few centuries (like the ghosts allegedly haunting English castles). But the claim for eternal survival is virtually unverifiable. How would we go about proving that a soul or something else that survives death will never become extinct? In any case, it would be difficult for us to date any souls we might uncover—even if we should manage to discover an incorporeal carbon-14 technique for dating surviving souls. If

15. Susan J. Blackmore, *Beyond the Body: An Investigation of Out-of-Body Experiences* (London: Heinemann, 1982).

we were to uncover a very old soul, we could not prove it to be eternal. It would have had a beginning (unless one believes in an infinity of prior existences), and most likely it would not have preceded the origin of the human species. But, more importantly, we would have no guarantee that it would continue into the infinite future, unless by eternity we mean beyond the categories of time altogether. Such claims would at best be in the form of postulates or conjectures, not proofs. Thus the *immortality* thesis remains to be tested.

In the last analysis belief in immortality is only an article of faith, an inference from an unsubstantiated belief in a divine universe, not a scientifically demonstrated truth. It is an expression of a person's hope that a divine being will in some way enable us to survive death so that we can exist throughout eternity, and/or that the deity will resurrect the physical body and soul at some point in the future and reinstitute personal identity and memory.

The argument from value

Belief in immortality, in many religions, is central to belief in God and perhaps is even its chief psychological ground. But the doctrine of immortality is not so much a descriptive claim about an alleged reality as a normative ideal postulated to satisfy an apparent psychological hunger. Indeed, the God-idea takes on meaning because man faces death; God is introduced, along with immortality, as a solution to the problem of death. This is the existential-psychological argument for immortality, which many earlier humanists such as William James and Santayana defended. It involves three factors: (1) a response to the problem of death, an attempt to explain away death and overcome it; (2) some moral direction and focus in what otherwise seems to be a random and purposeless universe; and (3) psychological sustenance and support, giving courage and consolation to the bereaved and fearful, helping individuals overcome anxiety, loneliness, and alienation. Kant thought that God, freedom, and immortality were unproven postulates of the moral life and that they were necessary in order to complete the foundations of ethics and to overcome the antinomy between happiness and obligation.

In this ethically pragmatic sense, the doctrine of immortality is interpreted not as making a descriptive truth claim about the world but as providing normative guidance. To say that one believes in immortality as an ideal is not the same as saying that immortality of the soul exists in a literal sense.

The secular-humanist critique of the ethical argument for immortality is telling, and on several grounds. The humanist considers the doctrine of immortality to be basically unrealistic and even morbid. It grows out of both fear of and fascination with death. The immortalist is fixated on death, yet he endeavors to deny its awesome reality. For the humanist this means a failure to face the finality of death and an inability to see life for what it really is. This attitude has all the signs of pathology, for a person holding it is out of cognitive touch with reality. It is an immature and unhealthy attitude, a form of wish-fulfillment. It exacerbates an illusion in order to sooth the heart aching over the loss of a loved one, or to avoid accepting one's own impending termination. Death is a source of profound dread. There is an unwillingness to let go. One hopes for an opening to another life in which all one's unfulfilled aspirations are realized. The eschatological myth enables one to transcend the pain.

This mood of denial expresses a basic lack of courage to persist in the face of adversity. Immortality is a symbol of our agony before an unyielding universe and our hope of some future deliverance. It is the tenacious refusal to confront the brute finitude of our existence, the contingent and precarious, often tragic, character of human life. Those who believe in immortality trust that somehow someone will help us out of our misery, however long we have to wait; that this vale of tears can be overcome; and that in the end, despite our present suffering, we will have a reunion with our departed loved ones. Immortality offers therapeutic solace. In the past, when disease was so prevalent and the lifespan so brief for the mass of people, the idea would have made some sense. Life was often "nasty, brutish, and short," and three score and ten was not the norm but the exception. Thus, the immortality myth functioned as a tranquilizer. Today, other attitudes dominate. Armed with modern medicine and technology, we can combat death in other ways. We attempt to prolong life and to make it abundant and enjoyable as far as we can— that is, until the dying process sets in, at which point many are willing to hasten its onslaught if there is great pain, and to accept the inevitable with stoic resignation.

Thus existential courage is a key humanist virtue. So essential is courage to life that Kierkegaard and Tillich recognized that without it life becomes difficult and unendurable. For the humanist, it is not enough to muster the courage to be—merely to survive in the face of adversity; rather one must cultivate the courage to become. In other words, the problem for each of us is to remake our lives constantly in spite of all the forces in nature and society that seek to overwhelm us.

It seems to the humanist that the immortalist has forsaken full moral responsibility, for he is unwilling to take destiny into his own hands. There are surely some limits on human achievement and independence

We need to make a distinction between things within our power and those outside it, as the Stoics recognized; but with those within our power we can tackle the great challenge to make the most of life, to re-create and redefine it, to extend the parameters of human power.

Here, freedom of human choice and action is pivotal. The humanist is not content with simply discovering and accepting the universe for what it is, in an act of piety; he seeks to *change* it. Nor is the task of life to discover what our nature is (whether God-given or not) and to realize it, but to *exceed* our nature. Thus man invents culture. We are post-Prometheans, stealing fire and the arts of civilization from the gods, continually tempting and recasting our fate.

A key objection which the humanist levels against those who cling to the doctrine of immortality is that it undermines ethics. One is unable to be fully responsible for himself and others, to be creative, independent, resourceful, and free, if he believes that morality has its source outside man. The reflective, deliberative, probing moral conscience is too vital to be deferred to the transcendent. We are responsible for what we are, and out of our compassion for other human beings and our desire to see that justice be done, we can achieve the good life here and now if we work hard enough to bring it about. It is not the hope of salvation or fear of damnation that moves us to seek a better world for ourselves and our fellow humans, but a genuine moral concern without regard for reward of punishment. Morality is autonomous.

In the humanist view, the believer has committed a grievous mistake: he has wasted much of his life. Life is short, yet it is rich with possibilities, to be lived fully with gusto and exuberance. Those who are morbid, fearful, timid, unwilling to seize destiny are often unable to experience fully the bountiful joys of life.

All too often those who believe in immortality are full of foreboding; laden with excessive guilt and a sense of sin. All too often the pleasures of the body, sex, and love are repressed, and a variety of opportunities for creative enjoyment are denied. Many such individuals have thus bartered this life for a future life; but if the promissory note is unfulfilled—as I think it is—then they have lost some important values in life. In retrospect, their lives have often seemed barren; they have missed many chances, failed to do what they really wanted; they could not seize the opportunities because of deep-seated fear and trembling.

Many theists believe that without immortality life would have no meaning. How puzzling and contradictory to so argue. It is a confession of their own limitations as persons. For it is, the humanist contends, precisely the doctrine of immortality that impoverishes meaning. If one believes in immortality, then little counts here. It is all preparation; life is but a waiting room for transcendent eternity. This life often is not fully cultivated, for only the next one counts. But life has only the meaning we choose to invest it with. All it presents are opportunities, which we may choose to capitalize upon or let pass by.

Does belief in immortality satisfy a necessary psychological need? Interestingly, many individuals without belief in immortality fear death less and are able to accept it and face it with greater equanimity than those with such a belief. Such unbelieving individuals are able to develop confidence in their capacities as humans, an independent moral conscience, a commitment to social justice or species welfare on this planet. Such individuals need not be without "transcendent" ideals, ideals larger than they. They may believe in contributing to a better world and be deeply concerned about the future of humankind; and they have a sense of obligation to ideals that are as powerful as any the immortalist possesses.

Humanists may indeed believe in "immortality" in a metaphorical sense: they are devoted to the good works that will outlast them. But they strive for them not in order to be rewarded or punished, but because while they live they find their goals worthwhile. They are moved now to do what they think will contribute to a better future for their children's children's children, even though they may never live to see it. They need nothing beyond that to support or sustain their moral dedication. Thus a thoroughgoing secular humanism does not need a doctrine of immortality to give life meaning or to provide morality with a foundation.

Nevertheless, it is apparent that the quest for transcendence expresses a passionate desire within the human breast for immortality and permanence. This impulse is so strong that it has inspired the great religions and paranormal movements of the past and the present and goaded otherwise sensible men and women to swallow patently false myths and to repeat them constantly as articles of faith.

XV: Space-age religions: Astrology and UFOlogy

The transcendental temptation expresses itself in ways other than traditional religious forms. We have seen its manifestations in the spiritualism movement and in paranormal beliefs. Of considerable interest is the fact that a similar form of the impulse is seen in the emergence of what I term space-age religions. There are two underlying premises of these quasi-religions. First is the belief that the universe manifests an overall order or plan and that this has some reference to every individual within it, whose fate is determined or influenced by it. Second is the belief that powerful, quasi-divine intelligences exist in space who are aware of our existence and who observe and perhaps even influence us. Two paranormal belief-systems represent either the first or the second assumption.

The first is astrology, an age-old system of beliefs that goes back an estimated 4,000 years to ancient Babylon and Egypt, and still has a powerful hold on the public imagination today. Astrology was first formulated and codified for the Western world by Claudius Ptolemy, in his book *Tetrabiblos* (around C.E. 140). Different astrological systems have been developed in India and China. It is a surprise to many scientists that astrology persists in the postmodern world, for the natural sciences and astronomy have largely refuted its basic premise. The second system is UFOlogy, the belief that earth is visited by extraterrestrial beings possessing semidivine powers, whether beneficient or demonic. As a projection of human needs and wishes, UFOlogy is a modern equivalent of Greek mythology's gods on Mount Olympus. Since it is a recently developed system of belief, we can observe its genesis and elaboration.

Astrology and UFOlogy combined may have a greater number of adherents than many of the traditional religious sects or faiths, though the level of faith and commitment is no doubt much less intense, and institutionalized churches have not yet developed to transmit the belief-systems and to codify and enforce orthodoxy. Both systems are curious byproducts of the space age, involving the speculative, fictionalized, or mythological aspects of a technology that enables human beings to leave our planet and embark on space travel. We may say that 1967 marked the beginning of the Age of Space, for that was the year man was able to break out of the earth's gravitational field. Perhaps 1967 should be noted in the future as Year One and 1968 as After Space One (AS 1). Everything before 1967 should be numbered retrospectively as Before Space.

Astrology

Historical assumptions

Human beings have always looked with awe at the magnificent heavens. They have worshiped the sun as the source of all light and life, and the moon as the heavenly body of the evening. The breathtaking spectacle of the stars and the galaxies has always fascinated human imagination, particularly when humans lived outside at night under the starry heavens, wondering what the celestial vault might portend about their origin and destiny. The first religions worshiped the heavenly bodies, and they expressed the conviction that these extraterrestrial bodies were ultimately related to us on earth, our needs and aspirations, our yearnings and sufferings, and that they offered providential signs and omens as well. It was thought prudent to supplicate these powerful forces in the universe. The creation myth in the Old Testament clearly demonstrates our cosmic wonder. In the beginning, God created Heaven and earth. As part of this plan of creation, he gave life to Adam and Even in the Garden of Eden. Even Jesus was born under a sign (the star of Bethlehem), and he was visited while in the manger by three wise men (astrologers) from the East, thus attributing cosmological significance to his time and place of birth.

The anthropocentric tendency is to invest the stars and planets, the sun and the moon with divine qualities, as signs of some divine order. The two most momentous events in a person's life are birth and death. Christianity is concerned with what happens to a person after death. Astrology, in contrast, is a highly personalized and cleverly contrived belief-system that emphasizes the moment of birth and seeks to give it some kind of transcendent meaning. It even has magical import, for at the moment of birth we are allegedly imprinted by the positions of the heavenly bodies, which provide signs marking our personality, physical characteristics, and future destiny.

The earliest astrologers claimed to ascertain the will of the gods. Human life and happiness were thought to be dependent upon what transpired in the heavens. The earth's fertility depended on the sun shining, the seasons, and the rain that fell from the sky. Great damage could be done by violent storms and floods, likewise traceable to the heavens, seat of the gods. If the sun and moon were especially worshiped, it was because they seemed to manifest orderly and regular movements. The luminous appearance of the celestial bodies suggested that they were made of a "heavenly" substance, unlike the kinds of substances found on earth. The same seemed true of the five known planets, each of which was equated with a god. The Babylonians identified Jupiter with Marduk, Venus with the goddess Ishtar, Mars with Nergal, Saturn with Ninib, and Mercury with Nebo. Mars appeared red, and thus represented the god of war, courage, and energy. Venus was white and denoted feminine qualities; she was the goddess of love and beauty. Astrological interpretations were placed upon what appeared to be the fixed stars. Various groupings of the stars in the firmament were designated as constellations. Twelve of them in the zodiac correspond to each month, and all but one are identified with living animals or people. The constellation Pisces looks like a fish, and thus it became a "water sign" by a process of magical correspondence. Taurus was a bull in the sky, and anyone born under the sign of this constellation behaved like a bull, was stubborn and strong. Thus astrology developed an elaborate system of signs and symbols and a literature; the heavens were interpreted anthropomorphically as having human qualities. The astrologer served as prophet-priest-magician. By his calculations, charts, and forecasts, he was able to cast horoscopes and predict the unique destinies of individuals and nations.

The key assumption concerns the relationship between the exact time and place of birth and the positions of the bodies in the heavens. The original astrological, cosmological system upon which it is predicated had the earth as the center of the universe, with the other planets, the sun, and the moon traveling about it. Ptolemy, who was also a great astronomer, contrived a system of epicycles, in which the heavenly bodies made circular loops. It was an ingenious construction, for it enabled ancient astronomers to plot the position of the planets, the sun, and the moon, and to make fairly accurate predictions. Astronomy and astrology were closely wedded until the 16th century. One dealt with the cosmic scene, the other with its effect on human affairs. The ancient astrologers compiled, over many centuries, elaborate observations of when and where people were born and their characteristics, and they attempted to correlate these with the positions of the celestial bodies as viewed from earth. A wise person would learn his horoscope and act accordingly, following the recommendations of his astrological advisors. We know from history that monarchs employed staffs of astrologers, who offered council as to when to launch an attack on the enemy or to pursue a fair maiden.

Some people consider astrology a science, others a humanistic study or art. Its skeptical critics consider it a classical pseudoscience. Attempts are made to make empirical predictions, and these presumably are grounded in detailed observations. Ptolemy's codification was supposedly based on written records and observations of what had taken place in the past, and the alleged correlation of the sun, moon, planets, and stars to persons and events on earth. At first these correlations were held to be of public importance, and of general concern for everyone. In time they applied primarily to the lives and destinies of rulers and the states they governed. Eventually they were extended to each individual, who had his own unique horoscope. The principles and rules of forecasting have become part of an astrological tradition that has been modified little in its essentials since Ptolemy. Today this tradition is set forth in a great number of astrological books, which are constantly published—much the same as the Bible repeats a legend. Most of the principles simply repeat what has been handed down, although there are great variations in interpretation. The original basis of the observations are unknown. Are they rooted in the recordings of ancient witnesses and were they tested for their accuracy? Or is much of the lore simply that—without foundation or fact? If one investigates the origins of a system of beliefs—as we have analyzed man's primary religions—we usually find that they rest on quicksand. Astrology seems to have had both religious and magical significance. An examination of the genesis of a belief has considerable value in understanding why it arose and what it means. One must not commit the genetic fallacy, however, for astrology may be true today, in spite of its dim origins. It will be useful, therefore, to give a brief outline of its main features.

There are different kinds of astrological systems. We will concentrate on natal astrology, which involves the development of a chart showing the positions of the planets in the sky at the time of a person's birth. The charting of a horoscope entails the use of celestial coordinates. First the sun, moon, and planets are located in the sky on the celestial sphere. Second, the celestial sphere (the apparent globe of the sky), containing the planets and the fixed stars is related to the earth at the time and place of a person's birth. The exact positions of the planets are determined with the help of astronomers, by locating their place on the zodiac. The zodiac refers to the belt on the ecliptic, the sun's apparent path among the planets. The horoscope describes the path that the sun appears to follow through the year, which represents a circle of 360 degrees. Astrologers divide the path into twelve sectors of 30 degrees each—somewhat arbitrarily, I might add. These are called the signs of the zodiac, or the "sun signs," since they indicate where the sun is at that period of time and hence at the time of a person's birth. Thus, for example, if you were born in early December, the sun would have been in the twelfth sign, Sagittarius.

The zodiac is generally determined by the vernal equinox. This refers to the position of the sun at the beginning of spring (about March 20th). Astronomers define this as the point in space where the earth's orbit around the sun, called the ecliptic, intersects with the plane of the earth's equator as it is extended into space. It occurs when the sun moves from south of the equator to north of it, and its apparent motion along the ecliptic as viewed from the earth. This is only apparent motion because, although traditional astrology positioned the earth at the center of the system with the sun moving around it, as we know since Copernicus, the reverse is the case. Even though the Copernican revolution completely transformed our knowledge of the solar system, it did not succeed in destroying astrology, which was based on an astronomical system that had to be abandoned. Astrologers merely kept on selling their services and adapting their ancient system as best they could.

Now what is distinctive about astrology, in contrast to astronomy, is that each sign of the zodiac represents certain unique positive and negative characteristics of human personality and behavior. Each individual has all twelve signs as part of his horoscope. His life and personality are determined by the position of the planets in the zodiac signs and the interpolation of the houses. For those born on the cusp, that is, the line between the two signs, there are conflicting positions.

Astrologers have also divided the horoscope into houses, each of which is related to a sign of the zodiac. The houses are based on the daily rotation of the earth on its axis. Like the sun signs, the houses are distinguished by the lines or cusps between them. The first house cusp is called the ascendant. This is calculated by the point in space where the eastern horizon intersects the ecliptic and the time and place of birth. Objects within the first house are the ones that will rise within the following two hours. Similarly for each of the other houses. The sign of the zodiac on its cusp is called the rising sign. It is the most important house cusp, for it designates an individual's basic reference toward life, how he or she responds, the basic personality and physical appearance. Also important to the astrologer's analysis is the nadir (fourth house), the descendant (seventh house), and the midheaven (tenth house). The various houses govern various human interests and activities, including financial affairs, marriage, profession, the domestic scene, and health. The horoscope of a person also includes various other planetary aspects, that is, the relative positions of the sun, moon, and planets. It ascertains whether the celestial bodies are in conjunction or opposition to each other in the various houses, and whether or they are 60, 90, 120, or 180 degrees apart. All of this has significance in casting a horoscope.

Given these positions in a detailed horoscope, the task of the astrologer is to interpret their meaning for the subject (called the "native"). There are more or less standard rules for interpreting the horoscope of an individual, but this can become extremely complicated. For how the rules are applied is often a matter of the astrologer's subjective judgment, and it may vary from case to case. Hence astrology has been called an art rather than a science. To illustrate, if a person was born under the sign of Aries (between March 21st and April 19th), he or she is supposed to be dominant, inventive, and impatient. If, for example, the moon was ascendant at the time and place of birth, he would tend to be moody and procrastinate. If Mars were in the sign of Virgo, he would be shrewd and appear to be impulsive. In regard to aspects, if Mercury were in conjunction with Jupiter, this would be a good omen, unless at the same time Venus were in equal aspect from Saturn (90 degrees from it), which would be bad. It becomes difficult, therefore, to interpret the exact meaning of charts; it may depend on how the astrologer and the native interact, especially if there is a personal encounter.

Astrological charts have been developed not only for individuals, but for nations, organizations, even domestic pets! Even the date of the signing of the Declaration of Independence, July 4, 1776 (the exact time is in dispute), has led astrologers to draw charts predicting the political future of the United States, as was done for ancient principalities.

Scientific criticisms

Astrology is a cumbersome though enormously fascinating system of belief, and it continues to captivate the human imagination. With the advent of modern astronomy, which displaced the Ptolemaic scheme, astrology was given a rude shock, and during the eighteenth and nineteenth centuries it seems to have been thoroughly discredited, at least among the educated classes. Yet it never disappeared entirely, and it has been revived in the twentieth century, with great popularity in even the most advanced, affluent, and highly educated societies of the Western world. Today, horoscopes appear in large-circulation newspapers and magazines, and professional astrologers outnumber professional astronomers. Why is this so? Is it because astrology is true? Are its empirical claims valid? Does it work? (This depends on one's definition of *workability*).

Yet there has been a devastating scientific critique of astrology's basic principles. I will outline some of the more important criticisms.

1. The earth is not the center of the solar system, as earlier astronomers held. Only the moon orbits the earth. All the planets orbit the sun. Thus the horoscope is rooted in a fundamental astronomical error. Classical astrology, which claimed to describe the planetary influences, only

knew of five planets, those visible to the naked eye, and was totally un-aware of Pluto, Uranus, and Neptune, which were discovered later. Astro-logers have since attempted to bring them into their calculations. Mean-while, all of the classical charts are mistaken. In addition there are thou-sands of asteroids in our solar system that are largely uncharted by the astrologers, but since they are part of the celestial scene, it would seem that their ability to "influence" events on earth should be calculated also.

2. Ancient astrology mistakenly thought that the stars were fixed in the celestial sphere, and they marked the sun signs by means of various constellations. In the more than 1,800 years since Ptolemy codified astro-logy, there has been a shift in the equinoxes due to the earth's spin, such that the position of the "fixed stars" in relation to the earth have changed by approximately 30 degrees. Hipparchus, in the second century, had already discovered that there is a gradual change in the direction toward which the earth's axis points. There was a precession of the equinoxes, which results in the vernal equinox moving westward about the ecliptic. This meant that the signs of the zodiac, which originally had the same names as the constellations they coincided with, have since altered their positions. The signs no longer line up with the constellations of the same name. For example, the sign Virgo is now in the constellation Leo. The astrologers of antiquity were aware of the phenomenon, and based astro-logy on the moving signs, not the constellations. This is called tropical astrology; another school of astrology, called sidereal astrology, is based on the positions of the planets in the constellations rather than the signs. Over a period of time, the signs will continue to shift further away from the constellations. There are thus alternative systems of astrology based on this fact.

3. What influences do the heavenly bodies have on earth at the time of birth? What are the forces at work—gravitational, electromagnetic, nuclear—or is there some unknown force? Both electromagnetic and gravitational influences from the planets have been measured and found to be weak. The sun and moon, of course, exert important gravitational and radiation influences on the earth, but those exhibited by the planets at the birth of an infant seem relatively small. So the question has been raised as to what kind of forces are operative in the delivery room. It has been estimated that the obstetrician or nurse hovering over mother and child exerts a far greater influence than the distant planets. Moreover, if physical forces are at work, why not while the infant is in the womb? Would the flesh and skeleton of the mother provide a better shield against prenatal influences than a modern building? This is questionable.

Scientists have thus been puzzled by being unable to determine the kind of mechanism to account for the imprinting of the infant at the moment of birth. This does not, of course, deny the vital influences of the sun and moon on things on earth—the tides, to take one example, or plants' need for sunlight. This is not at issue. What is at issue is the basic hypothesis that the positions of the sun, the moon, and the planets at the moment of birth exert a crucial influence over the individual.

4. The definition of the "moment of birth" is also a matter of dispute. Is it when the infant begins to emerge from the womb, which may take a period of time during labor? When the first breath is taken (which is what astrologers maintain), or when the umbilical cord is cut? Is labor necessary? In recent decades, obstetricians have performed many caesarean operations. Moreover, doctors have either induced or delayed the time of birth by the use of drugs. So is natural childbirth essential? Do astrological horoscopes operate only for those whose delivery is natural? A further problem is that many people do not have the exact time of birth recorded. Perhaps it is only approximated, so it is often impossible to get a precise time.

5. But even if one knew the exact moment of birth, a further question is raised as to what is imprinted when one is born. What role do genetic or biological factors play? Astrologers have maintained that in addition to personality traits, the physical constitution—skeletal build, hair and eye color, and tendencies for disease later in life—is correlated with one's horoscope. But biology teaches us that our physical traits are a function of chromosomes and genes transmitted at the moment of conception— not at the moment of birth—and that one's personality traits and most physical ailments depend upon environmental influences, generally well after birth. Perhaps at the moment of conception, when the sperm penetrates the egg and two sets of chromosomes are transmitted, the position of the heavenly bodies could make a difference. At least there would be a mechanism. But first, this has not been confirmed by any experimental evidence, and second, all astrological systems based on the moment of birth would still be in error and have to be discarded. If the moment of birth is decisive, then presumably, two persons born at the same time and place would be "astral twins," even though they have different parents and environmental influences. The two should look alike and lead parallel lives. St. Augustine became a critic of astrology when he noted that a slave and his aristocratic master were astral twins but led entirely different lives, given their different social roles and circumstances.

6. One can also ask: What is the "influence" of the celestial bodies? Does astrology involve a theory of determinism, predeterminism, or fate? One can read in different metaphysical conceptions, no doubt due to the ambiguity of the theory. Astrology in fact lacks a coherent conceptual framework, and astrologers have not thought deeply or clearly about its

foundation. There are various forms of determinism. Does astrology postulate predeterminism or fate, in which whatever will be is fixed? This hardly seems to fit the case, since astrologers constantly offer counsel and advice. They seem to imply that there is a kind of cosmic weather, and that certain times are more propitious for one to embark upon a course of action. Thus we often hear it said that the stars only "impel," not "compel," which suggests a form of weak determinism. I would have no objection to this kind of deterministic theory, if this is what they infer, but it is uncertain as to what they do maintain since in astrology the planetary bodies are supposed to imprint individuals and in some sense outline their future destinies, even though there is some room for human beings to maneuver within these parameters. Is there a direct causal relationship between the heavenly bodies and everyone who is born? This has been difficult to ascertain, unless there is an "unobservable mystical cause" emanating from the planetary bodies. Some have suggested that these bodies only trigger mechanisms in the infant (although this has not been demonstrated). Still others talk about a cosmic synchronicity, with the planetary positions playing only a symbolic role in correlating relationships. The difficulty with astrology is that it is so vague and ambiguous that it does not lend itself to precise interpretation or corroboration.

Sun-sign astrology

We may evaluate astrology from still another perspective, by examining the concrete empirical descriptions and prognostications of astrologers in order to ascertain whether or not they are accurate. We will do this, first, by evaluating the accuracy of sun-sign descriptions and, second, by analyzing horoscopes. Most people are familiar with sun-sign astrology, for this is what appears in popular newspaper columns. No idea is more influential in astrology than the notion that a person's traits are determined by the position of the sun in his or her chart. Linda Goodman summarizes the effect of sun signs as follows:

> The sun is the most powerful of all the stellar bodies. It colors the personality so strongly that an amazingly accurate picture can be given of the individual who was born when it was exercising its power through the known and predictable influences of a certain astrological sign.[1]

Following is a list of these signs and a brief description of their meanings:

1. Linda Goodman, *Sun Signs* (New York: Bantam Books, 1968), p. xviii.

THE CONSTELLATIONS OF THE ZODIAC, THEIR SYMBOLS, NAMESAKES
AND SELECTED SUN-SIGN CHARACTERISTICS

Constellation and symbol	Animal namesake	Selected characteristics
Aries	ram	headstrong, impulsive, quick-tempered
Taurus	bull	plodding, patient, stubborn
Gemini	twins	vacillating, split personality
Cancer	crab	clinging, protective exterior shell
Leo	lion	proud, forceful, born leader
Virgo	virgin	reticent, modest
Libra	scales	just, harmonious, balanced
Scorpius	scorpion	secretive, troublesome, aggressive
Sagittarius	archer/horse	active, aims for target
Capricornus	goat/fish	tenacious
Aquarius	water carrier	humanitarian, serving mankind
Pisces	fish	attracted to sea and alcohol

Exhaustive scientific testing of sun-sign astrology has been done. The results are invariably negative. R. B. Culver and P. A. Ianna, in *The Gemini Syndrome: A Scientific Evaluation of Astrology,* examined the claims in detail.[2] For example, people born under the sign of Aries are supposed to have the following physical characteristics, according to one astrologer: "A longish stringy neck like a sheep; a look of the symbol in the formation of eyebrows and nose; well-marked eyebrows; ruddy complexion; red hair; active walk."[3] Focusing on red hair, Culver and Ianna point out that this would exclude blacks and Orientals, and most Hispanics, who have black or dark brown hair. They conducted a survey of 300 redheads to find out the sign under which they were born, and found (1) that only 27 of them were born under Aries, and (2) that no one sign showed a predominance of red hair.[4] Natural red hair, which is a genetic trait, is not determined at the moment of birth.

2. R. B. Culver and P. A. Ianna, *The Gemini Syndrome: A Scientific Evaluation of Astrology* (Buffalo: Prometheus Books, 1984).
3. M. Hone, *The Modern Textbook of Astrology,* rev. (London: L. N. Fowler & Co., 1972), p. 49.
4. Culver and Ianna, *Gemini Syndrome,* p. 124.

The Astrologer's Handbook describes many physical characteristics. Scorpios are depicted as follows: "In appearance, they are generally of robust and strong build. They often possess keenly penetrating eyes and a strong aura of personal mystique and magnetism."[5] Culver and Ianna provide an evaluation of other physical characteristics, such as neck size, skin complexion, body build, height, weight. Again they find there is no correlation with any of the signs, so that one characteristic or type predominates over the others.[6]

An important and often dangerous element in astrology is its diagnosis of medical conditions and diseases. Yet Culver and Ianna find that there is no correlation between the sun signs and illnesses, such as diabetes, heart disease, rheumatism, or even acne. They find these distributed equally across the signs.[7]

Much less is said today about the physical aspects of astrology because of the patent quackery associated with it, yet that is an essential part of its tradition. Most emphasis now is on personality traits, occupations, and compatibility between people. Concerning personality traits, we find the following account of those born under Capricorn:

> Because Saturn rules Capricorn, the natives have a tendency to [be] melancholic and, at times, lonely . . . They have sensitive personalities and want very much to be appreciated . . . They are neat and methodical in their work and tend to be slave drivers at home . . . Capricorns are excellent executives and remain in subordinate roles for a short time only.[8]

Those born under Libra "are intellectual, and actively seek knowledge, new ideas, and mental stimulation. . . . they frequently play the role of peacemaker. . . . Libras are ruled by the planet Venus which gives them charm and grace . . . with a desire for popularity."[9]

The problem with these descriptions of traits is that they are so general they can be applied to almost anyone, especially since they tend to focus on positive qualities. Of Sagittareans we read the following: They are "naturally serious thinkers, concerned with the well-being of society as a whole as well as with their own lives. . . . they are honest, just, and generous . . . they are energetic and naturally outgoing."[10]

It is difficult to measure abstract traits; they tend to be elusive. In any case, tests have shown no correlation between the predicted personality

5. Francis Sakoian and Louis S. Acker, *The Astrologer's Handbook* (New York: Harper and Row, 1973), p. 46.
6. *Gemini Syndrome,* p. 126.
7. Ibid., p. 135.
8. Sakoian and Acker, *The Astrologer's Handbook,* pp. 48-49.
9. Ibid., p. 44.
.0. Ibid., p. 47.

traits and the signs under which a person is born. Culver and Ianna examined traits from aggression and ambition to understanding and wisdom and found no apparent correlation.[11]

One way to measure the accuracy of astrology is by correlating occupations and careers with sun signs. Studies have been made of sports champions, actors, army officers, men of science, politicians, artists, etc. by consulting various reference works that list occupation. Thus John D. McGervey examined the birthdays of 16,634 scientists in *American Men of Science* and 6,475 entries in *Who's Who in American Politics* and found no basis for the assertion that occupations tend to predominate under certain signs.[12] Other thorough statistical studies done by E. Van Deusen[13] and Culver and Ianna likewise show no correlations. Although rare correlations are sometimes cited, as in a study published in the *Manchester Guardian,* these can be attributed to statistical fluctuations.[14] Similarly for the alleged compatibility or incompatibility of people born under certain signs. For example, the marriage of a Virgo man and a Scorpio woman, according to astrologers, "has far too much against it for it to succeed," while the marriage between a Scorpio woman and a Cancer man is "a splendid match."[15] Divorce would seem to be a sure sign of incompatibility, yet Professor Bernie Silverman looked at 2,978 couples who were married and 478 couples who were divorced, in Michigan, compared their signs and marital status with the predictions of astrologers, and found no correlation.[16]

Horoscopes

So much for sun-sign astrology, which has no discernible basis in fact and yet enjoys tremendous popularity with a vast public. Most professional astrologers will admit, however, that sun-sign astrology is only part of their diagnosis, and hence unreliable by itself. They maintain that although the sun is the strongest influence on a person, the positions of the planets must be charted for a completely accurate account. Here it is not simply the monthly range, but, as we have discovered, the exact time and place of birth that must be recorded. Each horoscope is like a personal

11. Culver and Ianna, *Gemini Syndrome,* p. 130.

12. John D. McGervey, "A Statistical Test of Sun Sign Astrology," *The Zetetic (Skeptical Inquirer)* 1, no. 2 (Spring/Summer 1977): 49-54.

13. E. Van Deusen, *Astrogenetics* (Garden City, N.Y.: Doubleday, 1976).

14. G. A. Dean, I. W. Kelly, et al., "The Guardian Astrology Study: A Critique and Reanalysis," *Skeptical Inquirer* 9, no. 4:327-38.

15. Teri King, *Love, Sex, and Astrology* (New York: Harper and Row, 1973).

16. Bernie Silverman, *Contemporary Astronomy* (Philadelphia: W. B. Saunders, 1977), p. 437.

fingerprint or signature. Only a trained, professional astrologer can give a precise diagnosis. So the question can be raised: Are detailed horoscopes accurate? I am afraid that the same criticisms we made of sun-sign astrology can be raised against the more elaborate horoscopes and that they also fail the test of adequacy—but for a variety of subtle reasons.

How accurately do horoscopes analyze personality? A number of researchers have randomly distributed horoscopes in order to test their fit. The French astronomer Paul Couderc advertised in a French newspaper that free horoscopes were available for those who wished them. Every respondent was sent the same bogus horoscope. It included such phrases as "You have inner conflicts . . . life has many problems . . . you sometimes upset people," etc. He asked for comments and received them from 200 persons. A large number of people claimed that the account fit their personalities perfectly.[17]

The next case concerns the horoscope of a person born on January 17, 1898, at 3:00 A.M. in Auxerre, France. It reads in part:

> As he is a Virgo-Jovian, instinctive warmth of power is allied with the resources of the intellect, lucidity, wit . . . He may appear as someone who submits himself to social norms, fond of property, and endowed with a moral sense which is comforting—that of a worthy, right-thinking, middle-class citizen . . . The subject tends to belong wholeheartedly to the Venusian side. His emotional life is in the forefront—his affection towards others, his family ties, his home, his intimate circle . . . sentiments . . . which usually find their expression in total devotion to others, redeeming love, or altruistic sacrifices . . . a tendency to be more pleasant in one's own home, to love one's house, to enjoy having a charming home.

An advertisement was placed in a French newspaper by Michel Gauquelin inviting people to send in their name, address, birthday, and birthplace. About 150 replied; each person was sent a full, ten-page horoscope, from which the quotation above is taken, a return envelope, and a questionnaire. Of those who answered the questionnaire, 94 percent said they were accurately portrayed in the horoscope and 90 percent said that this judgment was shared by friends and relatives. It was the horoscope of Dr. Marcel Petoit, a mass murderer. Dr. Petoit posed as an underground agent who would help refugees fleeing from the Nazis. Instead, he lured them to his home, robbed them, and dissolved their bodies in quicklime. Indicted for twenty-seven murders, he cynically boasted of sixty-three. So much for the accuracy of horoscopes!

Three recent scientific studies have reached the same conclusion about

17. Quoted in M. Gauquelin, *Astrology and Science* (London: Peter Davies, 1969), p. 149.

the accuracy of astrology. In *Recent Advances in Natal Astrology: A Critical Review 1900-1976,* Geoffrey Dean and Arthur Mather (under the auspices of the Astrological Association) reviewed seventy-six years of literature and reached the following negative conclusion:

> Astrology today is based on concepts of unknown origins, but effectively deified as "tradition." Their application involves numerous systems, most of them disagreeing on fundamental issues, and all of them supported by anecdotal evidence of the most unreliable kind. In effect, astrology presents a dazzling and technically sound superstructure supported by unproven beliefs. . . . there is not the slightest evidence of any empirical basis. . .[18]

Equally negative were the studies of Culver and Ianna already cited and that of Eysenck and Nias,[19] who state at the outset: "We feel forced to conclude that most of these beliefs (of astrology) do not stand up to rigorous examination."[20]

The sole possible exception to this indictment concerns the work of Michel and Françoise Gauquelin, already cited. Michel Gauquelin began as a believer in astrology, spent years investigating horoscopes and astrological charts, and found there was no evidence to support them. Indeed, the work of Michel Gauquelin emerges as the most important body of statistical data negating the popular version of classical astrology.

But in the course of his studies, Gauquelin believed that he had found a relationship between the positions of planetary bodies at the time and place of birth and personality traits and careers. He labeled this *cosmobiology* and *astrobiology.* He maintained that Mars, for example, is correlated with being a sports champion, Jupiter with actors and politicians, and Saturn with scientists and physicians. He divided the heavens into twelve sectors and found, for example, that Mars appears in the first and fourth, higher than chance (22% versus 17%). Gauquelin's work is based on a sample of European sports champions. I will not review his claims here in detail. Both Mather and Dean, and Eysenck and Nias believe that Gauquelin's work is significant, and that if true, it will vindicate the basic premise of astrology that there are planetary correlates.

A group in Belgium known as the Committee Para attempted to replicate Gauquelin's findings. Although Gauquelin claimed that the group replicated them, Committee Para was skeptical, for it disputed the theoretical frequency of nonchampions. This was finally resolved by the

18. Geoffrey Dean and Arthur Mather, *Recent Advances in Natal Astrology: A Critical Review 1900-1976* (Southampton, England: The Astrological Association and Camelot Press, 1977), pp. 1, 23.

19. H. J. Eysenck and D. K. Nias, *Astrology: Science or Superstition?* (New York: St. Martin's Press, 1978).

20. Ibid., pp. ix-x.

Zelen Test, a study conducted by Professor Marvin Zelen of Harvard University, which showed that if the Gauquelin data was accurately assembled, then the frequency for nonchampions being born in the first and fourth sectors was 17.1 percent, as Gauquelin maintained. However, the replication was not an exact fit, for although sectors one and four were higher than expected, other sectors were higher or lower than predicted.

Three researchers, who were members of the Committee for the Scientific Investigation of Claims of the Paranormal—George Abell, Professor Zelen, and I—attempted a replication of Gauquelin's hypothesis in the United States. We compiled a sample of 408 sports champions, and the results were negative, indeed, slightly below chance. The American sample was derived by drawing sports champions from various *Who's Who*s of sports. Gauquelin has disputed our negative interpretation, claiming that the "Mars effect" applies only to the "most famous" champions, that they must be selected from these volumes, and that it works only for those of international reputation. Our response was that it is curious that the Mars effect (since it supposedly points to the personality traits of competitiveness and aggressiveness) does not apply to famous sports champions listed in *Who's Who* but only to the *most* famous, internationally known ones and that those who fit into the latter category are a very small class. Moreover, it is often a question of subjective interpretation as to who are the "most famous." Thus, at present, the theory of Gauquelin has not been corroborated. Thus far all we have are data largely assembled by or with Gauquelin and interpreted by him, using his criteria, which he has changed in different studies. In science we need independent and impartial replications of hypotheses. Until we have that, we must be open-minded yet skeptical about his results. In any case, even if Gauquelin's work is replicated eventually, this would in no way validate the entire system of traditional astrology, which continues strong, in spite of the fact that no basis in empirical fact has been demonstrated for it.

Conclusion: Why does astrology persist?

The question that can be raised is why, in spite of a lack of evidence for its claims, and considerable negative evidence to the contrary, does astrology persist as a belief-system? This is the same question we have raised about religious belief-systems. Similar considerations apply.

One explanation is that the general public has not been exposed to the benefits of scientific criticism. Generally, proastrology propaganda is far greater than any negative criticism. If many were given scientific studies, they might be persuaded to reject the claims. There is no guarantee, however, that even if many people were aware of the negative arguments, they would reject astrology.

A second explanation is that gullibility is often very strong in the human breast, and that many people tend to accept claims that appear authoritative. They are often deceived by con men and women. Here, the guru is the astrologer, who is speaking mumbo-jumbo but is quite persuasive. There is still another element that enters in, however, and that is self-deception; this applies to both the client and the practitioner.

Astrology, like religion, is a faith; it is based upon, and reinforced by, subjective validation. In other words, the prognostications and evaluations of the astrologers appeal to a kind of inner intuition. The readings are so general that they can be applied to almost anyone. The truth is in the eye of the beholder, who stretches the diagnosis to meet his own case. "You are outgoing, yet often want to be alone," or "you are sympathetic, but often not appreciated by others." Where a horoscope is cast and the astrologer meets the client on a one-to-one basis, the former is often able to adapt the reading to the context of psychological interaction, and the reading may be facilitated by what the reader observes about the person. The reading depends upon common sense and shrewd intuition. Here, the effectiveness of a cold reading is evident, for visual and verbal cues are able to focus the reading on the interests of the client. These may be both conscious and unconscious. Thus, a personal fit of a general scheme is facilitated; for there is already sufficient validity in a horoscope that it can be modified to the case at hand. A horoscope is like stretch socks, which can be made to fit sizes 8 to 12.

There is a deeper and more profound lesson from our study of astrology, and that is that the transcendental temptation is probably influencing a person's attitude toward what he is being told. In his conscious or unconscious is something like the following: The universe has a plan, and this includes the heavens. The plan encompasses each and every individual, including myself. Thus, my personal character and destiny is tied up with the stars. There is something beyond, and what I am or will become is synchronized with that.

Astrology thus speaks to a person's desires and aspirations and gives a cosmic explanation for even the most trivial, inconsequential, or idiosyncratic happening in the lives of individuals. The temptation to believe in a transcendental force tends to reinforce people's hopes and wishes, and allows for subjective validation. Whether astrology is empirically true or false is not the central issue. Whether it works—or is *made* to work—by fulfilling a hunger for meaning is of vital significance. And that, I submit, is its psychological function, and the key to its continuing hold on the human imagination through the centuries. It reinforces the longing for each person to be, in some mysterious way, related to the cosmic scene over and beyond his comprehension, so that every person's life has some ultimate significance. Herein, I submit, lies the key to the persistence and success of astrology, in spite of negative evidence to the contrary about its truth.

UFOlogy and extraterrestrial life

UFOlogy is another paranormal belief-system that may very well express the transcendental temptation; at least all of the signs at the present time seem to point in that direction. But this paranormal belief-system is of an audacious and fascinating kind: the idea that earth is being visited by highly intelligent and mysterious beings from afar—in this case, elusive entities from outer space. The similarity of this to the idea of revelation is striking; the difference is that this new religion of revelation is garbed in the latest forms of science fiction. Is UFOlogy simply a mass delusional system that has captivated millions of people on this planet? If so, it would not be the first such belief-system that has no foundation in proven fact, yet has strong support. Undoubtedly, it is given impetus by real and impressive scientific efforts to explore the universe for extraterrestrial life.

Modern UFOlogy began on June 24, 1947, when Kenneth Arnold reported that he saw a formation of nine disclike objects over Mt. Rainier, in the state of Washington. Arnold said that each disc resembled a "saucer skipping over water." His claims were given worldwide attention, and subsequently "flying saucer" sightings have been reported by tens of thousands throughout the world. Believers run into the millions. In some years, UFO reports became epidemic, and public interest increased enormously. These reports have come from most countries in the world, and from all strata of society. Are extraterrestrial beings and spacecraft visiting and observing life on the planet earth?

The accounts of UFOs have varied, but there seems to be a common syndrome: strange objects in the sky, cylindrical or saucer-shaped, glowing or flickering lights, making beeping noises, darting about at incredible speeds, taking off at odd angles, suddenly standing still, or disappearing. There have been many reports of human encounters with humanoid creatures who came out of these saucers, and even incredible accounts of having been abducted, examined, taken to other galaxies, and then returned to earth.

What is genuine in these accounts, and what is sheer fantasy? If we are committed to the empirical method, must we not take the testimony of eyewitnesses seriously? These reports were not rare or isolated in number, but have been numerous. There are at least two salient considerations: (1) Did the events occur as reported? (2) How should we interpret what people claimed to have seen? Can natural explanations be given?

UFOlogy divides into two major camps: (1) UFO believers who are convinced that at least some of the UFOs are extraterrestrial in origin, and that knowledge of them is being systematically withheld from the public by national governments (but for what purpose is unclear). (2) Skeptics who have examined the evidence and have offered prosaic natural interpretations of the phenomena. Most UFO reports, they say, if carefully investigated, become IFOs—that is, Identified Flying Objects, objects in the sky that are commonly misperceived or misinterpreted. Most of the accounts can be explained as bright stars, the moon, or planets that stand out in the sky (Venus, Mars, Jupiter), meteors, weather and other balloons, helicopters, passenger or military planes, missile launches, reentering manmade rockets and satellites, searchlights, flares, fixed ground lights, and other visual anomalies—even birds, bolt lightning, kites, or unusual cloud formations. In some cases, sightings have been a prank or hoax. Not all UFOs have been identified, primarily because there is no evidence corroborating the claims of the initial eyewitnesses. Thus an air of mystery attends those cases that still have not been fully explained to everyone's satisfaction.

Let us examine some of the classic cases. I will focus first on some of the most highly discussed instances of abduction, by witnesses who claim to have had direct encounters with extraterrestrials.

Abductions

Betty and Barney Hill. J. Allen Hynek, noted UFO investigator and astronomer, has labeled abduction cases "close encounters of the third kind." No doubt the most celebrated case and also the most thoroughly documented abduction on record is that of Betty and Barney Hill, residents of Whitfield, New Hampshire. According to their testimony, they were driving in northern New Hampshire on their way home from a vacation in Montreal on the evening of September 19, 1961. Sometime between 10:00 P.M. and midnight, their attention was captured by an erratically moving light in the sky, which seemed to be following their car. The sky was clear and the moon was shining brightly, low in the southwest. Betty said that she had seen a star or planet below the moon, but soon afterwards, she reported, she saw a second object in the sky, which was a bigger and brighter "star" above the first object. This, she said, was a UFO. Barney at first believed it to be an ordinary object, perhaps an airplane or satellite, but Betty claimed it was a flying saucer and insisted that Barney stop the car and look at it. "It's amazing," she exclaimed.[21]

Barney got out of the car, looked at the object with binoculars, and thought he saw a double row of lighted windows and aliens within, sneering at him and turning their backs to pull levers. The leader of the aliens appeared to be a "Nazi," he said. Terrified, he jumped back into the car and they hastily drove off. Betty reported that Barney was in a "hysterical condition," fearing that they were going to be captured.[22] The aerial craft, they alleged, continued to follow their car. It was at this point that their journey was supposedly interrupted. According to the story, they heard a short series of beeping sounds coming from the rear of the car, which caused the car to vibrate. They said that they lapsed into drowsiness. When they regained consciousness, they found that two hours had elapsed and that they were some thirty-five miles south of where the beeping sounds had begun.

The Hills were subjected to intense questioning by dedicated UFO proponents, who met with them within weeks following the alleged encounter. They were plagued by nervous disorders and frightening dreams and nightmares about being taken aboard an extraterrestrial spacecraft. After two years, Barney was so disturbed that he was forced to seek the help of a psychiatrist, Dr. Benjamin Simon of Boston. Under hypnosis, Barney and Betty Hill attempted to account for the "two lost hours." They told of being taken aboard the spacecraft and of being examined by the strange creatures. Moreover, Betty Hill claimed to have seen a "star map," which described the star system from which the UFO came. Fifteen stars on this map were allegedly matched by Marjorie Fish, an elementary school teacher, to stars in the sky, and this is often cited by UFO believers as independent confirmation that the Hills' story is true. Betty's tale is concluded when she and Barney were returned to the car some two hours later.

What is to be made of this famous case? Robert Sheaffer, a UFO skeptic, reports that if we consult weather reports and astronomical charts of that day, we can piece together certain facts: the moon was shining, the first planet they saw was Saturn, and the other bright object was probably Jupiter, which had reached the position they described. If one is riding along in a car, it appears that the planet is moving when the car does. "If a genuine UFO had been present, there would have been three objects near the moon that night: Jupiter, Saturn, and the UFO. Yet they reported seeing only two."[23]

21. John G. Fuller, *The Interrupted Journey* (New York: Dell, 1966), chap. 7.
22. Ibid., chap. 2.
23. Robert Sheaffer, *The UFO Verdict* (Buffalo: Prometheus Books, 1981), p. 35.

Were their imaginations responsible for the rest of their story? It is interesting that Dr. Simon, who put the Hills under hypnosis two years after the event, held that the entire event "was a fantasy . . . in other words, it was a dream. The abduction did not happen."[24] According to Dr. Simon, Betty had described many specific details of the abduction, whereas Barney seemed to remember almost nothing. Dr. Simon indicated that the abduction incident was *not* a common, shared experience, and it is suggested that Barney had derived his knowledge of the alleged abduction from hearing Betty recount her dreams. Dr. Simon told Philip Klass he did not believe the Hills had been abducted, but rather that the story was a fantasy they had come to believe and retold under hypnosis.

What about the star map Betty saw? Sheaffer points out that the map could match many areas of the sky, that there are many star patterns which would fit Betty's sketch. Some UFO believers maintain that radar sightings at that time corroborate the story. Pease Air Force Base in Portsmouth, New Hampshire, several miles away, reported an unidentified object on its radar at 2:14 A.M., but there was no corroboration by the Airport Surveillance Radar. Sheaffer maintains that radars are sensitive to many targets, including even birds and insects. Moreover, the radar anomaly occurred later, many miles away. Thus, there is actually no independent corroboration of the Hills' "eyewitness testimony" of the encounter.

Indeed, subsequent developments in the Hill case strain our credulity further. Betty Hill, in 1977, began talking about a "UFO landing spot" in southern New Hampshire, which she would visit three times a week to observe UFOs. Even UFO devotees have avowed that she is able to see UFOs when no one with her can. On one occasion, reports John Oswald, field investigator for a UFO group (who accepts abduction stories), Mrs. Hill was unable to "distinguish between a landed UFO and a street light."[25] This raises the interesting question of whether UFO abduction reports are like mystical or relevatory experiences, uncorroborated but built up out of firmly held psychological convictions.

The Travis Walton Case. Another famous incident referred to in the annals of UFOlogy is the abduction of Travis Walton aboard a UFO. This incident supposedly occurred on November 5, 1975, near Snowflake, Arizona. Walton, a 22-year-old woodsman, was working with a lumbering crew in the Sitgreaves National Forest. According to the story, Walton and six members of the crew were riding in a truck when they encountered a saucerlike object hovering over the ground about 110 feet away. The

24. NBC "Today Show," October 20, 1975; see also Philip J. Klass, *UFOs Explained* (New York: Random House, New York, 1974), pp. 252-54.
25. Robert Sheaffer, *The UFO Verdict*, p. 43.

strange object was emitting a high-pitched buzzing sound. Walton became excited about the object. He got out of the truck and approached it. When he was under it, a beam of light suddenly appeared from the bottom of the craft and knocked him to the ground, apparently unconscious, and he claims to have experienced a kind of electric shock. The other six men were frightened and hastily sped off in their truck.

Meanwhile, Walton disappeared, and efforts by the police and others to locate him proved to no avail. After five days, he returned and called his folks from a pay phone. He was found in a dazed condition.

According to Walton's account of what happened, after he had been knocked to the ground, he woke up in an all-metal "hospitallike" room. He was being observed by three strange creatures with bald heads, no more than five feet high. He tried to do battle with them, but they left the room unscathed. Next, a human entered the room. He was about six feet tall, with brownish-blond hair and hazel eyes, and wore a helmet. Not talking, he led Walton down a corridor where he met three other humans, who put a clear plastic mask over his face. Walton blacked out. The next thing that he recalled was walking along a highway with the flying saucer departing straight up. He was only able to remember a few hours of the time that he had disappeared.

Was the story true? Was his testimony reliable evidence for the existence of extraterrestrial beings? When word of the incident was revealed, it became instant news, and the world press descended upon Snowflake. Believers and skeptics alike came. One sensationalist tabloid, the *National Enquirer,* covered the incident, and a $5,000 prize was awarded to Walton and the woodsmen for the "best UFO case of the year." It was alleged that they all passed lie-detector tests. UFO skeptic Philip Klass was able to uncover many discrepancies in their accounts. First, the polygraph tests they had taken were poorly administered. Further, Klass said that Walton had failed an earlier, unpublicized lie-detector test given by an expert, John J. McCarthy, who concluded that the case involved "gross deception." McCarthy's report concluded that "based on his [Travis Walton's] reaction on all charts, it is the opinion of this examiner that Walton, in concert with others, is attempting to perpetrate a UFO hoax, and that he has not been on any spacecraft."[26]

Klass found interesting sidelights to the case, notably that Travis Walton, his brother, and mother all believed in UFO phenomena, and that Walton had told his mother a few weeks before the incident not to worry if he were ever abducted—that he would return safely. Klass also found that the work crew, and especially its chief, Mike Rogers, had a

26. Klass, *UFOs: The Public Deceived* (Buffalo: Prometheus Books, 1983), p. 186.

possible economic motive in helping to concoct the hoax. It is clear that the actual abduction aboard the spacecraft and what transpired inside it is attested to only by Travis and not the other members of the work crew, who may or may not have seen a strange object in the sky. There is no corroboration of Travis' story, and so it remains a subjective account.

The Hickson-Parker Encounter. A third highly publicized event occurred on October 11, 1973, in Pascagoula, Mississippi. Two shipyard workers, Charles Hickson, 42, and Calvin Parker, 19, claimed to have been kidnapped by a UFO while fishing on the Pascagoula River. They maintained that they heard a buzzing or zipping sound, and saw a round flying saucer hovering over the ground. The vehicle was flashing blue lights. Hickson's accounts of the event were contradictory. On one occasion he said that the vehicle was eight to ten feet wide, and on another he described it as twenty to thirty feet long. According to his account, three humanoids emerged from the UFO and floated toward Hickson and Parker. They were described as about five feet tall, with grey crumpled skin, elephantlike, egg-shaped heads, no necks, and clawlike arms. Two of the creatures took hold of Hickson and floated him aboard the UFO, where he was taken into a brightly lit chamber. A third creature grabbed Parker, who fainted, and also floated him into the craft. The humanoids examined Hickson with an object that resembled an eye but was not attached to anything. He felt completely paralyzed when they did so. About twenty minutes later, they floated Hickson outside, where he rejoined the greatly disturbed Parker, who had regained consciousness. The UFO then shot straight up and disappeared. Parker, having passed out, was unable to supply any details of the event. After several hours, the two men got up enough courage to relate their tale to the sheriff, and this miraculous incident rapidly became public news.

What are we to make of this incredible story? Was it verifiable? Should it be accepted as a genuine encounter? The UFO buffs—a group of committed believers—seem to have almost a quasi-religious desire to accept the E.T. hypothesis. Hickson and Parker were interviewed by two well-known UFO experts, Professor James A. Harder and Dr. J. Allen Hynek. Harder concluded that "there was definitely something here that was not terrestrial . . . Where they came from and why they were here is a matter of conjecture, but the fact that they are here is true, beyond a reasonable doubt." And Hynek cautiously maintained, "There is no question in my mind that these two men have had a very terrifying experience."[27] Additional credibility was given to Hickson's testimony when it was revealed that he had taken a lie-detector test. Meanwhile, the case

27 Klass, *UFOs Explained*, p. 298.

aroused a great deal of interest, especially in the Mississippi area, where there were a number of similar reports of UFOs.

Fortunately this case, like the two previous cases, has been examined by skeptics, who can provide alternative explanations. Klass maintains that the evidence points to the strong possibility of its being a cleverly contrived hoax. According to Klass, the lie-detector test was given to Hickson by an inexperienced operator. Efforts to get Hickson to take another test or to test Parker were unsuccessful. Klass discovered that Hickson had filed for bankruptcy a few months earlier. He speculated that Hickson hoped that his story would lead to film or TV offers, motives that one can only infer and not confirm.

What are we to make of these three abduction cases? Are they all hoaxes? Were the accounts of the abductees simply hallucinations, or was there a combination of motives? Some commentators have found similarities between abduction stories, near-death experiences, and drug-induced hallucinations. The common threads runing through these stories are similar: the seeing of a bright light, a buzzing sound, a sense of floating out of one's body, moving through a tunnel or corridor, encountering a being or beings bathed in light, undergoing examination, and then returning to real life. This leads one to speculate whether the transcendental temptation is also an important factor in the UFO phenomenon—not only for the 200 or more reported abductions but also in the general population's willingness to believe in the phenomenon. UFOlogy, at least for some, seems to function as a quasi-religious phenomenon.

Millions of people claim to have seen strange things in the sky that they cannot explain by natural means, and which they attribute to an extraterrestrial, out-of-this-world source. Unable to give a causal explanation, they read in a magical account, reminiscent of spiritual, psychic, paranormal, supernatural, or other miraculous events. Is a similar, subtle, psychological process at work, with only the content of the beliefs differing but the origin and function the same?

The extraterrestrial hypothesis and UFOs

The extraterrestrial hypothesis is perfectly plausible. The possibility that life may exist in other galaxies in the universe is a meaningful scientific hypothesis. Some astronomers have postulated the probabilities of life having evolved on other planets. The argument runs as follows: there may be at least one million planets in our galaxy alone. There are tens of billions of galaxies throughout the universe. If the conditions for life are present in these planetary systems, presumably higher forms could have evolved, and there may be intelligent forms which have developed advanced technological civilizations. It is reasonable to assume that life will spontaneously emerge on suitable planets, given adequate surface temperature and other conditions. Organisms capable of photosynthesis can then develop, and it is estimated that after three to four billion years, other higher forms of life will or have evolved. This presupposes the presence of carbon, oxygen, nitrogen, hydrogen, and other elements. It is conceivable that such civilizations have advanced far beyond us and have conquered the technology to make space travel over infinite distances possible.

Some astronomers have denied that this is the case, however, maintaining that the probabilities are very low. There may be far fewer planetary systems than was earlier believed; indeed life on the planet earth may be a relatively rare event. Moreover, for a planet to be propitious for life, the necessary basic elements must be present, at least for the kinds of life with which we are acquainted, and many planets may contain far fewer elements. For life to be present, the planet must be at the right distance from its sun, that is, neither too far nor too near. If the earth were closer to the sun, it would be too hot to support life, and if too far, glaciation would have taken place. The zones that are habitable may be relatively limited. Thus the probabilities for life are, we are told, much smaller than those extrapolated by the E.T. optimists. Some astronomers differ from this pessimistic appraisal and believe that organic matter need not be based on carbon and water, as on our planet; they think that life grounded in silicon or other forms of chemistry might be possible, and that these forms of life might be able to survive extremely high or low temperatures. To assume that life on the planet earth is an entirely unique phenomena, they say, would be surprising.

Still another speculative suggestion is that there may be cosmic clouds of organic material able to survive over long periods and/or that organic matter in spoors of life have been transmitted by meteors from different galaxies, enabling the evolutionary process to be repeated throughout the universe.

Whatever the probabilities of these processes are is still uncertain at present. What is of momentous significance to the human species is whether intelligent life exists elsewhere. If so, we should attempt to make contact and to communicate with these forms of life—as Carl Sagan has urged. Radio telescopes have been monitoring the heavens and transmitting messages, but thus far, no identifiable messages have been received. One can neither confirm nor deny the E.T. hypothesis on a priori grounds at this stage of scientific inquiry. But this is all rather far removed from the question of whether the planet earth has been visited by ancient astronauts, as Von Däniken has maintained, or is being visited today by UFOs. Aside from the question of probabilities is the empirical question of confirmation, and this can only be resolved by reference to the evidence.

What is the evidence in support of the UFO-E.T. hypothesis? I have referred to three of the most famous abduction cases, but actually the number of sightings of UFOs is enormous. In the United States, something like 9 percent of the population, millions of people, claim to have seen a UFO (according to a Gallup poll of 1978). Fifty-six percent of the population say that they believe in UFOs. This is repeated worldwide, so that we are dealing with phenomena of mass proportions. Generally, the "sightings" are of strange lights or objects in the sky behaving in bizarre ways, hovering and darting about at unexpected angles. This is the age of air travel, so people are accustomed to look skyward and see balloons, helicopters, propeller and jet airplanes, rockets, and missiles— all phenomena that would have seemed strange in earlier centuries. But they are also puzzled by other things they think are anomalous and inexplicable.

In 1952, the U.S. Air Force launched a special inquiry, Project Blue Book, to investigate such phenomena. It dealt with approximately 13,000 cases of alleged sightings. After an extended study, the Air Force concluded there was no evidence for the supposition that the phenomena were extraterrestrial. It maintained that it was able to explain approximately 94 percent of the sightings as due to natural causes. The overwhelming majority of cases were based on simple misperceptions and misinterpretations, even by highly trained airline pilots, engineers, and scientists. A lengthy study conducted at the University of Colorado, known as the Condon Report, recommended in 1969 that no further study be made by the Air Force, even though some cases remained "unexplained." UFO believers often point to that fact as significant, but this does not necessarily confirm the extraterrestrial hypothesis, but simply indicates that there is not sufficient data to determine what was present or to corroborate the testimony of eyewitnesses. Not every murder has been solved by police departments, but we are not entitled to suppose that a paranormal agent committed unsolved murders. The burden of proof rests with the advocates of the extraterresterial hypothesis. And they have not produced sufficient evidence.

In another important study,[28] Allan Hendry, associated with the Center for UFO Studies, reviewed over 1,300 UFO reports that occurred

---

28. Allan Hendry, *The UFO Handbook: A Guide to Investigating, Evaluating, and Reporting UFO Sightings* (Garden City, N.Y.: Doubleday, 1979).

in a fourteen-month period (August 1976 to November 1977) in the United States. He interviewed, largely by telephone, witnesses who had reported UFO sightings. He concluded that prosaic explanations could account for the great majority of such reports. Hendry found that the reports were global in nature, came from all parts of the United States, and were made by a cross-section of individuals representing all ages, occupations, and educational backgrounds. Reports were of various kinds: those that were "close" (that is, a UFO which was said to appear within 500 feet of the witness), those that influenced the environment (and allegedly left some physical trace), and those that involved meeting occupants or entities. In the 1,307 cases that Hendry investigated, 1,158 were readily translated by him into IFOs (88.6%), 36 cases were excluded as not providing appropriate data (2.8%), and only 113 cases (8.6%) remained as unidentified phenomena. Hendry confesses at the end of his case-by-case study that he was "*still* no closer to the nature of this complex beast" than when he started.[29] He says that much or most of the data is anecdotal testimony of excited witnesses. "*Never* does the evidence suddenly allow a burst of approval for even one UFO,"[30] that is, for the extraterrestrial explanation.

What Hendry has uncovered is surely significant. For it is clear that after decades of searching, we still do not have one incontrovertible case that stands up under careful scrutiny. We have no decisive proof, no hard corroborating evidence, that UFOs are extraterrestrial. Perhaps by continuing to investigate such sightings, we will some day discover sufficiently hard and rigorous evidence to corroborate the claim unambiguously. But until we do, an alternative explanation is available, namely, that the UFO phenomenon tells us something about the psychological and sociological behavior of the human species, of the fascination with the unknown, and the hunger for belief in the existence of realities beyond.

Here we find the reappearance of a common thread. UFOlogy is the mythology of the space age. Rather than angels dancing on the heads of pins, we now have spacecraft and extraterrestrials. It is the product of the creative imagination. It serves a poetic and existential function. It seeks to give man deeper roots and bearings in the universe. It is an expression of our hunger for mystery, our demand for something more, our hope for transcendental meaning. The gods of Mt. Olympus have been transformed into space voyagers, transporting us by our dreams to other realms. The transcendental temptation has again overcome us. And so we see what we want. We fashion a universe to our liking.

29. Ibid., p. 283.
30. Ibid.

Typical examples of UFO sightings

1. On July 17, 1957, at 4:10 P.M. the crew of an RB-47 aircraft from Forbes Air Force Base in Topeka, Kansas, saw an intense blue-white light over Louisiana. A radar signal was picked up by the radar operator aboard the plane. Although the RB-47 increased its speed, it could not overtake the UFO. The UFO was also picked up by radar over Texas, as the plane pursued it. The UFO disappeared and reappeared to the plane and on the radarscope. At one point, the UFO disappeared from sight but then reappeared on the RB-47's tail. The plane followed it for ninety minutes over four states, for a distance of 700 miles.

2. On the night of April 29, 1978, the police department in Aurora, Illinois (on the outskirts of Chicago) received ten reports of a UFO sighting, between 10:30 P.M. and midnight. Eyewitnesses claimed to see a saucerlike object, with lights circulating around its perimeter. There were varying reports as to its exact size, with estimates anywhere from twenty-five feet to the width of a football field. Some suggested that there may have been a mother ship and smaller craft as well. Mr. and Mrs. S. reported a close encounter with the craft. The saucer hovered over their car, and as they drove to alert their neighbors it followed them, flying at treetop level. The saucer was completely silent. After hovering motionless above them for a time, it abruptly took off to the east.

3. Late in August 1984, numerous residents of Brewster, New York, a Hudson Valley town forty miles north of New York City, began to report UFO sightings. They said that they saw high objects, circular or V-shaped, emitting red, green, and white lights, and hovering noiselessly or with a faint hum. They performed erratic maneuvers, then suddenly disappeared. Monique O'Driscoll related that as she was driving home one evening, the voices on her CB radio were suddenly drowned out by static. She saw something approaching her vehicle very slowly, and she followed the craft as it veered away and hovered over a nearby house. Looking at the belly of the craft, she noted that "it was a dark grey metal, like the framework of a bridge. After a few minutes it started going away . . . and then, zip, it was gone."[31] A commercial airline pilot, Michael Faye, spotted a strange object shortly after taking off one night from Newark Airport. It had six or eight lights that went out all at once as he began to come close to it. "It had disappeared," he asserted in amazement.[32] Such a furor was raised about the Hudson Valley sightings that

31. *Discover* magazine, November 1984, p. 19.
32. Ibid.

scores of well-known UFO investigators descended on the area. Several attested to their authenticity.

What are some possible explanations for the preceding sightings? In regard to case 1, Philip Klass has suggested that the disappearing and reappearing UFO most likely could be accounted for by a variety of different natural objects. Moreover, the radar signals which the crew of the RB-47 thought came from the UFO had the characteristics of those used by air-defense radar installations that were operating in the area at that time. Case 2 is explained by Allan Hendry as due to a brightly lit advertising plane, which he found was in the area at that time. Viewed from underneath, it appeared noiseless. *Discover* showed case 3 to be a hoax perpetuated by pilots of private cub planes. Flying in circular formations, they would suddenly turn off their lights, which made it seem to those watching from the ground that they had disappeared!

I have cited only three cases as typical of what occurs, but the list of faulty observations and hasty generalizations is endless. No doubt reports are colored by the widespread publicity given to previous reports and the belief, and hence expectation, that UFOs are extraterrestrial in origin.

Other monsters, other seas, other galaxies

Although thus far the UFO field has proven to be a blind alley full of hopeful anticipation and the will-to-believe, this still leaves unanswered the broader question of whether there are other beings and forms of life yet unknown to man. The great range and variety of species on this planet is stunning. It has been only recently that virtually the entire earth has been explored. For centuries there were reports from explorers of strange flora, fauna, sea and mammal life, from kangaroos in Australia to Darwin's remarkable discoveries on his voyage on the Beagle to the Galapagos Islands. There may be species yet to be discovered and classified, for example, new varieties of ocean-floor life. Indeed, it was only in the last century that the okapi and pygmy hippo were seen by the first European explorers of Africa.

There is still a great fascination with unknown monsters, whether dragons of the deep or other potentially ferocious animals. All of this is no doubt understandable, given the history of the human species stretching back to its primeval origins. Today the Loch Ness monster and other creatures still captivate the imagination. Perhaps other primates will be encountered: Yeti, Sasquatch, or Bigfoot, as he is called in various parts of the world where people claim to have seen a large human-gorillalike creature. Dinosaurs have been investigated by cryptozoologists in the

African Congo and mermaids (dugongs) off the coast of New Caladonia and New Ireland in the South Pacific. The entire history of the human species is an encounter with strange wild beasts, ever dangerous and threatening, defying our efforts to tame them. This belief is so deeply engrained in the subconscious of the race that it persists today. Paleontologists have unearthed the remains of mammoths and dinosaurs that inhabited the earth in the past, and novelists have contrived strange Frankenstein-like monsters to intrigue and beguile us.

What about the universe beyond? There may be somewhere an infinite variety of forms of life besides the giraffe, the katydid, and the myriad fauna on earth. The extraterrestrial imagination can create multiplicities of possible beings: egg-headed or furry, clawlike or withered, demonic or beneficient. Will they be like us, or capable of telepathy and psychokinesis, conquering death and disease, of infinite wisdom, even immortal? If and when we make the first contact, what will we be able to learn from them about the mysteries of the universe and life?

The imaginative constructs of science fused with science fiction contains black holes and time warps, the romance of being able to go backward and forward on the time scale. What a dazzling feast for the transcendental temptation! How startling will the religions of the future be, fed by the daring ideas of scientific imagination? How pale and insignificant by comparison will the religious mythologies of the past become! Astrology, an age-old faith, and UFOlogy, a new one, turn human attention skyward. What lies beyond the dark reaches of outer space, punctuated by the flickering bright lights of distant galaxies? The transcendental temptation is just at the point of taking its greatest leap forward, soaring this time not on the wings of angels but on the gigantic telescopes and radar antennae of astronomers and the Promethean promise of the new technologies of space travel.

Prometheus stole fire and the arts of civilization from the gods and bequeathed them to man. Today man threatens the gods in their heavenly abode itself. He seeks to leap out, populate other planets beyond the earth, to unravel and conquer the mysteries of the cosmos. Will it be too much for man? Will his courage collapse? Will the transcendental temptation again overwhelm his audacity with new forms of space-age religious myths?

Part Four

BEYOND RELIGION

XVI: The transcendental temptation

Magical thinking

We have taken a long journey in this book, from the ancient religions that persist all the way to the newer forms of space-age mythology. We have found a common thread running throughout, and so we may ask: Why have countless generations of men and women down through the centuries undertaken the quest for transcendental realities? Why do they continue to believe in unseen powers and forces that guide the destiny of the universe?

There are no doubt a variety of explanations beyond those that I have already suggested, and these are complex and diverse. A chief source of the transcendental temptation is the propensity for magical thinking and the ready willingness to accept its efficacy. The term *magic* is commonly identified today in the public mind with conjuring tricks and entertainment. Magicians are viewed as performers who can both mystify and delight us. But this is not the original meaning of the term. In the long history of human culture, magicians were thought to be able to cast spells and untap mysterious forces in the world beyond. In this regard, magic was similar to other forms of religious behavior. The magician performed ingenious acts that were thought capable of influencing events; he could evoke supernatural powers beyond our ordinary comprehension. This was akin to the behavior of the prophet, priest, rabbi, or shaman, who also claimed to have special powers and could conduct sacred rites and rituals that enabled him to function as an intermediary between supernatural powers and human beings.

Belief in magic per se is mistakenly thought to apply only to primitive cultures, and is viewed today as a form of superstition. The medicine man or sorcerer of old evoked spells, used amulets, was skilled in brewing concoctions, and in chanting formulas that could ward off illness, harm enemies, and influence nature, fertility, disease, even death. He could summon a spirit or force by means of incantation or invocation. Magic is also confused with witchcraft, often a form of black magic, which is intended to achieve nefarious ends. In some cultures, both countersorcery and white magic were developed not only to ward off the malevolent effects of sorcery but also to guarantee beneficent ends, such as victory in love or athletic contests and success in other pursuits of life.

Yet the similarity between the magical and the religious mode of response is pervasive even today. The priest is like a magician in that he too engages in rituals and ceremonies to propitiate the unknown powers and gain their favor. The priest also uses amulets and talismans (such as the cross or tefillin), and he follows prescribed rituals, enlisting the aid of a deity. Instead of many gods or spirits, monotheism postulates one God, the Father, who though hidden from our view, cares about us, intervenes in history, and rewards or punishes us for our faith or transgressions. The priest, rabbi, or shaman is supposedly privy to a secret knowledge, and he can employ the proper modes of piety to entreat God and guarantee salvation. The rites of baptism, circumcision, confirmation, and communion provide divine blessings for the passages of life. The last rites, if properly administered by the magician-priest, will help ensure a positive reception by God at judgment day.

The magician-sorcerer, religious prophet, or astrologer-psychic is also thought to be capable of divination, and hence is endowed with extraordinary powers. He or she is able to peer into the future and can precognize or foretell what will happen. He can also read omens and signs for what they portend. And there are other psychic gifts that he possesses: clairvoyance, telepathy, and, if a miracle worker or sorcerer, he can exert psychokinetic powers on the physical world—part the seas, cause the heavens to rain, or heal the sick and dying.

Primeval magic and religion is the soil from which present paranormal beliefs spring The religious and paranormal beliefs that persist today are only the most recent manifestations of practices that have deep roots in human culture. Anthropologists and archeologists who have studied the varieties of human culture have found that the magical mode of thinking is a perennial phenomenon.

Sir James G. Frazer has provided some empirical documentation of the close affinity between religion and magic. Religion, he maintained, was simply a more developed refinement of the more primitive mentality; both shared a common outlook on the world. Frazer postulated two principles that governed magic: (1) homeopathy, or the law of similarity, the principle that like produces like, and (2) the law of contagion, the idea that persons or things that had been in contact once can influence each other forever.

Homeopathic magic is illustrated by Frazer as the common belief that a man could injure or destroy his enemy by damaging or destroying an image of him. For thousands of years this was practiced by the sorcerers of ancient India, Babylon, Egypt, Greece, and Rome; and, according to Frazer, could still be found at the time that he wrote (late nineteenth century) among primitive peoples in Africa, Australia, and America. Frazer relates that the Ojibway Indians attempted to inflict evil by creating a wooden image of their foe and poking a needle into his heart or head.[1] If the Indian intended to kill the person, he burned or buried the doll, reciting magical incantations as he did so. In Burma, a jilted lover resorted to sorcery by making a small image of the loved one, containing a piece of her clothing, and casting it into water to the accompaniment of chants. If the revenge was effective, the girl was supposed to go mad.[2] Similarly, the Ainu of Japan destroyed an enemy by making a likeness of him out of mugwort or guelder-rose and burying it under a rotting tree. Analogous devices were employed by the Muslims of North Africa, the Chinese, and other peoples. Magical images have often been used to secure the love of another. The ancient Hindu method was to pierce the heart of an image of the beloved with an arrow—hopefully with amorous results.[3]

Even today we can see vestiges of these beliefs. For example, voodoo is practiced in Haiti. We are told by many that this form of magic is very effective. Thus, if a doll image of a person is injured, then evil will befall that person, particularly if he is aware of or is sent the damaged doll. Burning someone in effigy is the modern counterpart of this ancient belief; it is practiced in even the most advanced societies to indicate hatred of or profound displeasure with a leader.

There is no evidence that homeopathic magic works. There is no underlying causal relationship between the spells and incantations of the sorcerer and what may befall the targeted individual. There may, of course, be a psychological relationship, such that if the targeted person truly believes in the efficacy of black magic, he may become so frightened by what will happen that his fears may provoke an unfortunate or even deadly reaction. But if so, it is not the magic but the *belief* that it works that brings about the desired effect.

Homeopathic magic has been used in a positive way also—and apparently still is—to heal the sick or ward off illness. In ancient Greece, when a person died of dropsy, his children were supposed to sit with their feet in water until the body was buried. The natives near Rajamahall, in India, threw a person who had died of dropsy into a river, believing that otherwise the disorder would return and afflict others. The ancient Hindus

1. Sir James G. Frazer, *The New Golden Bough,* ed. Theodore H. Gaster (New York: New American Library, 1959), p. 35.
2. Ibid., p. 36.
3. Ibid., pp. 36 ff.

performed an elaborate ceremony for the cure of jaundice: the task was to confine the color yellow to yellow creatures or things to which it belonged, such as the sun, and to secure for the patient a healthy red color from a living source, such as a red bull. Today people carry charms and amulets to ward off diseases and bring them luck, and old wives' tales persist for the cure of illnesses by homeopathy.

Contagious magic is also widespread. It can be seen in the magical symmetry that is believed to exist between a person and any of his severed portions, such as nail parings or hair. Whoever gets possession of a person's hair or nail clippings can influence him from a distance. In the Marshall Islands the navel string of a boy was cast into the sea in order to assure that he would become a good fisherman; the navel string of a young girl was inserted into a leafy pandanus tree so that she would be skilled in plaiting the fiber from the tree. In the Marquesas Islands, when a birth had taken place, the placenta and rest of the afterbirth were buried under a frequented path so that the women walking over the spot would become fecund.

Frazer illustrates the application of contagious magic in the relationship between a wounded man and his assailant. He quotes Pliny to the effect that if you wound a man and are sorry and if you spit on the hand that afflicted the injury, the suffering man will be relieved.[4] The Kagoro of Nigeria believed that if a man was wounded by a spear or sword and it did not heal, then he must obtain the weapon, wash it with water, and drink the water to be healed. In Prussia, according to Frazer, it was believed that if one could not catch a thief, the next best thing was to get hold of one of his articles of clothing, which he might shed in flight. If the garment was beaten stoutly, the thief would fall ill.

Again, there is no evidence that any of these forms of magic had any genuine effect in the real world, though undoubtedly they played some kind of psychological and/or sociological role. They served as a sort of relief valve—one is not helpless against adversity; one could do something. These are pragmatic side effects, byproducts of an acausal process. A good illustration of this is the rain dance, which numerous tribes have performed. During periods of intense drought, as primitive men and women awaited the rains to replenish the earth, there was very little that they could do to bring an end to the drought. And so the medicine man was called in. If one wanted to produce rain, the best way was to simulate it by sprinkling a bit of water or mimicking the clouds. The Hopi Indians engaged in the rain-dance ceremony. If they did not succeed in causing rain, at least the dance enabled the members of the tribe to share their common yearnings and do something in concert. If the goal is to stop the

4. Ibid., p. 65.

rain, then one should avoid water and use fire and warmth to dry up the excessive moisture. This form of magic thus is evocative and expressive—and it has been pointed out that if enough people perspire a sufficient quantity of moisture over a period of time, this may lead to some condensation! Eventually no matter what they do, it will rain, for which they can then take credit. Rainmaking has its counterpart today in belief in psychokinesis; that is, that the mind by mere willing can cause physical changes in the universe without an intermediary physical cause. For modern man the most effective way of dealing with times of drought is to build dams and irrigation streams, but this is to rely on science and technology rather than magic and religion.

Prayer is surely a form of magic; it is the residual belief of a primitive mind that one can influence the deity who controls nature and one's destiny. It is hoped that the deity will respond to our prayers and sacrifices. The priestly class uses holy objects and symbols: the Torah, bread and wine as the body and blood of Christ, baptism by water; and they don special robes and garments, all of which are vested with sacred qualities.

Those who pray are convinced that their prayers will be answered by God in this life and/or the next. As we have seen, it is difficult to know whether sinners will be punished and saints rewarded in the afterlife or even whether such a felicitous state exists. But perhaps we can judge the empirical efficacy of prayer in this life. In a trenchant study, "Statistical Inquiries into the Efficacy of Prayer," the English scientist Sir Francis Galton (1822-1911) did not find any evidence for prayer's positive effects.[5] It seems apparent that the clergy pray more than anyone else. Yet Galton found that the death rates of eminent clergymen were not any lower than those of other professions, such as lawyers and doctors. Moreover, the efficacy of prayer may be tested further by examining the numbers of infants who died at birth among the praying versus the nonpraying classes. Surely, Galton said, the solicitude of pious parents would be directed toward the health of their expected offspring, especially since a death before baptism is considered by Christians to be a great evil. An analysis of stillbirths, however, showed no appreciable difference in death rates between the infants of people who prayed and those who did not. In any case, it is surely questionable as to whether believers who pray are any healthier, wealthier, or happier than nonbelievers. Both the German and French armies prayed to the same God in the world wars as they went off to the slaughter singing: "Gott mit Uns" and "Mon Dieux, Mon Dieux!"

5. Sir Francis Galton, "Statistical Inquiries into the Efficacy of Prayer," *The Fortnightly* (Aug. 1, 1872); reprinted in Paul Blanshard, *Classics of Free Thought* (Buffalo: Prometheus Books, 1977).

Is magical thinking acausal?

The fundamental premise of those who believe in a magical-religious universe is their conviction that there are hidden and unseen powers transcending the world, yet responsible for what occurs within it. This belief is also endemic to belief in paranormal phenomena. It presupposes that there are exceptions to materialistic or physical causality. Strange spiritual agencies and forces, unfathomable to human understanding, are responsible for what occurs in nature. Early anthropologists considered magical thinking to be a prelogical mode of response to natural events. Ancient peoples were troubled by seemingly inexplicable occurrences. At first they attributed them to animistic causes, believing that material objects and animals have an inner spiritual consciousness like ourselves, which was separable from the body and had causal efficacy.

Magical thinking from one frame of reference appears to be acausal, for those who appeal to it are unable to find explanations for what happens. A sudden lightning storm may spark a forest fire. "What causes the lightning and high winds?" the primitive man ponders. There is an intense drought and the crops fail. Are these terrible events due to the wrath of the gods? A child is ill with a high fever and appears to be dying. Why does this sickness rage? What can be done about it? Knowledge of meteorology or of bacterial infection is absent. Or conversely, there is a bumper crop, or a tribe is blessed with many healthy cattle, which provide it with abundant milk and meat. Whom shall the tribal elders thank for this good fortune?

On the one hand there is a presumption that there are no known causes. Things seem to happen mysteriously and by happenstance. The primitive mind cannot find any connections and is unaware of the natural antecedent or concomitant conditions that are present and could explain the observed effect. Does this mean that magical thinking is *acausal?* In one sense it appears to be contracausal, for it involves a break in our normal patterns of expectation. If we cannot explain something, we say that it is "magical" or "mysterious," for it defies our powers of comprehension. But in another sense it is difficult for humans to accept a totally chaotic universe. The mind seeks order and regularity, and so primitive man supposes that there is a cause, though it is occult or unseen. He believes that he can influence it and cause it to modify its otherwise capricious course. Thus magical thinking invokes a sequence of causes, but they are hidden, unknowable, or unknown. The effect is not deemed causeless: a supernatural, supernormal, or paranormal cause is interpolated to account for it.

Magical and religious thinking involves a rupture or break in natural causality, which is based on normal habits of expectation. It thus reads in an occult cause to make sense out of the fragments of experience. Magic is a synonym for the unknown, that which is not readily discernible or understood. *But it is not acausal,* for one assumes that something or some power is at work and is responsible for what happened.

When we watch a magician entertain us by escaping from chains, sawing a woman in half, or guessing the card that someone has selected, we ask, "How did he do it?" We assume that there is a clever trick, and once we find out how it is done, we may be surprised at how easy it was. But until we do, we are befuddled, for we can find no known cause; though we believe that there is one.

Supernormal, magical, religious thinking, which invokes miraculous events, has two ingredients: (1) Our ignorance of the real or natural causes at work, and (2) our supposition that there must be an unknowable cause, which we deem miraculous or supernormal. If it involves a miracle worker or godman, the first part of the equation is present: we infer the second. Spiritualist and psychic believers adopt a similar mode of thinking. They find that strange things happen—raps on tables, levitation, precognition. Being ignorant of the cause, they assume that a true paranormal event has occurred. But this kind of phenomenon is not without a causal basis. The causes are simply transposed out of the natural world to another realm, which is deemed to exist on a parallel plane. There is a rupture in the process of inference and a move from naturalistic-materialistic to paranormal causality. The difference between these two modes of explanation is that the former can be isolated under controlled conditions and tested; the latter cannot and is mired in magical thinking.

Today we attribute serious illness to a bacteria or a virus, which is unobservable by the ordinary man. A complicated theory of disease introduces microorganisms to account for symptoms, and antibiotics to defeat them when they invade the body. When the germ theory was first proposed by Semmelweis, this thinking was ridiculed by the medical profession—it seemed occult. But the difference was that it was testable. Naturalistic explanations, unlike supernormal ones, can be corroborated experimentally.

Is magical thinking irrational? Does it defy the principles of logic? Is it antilogical? In being unable to find proximate natural or material causes for phenomena, the primitive mind invoked occult explanations. It was not unreasonable to do so given the level of knowledge. Today it is no longer permissible to do so, though vestiges of occult thinking persist in religious and paranormal areas, so deeply are they embedded in our cultural habits. Occult thinking is illegitimate in virtually all fields and disciplines of human inquiry today. It would be inadmissible for economists, physicians, or physicists to invoke occult explanations for phenomena that are currently inexplicable in terms of known hypotheses and theories. If we don't fully understand all the causes of inflation, we are not entitled to say that these are due to occult forces at work. Similarly, if the cause of a new disease is unknown, we cannot invoke a supernatural explanation—spirits, demons, or other satanic agencies—to account for the malady. We are not permitted to postulate a spiritual agency to explain any phenomenon in the laboratory of natural science. If we do not know the cause of something, we suspend judgment until sufficient evidence has been marshalled and adequate tests for the explanatory hypotheses introduced.

Similarly, magical and occult thinking must be rejected in ordinary life. If we misplace a set of keys or if our car engine suddenly stops and we cannot get the motor going again, we are not justified in postulating occult or magical causes. To do so would usually bring a charge of insanity. Similarly, it is illegitimate to postulate ESP (which functions as an occult cause) to account for alleged above- or below-chance runs. To read in ESP is to postulate an unknown entity, a factor that cannot be observed. Some look for a theory to explain ESP, such as hidden relationships in the universe (Arthur Koestler) or an underlying synchronicity of all events (Carl Jung). But these are untested conjectures.

The basic methodological principle of science is that we should seek natural causal explanations for phenomena. The occult or transcendental temptation is antiscientific. Where there is uncertainty, the most sensible response is agnosticism or the withholding of judgment. But this is often very difficult. Magical, occult, religious thinking persists in many areas of life, particularly when we are beset by quandaries. It is on the borderlands of knowledge and in areas concerning human meaning and purpose that the transcendental leap is especially tempting. The dissatisfaction with ambiguity and the quest for order often tempts us to invoke unknown occult or magical causes.

How are w . 'o explain the persistence of magical religious thinking today, not only among ordinary and uneducated people but also among sophisticated scientists and scholars? We have seen that magical-religious systems of belief persist in spite of refutations by critics and abundant evidence to the contrary. Celsus and others in ancient Rome denigrated Christianity, yet it survived its critics. In the nineteenth century, biblical scholars performed the same task by submitting the Bible to rigorous critical analysis, seemingly to little avail. Why do powerful mythic systems seem to withstand criticism? Why do magical-religious systems of thought seem to have a life and resilience of their own, impregnable to logic and evidence? Are there other reasons that can account for their tenacious hold on human beings?

The role of creative imagination

One obvious explanation for the persistence of the transcendental temptation is irrefutable: namely, that our knowledge of the universe is limited at any one stage of human history, and there is a tendency for us to wish to leap beyond it. We live our brief lives in a particular slice of space-time history and in a specific sociocultural context. A relatively minor planet in one galaxy among billions, we are only an infinitesimal part of the total cosmic scheme; and there are vastly more things that we do not know than we do. The world as viewed from our individual vantage point is often ambiguous. The entire universe confronts us with its enormity, inviting us to unravel its secrets, yet resisting any easy interpretation of its mysteries. Yet since the inception of philosophical reason and science we have made significant headway in this adventure. The sciences have enabled us to develop testable hypotheses by means of which we interpret nature. Astronomy and the natural sciences, biology and the life sciences eloquently testify to this impressive advance.

Still, our knowledge is constantly being revised, and we can have no confidence that what we have learned has reached—or will ever reach—its final or ultimate formulation. This is true of many corners of the universe we seek to penetrate. E. O. Wilson, the sociobiologist, has observed that if we turn from the cosmos to earth and pick up a handful of dirt, we find infinite variety and complexity in the biosphere as well. We have done much to understand the forms of life on our planet and to unravel the functioning of the intricate microstructures of the DNA molecule. But there is much more that we do not yet understand, not only about our own habitat but also about other parts of the universe.

Human beings are endowed with a sense of wonder and the creative capacity to wish for the stars. As young children we build castles in the sand, and as adults we extend them so that they soar in the universe at large. Our playthings fulfill our hopes and aspirations, satisfy our passions and needs. A little boy formerly collected toy soldiers and imagined that he was a great general; today he imagines spaceships to bombard alien invaders. A little girl plays house and imagines she is a fairy princess and her husband is a prince; tomorrow she is a member of a space mission. We all live by illusions: about ourselves, our lovers and friends, our society, the universe at large. A person may dream that he or she has won a huge prize in a lottery and think of all the things that he or she might buy with the money. At some point the reality principle intrudes; otherwise one could not go about the business of the day. Some illusions are a source of satisfaction; others, if we do not know that they are sheer fabrication, may land us in deep trouble.

Nevertheless, given the limits of our knowledge and our ignorance of the deeper recesses of nature, we allow our imagination to spill out and we invest the universe with powers and realities that meet our psychological needs. A provocative illustration of this is the cargo-cult phenomenon, a modern-day religion among the Melanesians in the South Pacific, who contrived a system of fanciful beliefs and rituals. According to their tenets, the white people, who had stolen from the gods the secret of producing material goods, would deliver such goods in ships, planes, and rockets to believers at no cost. A messiah would appear to make this possible. Believers would never have to work, and their deceased ancestors would accompany the goods back to earth. When the Germans arrived during World War I, the natives thought they were their saviors; this happened again when Japanese and Americans landed on the islands in World War II. How ludicrous, one thinks—but perhaps no more so than other belief-systems that are fabricated by the human imagination to cope with natural events and give vent to human wishes.

Imagination has a double function. First, it enables us to conceive of an infinite number of possible worlds. Here myth and fiction intervene, as well as poetry, drama, and religion. These imaginative constructions attempt to order the disparate pieces of our life-world and to give them some meaning and symmetry. Science also involves the use of creative imagination. But there is a profound difference. For although science creatively formulates hypotheses, these must be tested experimentally before they can be accepted.

In a sense, we live in an ideal world that we have constructed out of our reveries, expectations, and musings. We are entranced by the world of the possible: the Christ myth exemplifies this tendency. How incredible is the story of a god who took on human form or the belief that the divine word became flesh, whatever that means. How presumptuous to believe that the divine creator descended in order to save each and every person from suffering and death! The imagination draws a fanciful picture of a transcendental reality, some kind of celestial kingdom. Time and again the theistic myth appeals to the hungry soul; it feeds the creative imagination and soothes the pain of living. There *must* be something beyond this actual world, which we cannot see, hear, feel, or touch. There *must* be a deeper world, which the intellect ponders and the emotions crave. Here is the opening for the transcendental impulse. *Yes,* says the imagination, *these things are possible.* It then takes one leap beyond mere possibility to actuality. How can we explain this conversion? At what point does the ideal become the real? When is fiction transformed into truth?

Much of our daydreaming is focused on possible futures. We live for tomorrow, and the sands of time slip through our fingers. The present moment gives way to the onrushing future. Our future, in part, involves ideal possibilities being fulfilled. We make plans and dream of projects, and if we work hard enough at them, they may become true.

Simon Rodia, a poor Italian immigrant living in the Watts section of Los Angeles, spent thirty-three intense years building towers. He collected seashells, tiles, litter, and any scraps he could find; reinforcing them with railroad-track rods and working without blueprints, he built several high towers that had no utilitarian function. The Rodia towers have been designated a California state park, a tribute to his artistic talent and in recognition of his creative persistence. In this venture, Rodia typifies human imagination at work; though others may squander their talents on mundane projects, Rodia sought to build a cathedral. It is only the rare individual who can create a new religion or institution that attains historic grandeur.

A similar creative process is at work in our reconstruction of history. We take the materials and memories of the past and weave them into great fiction. We only remember those moments or years that we wish to, and we savor them in delight and enjoyment or fear and anguish. History involves the interposition of our imaginative reconstructions of what we think might have been. The great religious traditions are fictionalized monuments to individuals and events of the past that have been so transformed that they may have lost all resemblance to what actually occurred.

There is a constant battle in the human heart between our fictionalized images and the actual truth. We fabricate ideal poetic, artistic, and religious visions of what might have been in the past or could be in the future. But whether these idealized worlds are true is another matter. There is a constant tension between the scientist and the poet, the philosopher and the artist, the practical man and the visionary. The scientist, philosopher, and practical man wish to interpret the universe and understand it for what it really is; the others are inspired by what it might become. Scientists wish to test their hypothetical constructs; dreamers live by them. All too often what people crave is faith and conviction, not tested knowledge. Belief far outstrips truth as it soars on the wings of imagination.

The second function of imagination is that it provides a motivating fuel for life: it is a stimulus and impulse to action. Human civilization and culture are the creative constructs built out of human imagination. Nature is what it is; culture more often is what men and women have wanted it to become—though they often fail in their efforts. Our vessels and tools, cities and nations, poetry and art, science and philosophy, laws and institutions are all products of imaginative human creativity. We idealize an end; we act to bring it about. Not all of our possible worlds can be realized. We can dream of universes that are far beyond our capacities to reach. Some of our dreams do come true however—those that we have acted upon by finding the means and fashioning the tools to realize our purposes. We take inchoate matter and raw data and impute meaningful forms to them. But the forms that are brought into being are interposed by us.

Nature limits what we can and cannot do. It also suggests the ideal essences that we then resolve to implement, although the essences we finally interject are the products of our own inventive minds, for they outstrip bare nature. Men observe the moon and the sun and they invent the wheel. They find a rock or stick and learn how to use it as a tool. They gather branches and leaves and build a shelter. Looking skyward, they seek to emulate the hawk and the eagle by conceiving of and building an airplane. The Empire State Building and the rocketship take the form of the human penis as they protrude upward to penetrate the heavens. The possible always outstrips and precedes the actual—at least that is the splendid drama that unfolds throughout human civilization. Although implicit in the womb of nature, since humans and the materials they work with are all part of nature, culture is a product of art and artifice: it is both natural (since we are all part of nature) and artificial (since we bring into being something new). Like the nests of stray straws built by the bird or the dam of fallen logs cut by the beaver, these are the products of expended effort and labor.

Work is the most forceful expression of human ingenuity. The creative imagination desires an end, conceives of a plan, and then resolves to bring it into being by means of action. The realm of the possible outstrips past, present, and future realities, but we are forever called to account by reality. We need to understand nature and fathom its inner workings if we are to succeed. We need to distinguish our illusions from genuine truths. This is the cognitive-descriptive-theoretical function of imagination and language. Once an idea has been judged by practical common sense or science, it can be added to our store of knowledge; before it has been tested in the crucible of experience, it may be just so much balderdash.

Still, the imagination is essential for the human enterprise, for it is the means by which we seek to overcome obstacles to our efforts and to provide new goals for our achievements. But it also offers an escape route for our frustration; it provides both a crutch and a catapult to overcome our pain and torment. It is the latter that deserves our special attention, for the transcendental temptation is a kind of psychoexistential device that we resort to in order to cope with our unresolved ambiguities and unfulfilled projects, and to give vent to our deepest longings and tragic despairs.

The religious systems that human cultures have built are the creative products of an idealized imagination in which human beings have poured forth their dreams and tears. Though untrue, they persist because they satisfy needs that lie deep within the psyche. They are valid in this sense: our knowledge of what exists in the universe is only fragmentary and incomplete. There are many more things that will be discovered by the human mind that may excite, surprise, amaze, or even delude us. It may be the case that there are other beings that exist somewhere in the universe, with knowledge and powers far beyond our wildest aspirations. It may be that they care about us or even influence us. It may be that we will learn from them and that some of the mysteries will be unraveled and, more importantly, that some of our wishes will be fulfilled. It is the fulfillment of our unattainable hopes and desires that transcendental systems of beliefs prey upon. Just possibly, we will live again. Just possibly, we will not die. Just possibly, all our shattered plans will be reconstituted and we can be reunited with those whom we love and cherish. Just possibly, our deepest human hopes and aspirations will eventually be realized.

We talk a great deal about truth and knowledge, as we should, since we cannot live in the ideal realm alone but must descend into the concrete world of practical choices as we go about the business of living. Yet the world of fiction and reverie becomes real and is taken as true. We read books of nonfiction to learn the truth—biographies and histories, geographies and encyclopedias. But the fictionalized accounts in which we can indulge our fancies also fascinate and intrigue us. Novelists and dramatists, cinematographers and artists open up new worlds for us. They are not true, or genuine, or real, but only somewhat true, genuine, or real. Nevertheless, they tell us something about ourselves and others. The characters of Shakespeare and Balzac may be contrived, but they take on lives of their own. In reading a novel or viewing a play, we may be profoundly moved; there is excitement to the story, power in the tale. Even though we may label it as fiction and know that it is contrived, it nevertheless entrances our imagination.

Religious works are largely fiction. Invented by prophets and poets out of the creative fabric of imagination, superstitious longings, forebodings and dreams, they are eagerly devoured by a grateful audience. It is astounding to discover that the greater the untruth, the more unlikely the fairy tale, the more outrageous the myth, the more likely is the fact that large numbers of people will cleave to it.

What is so fascinating about the ancient religious systems that persist is that they do so in spite of the fact that they rest on shaky epistemological grounds. Though they appear to have little basis in fact, they continue to prosper in power and influence: Roman Catholicism and astrology are good cases in point. Obviously these belief-systems have internal functions—psychological and sociological—if no verifiable referents. Even though they began as figments of the human imagination, they soon assumed lives of their own. As artifacts of human culture, they have a factual basis in history, and extensive literatures and traditions are devoted to their preservation and promulgation. To challenge or overthrow an institutionalized system of beliefs and practices such as Catholicism or astrology becomes difficult, for they define a way of life, and hundreds of millions of people in every generation may be intensely committed to the maintenance of the delusional system. This is true of Catholicism far more than astrology, for the former's promises are more grandiose and all-encompassing, and its system of beliefs is well-defined and regulated by institutional authorities. Indeed, those who do not accept the institutional framework of beliefs are considered by society to be deviant and anomic to the culture. The purveyors of fairy tales have become the guardians of everyone's morals, and, unbelievably, heretics and dissenters are considered to be recalcitrant.

Is it possible to overcome the human tendency toward the transcendental temptation? Can we live without religious poetry and paranormal mythology? Can we face the world in cognitive-realistic terms? Voltaire remarked that religion began when the first knave duped the first fool. No doubt this is an oversimplification, for there are other motives and impulses responsible for religious belief.

Does religion have a biogenetic basis?

Because of the persistence of religious forms, even in highly sophisticated cultures, one needs to ask whether religion is biogenetic—that is, is there something within the species *homo sapiens* that demands it? Is there a biometaphysical craving, a specific gene or genes predisposing humans toward it? Is the human species *homo religious* by nature? Has this tendency been bred over the long history of human evolution, largely because it served an adaptive function for those members of the species who possessed it?

Sociobiologists have suggested that this is the case—namely, that religious belief-systems have profound biopsychological and sociocultural functions, and those groups and individuals that had religious tendencies have had an advantage in the struggle for survival over those that did not.

More specifically, they selectively influence who shall reproduce. If this is the case, then how or why and by what mechanisms it operates is a crucial question for investigation. One can imaginatively reconstruct a possible historical scenario for this argument, albeit a highly speculative one. Religion has encompassed many different aspects of human experience. I wish to focus here only on the mythic aspect, that which expresses reverential piety toward the transcendent. Primitive man was faced unremittingly with adversity and danger: disease, natural disasters, dangerous beasts, conflicts with marauding tribes, the imminence of early death. In this context of adventure and challenge, priest-magicians emerged who offered some kind of balm to the weary soul, enabling a person to persist in spite of the tragic dimensions of primitive existence. Various religious systems in different early cultures provided outlets for despair: one could offer sacrifices and propitiate the gods of the spirit world, in the hope that they might be mollified and that things would improve.

The death of one's cherished relatives and friends could be mitigated by burying them with ceremonies to ensure that their departed souls would be cared for as they were transported to the next world. They could provide solace for men and women who felt impotent before an unyielding universe and the inexorable march of events.

All of these sacrificial-propitial activities could lessen psychological anxiety, and help alleviate fear about the awesome character of brutal existence. Those social groups which provided mythic stories and incantation rituals helped their members overcome rude shocks, and they offered some means of coping with distress, reinforcing the will to live. Those overwhelmed by grief and dismay with nowhere to turn for answers, with no mythic or ritual system of release and hope, might fall more easily into disarray and disappear. They would be overwhelmed by the tides of fortune, unable to comprehend or adjust to them. They could not reproduce their kind nor endow them with their genes, for they lacked the "mystical glands." It is conceivable then, that these myth systems provided a basis for the group's solutions to the mysteries of existence. Those social groups which could transmit such belief-systems from one short-lived generation to the next might survive more readily than those which did not. In the most extreme case, in a relatively advanced civilization like ancient Egypt, the great pharaohs built huge monuments to assure their passage into the next world. But even the smallest and most primitive tribes, as they developed mythological belief-systems, could do something similar. Anthropologists have uncovered fairly elaborate systems of worship and burial of the dead back to the times of Neanderthal man in the remote past of human culture.

Of course, there are limits to the mythic constructions that can be fashioned. Mythic systems cannot be in total disharmony with the real world, since men and women still have to deal with the eternal practical problems: hunting, planting, and fishing, building shelters and baking bread, defending themselves from attack, etc. Still, mythic beliefs help soothe the aching heart so that men and women can go about the business of living.

The problem with the sociobiological hypothesis, however, is that these impulses seem to be absent in some humans, who seem to lack any sense of reverential piety and are unwilling to deify and propitiate unseen forces or beings. If *homo religious* is a universal characteristic of the species, why is it not present in everyone? Two possible answers may be given: first, that it actually is present but simply assumes different forms. Thus, new systems of religious piety replace the older mythologies. For example, with the collapse of traditional theistic forms of devotion, alternative religions emerge. Marxism is a new religion insofar as it worships the dialectical process and has faith in a vision of the ultimate Communist utopia. And in its extreme expression in totalitarian societies it deifies Marx, Engels, and Lenin, saints of the millennial ideology.

Some forms of humanism may also function religiously, that is, insofar as they have an inordinate faith in the powers of science or in the importance of democracy or the belief in world government. Religious humanists have argued that since religion is essential, humanism must perform all of the reverential functions of traditional religion. Is the excessive adulation of secular ideals a form of magic and idolatry, and is this not the moral equivalent of the genetic impulse? I do not think that this need necessarily be the case, because side by side with believers, polytheistic and monotheistic, are the skeptics, atheists, and agnostics, who deride systems of overbelief, and maintain that they can face the challenges of life without the need for sugar-coated pills. Clearly, there have been skeptics in the history of the human race who have stood against the times and rejected the sacred cows of their age. Socrates and Plato lampooned the myths of the reigning religions of the Hellenic world. The Skeptics and Epicureans in Rome rejected the faith of their countrymen. Spinoza, Hume, Voltaire, Mill, Marx, Dewey, Russell, Margaret Sanger and Simone de Beauvoir, Sartre and Freud, and countless other modern skeptics have done the same. Why, if the religious impulse is a constant of our biological nature, is it absent in some individuals?

William Blake's response to the skeptics was:

> Mock on, mock on, Voltaire, Rousseau;
> Mock on, mock on, 'tis all in vain!
> You throw the sand against the wind
> And the wind blows it back again.

A second possible answer to our question is that perhaps the skeptic is an anomic deviant, who, living against the norm, is casting sand against his biological nature. Much the same, for example, that even though homosexuality is the tendency of a minority of humans (anywhere from 4 to 10 percent), it is hardly the norm. This is not the time and place to enter into a discussion of homosexuality. Clearly the species could not survive if the opposite sexes did not mate, and finding pleasure and satisfaction in heterosexuality is essential to this process. Yet homosexuality is so widespread (and has even been observed in animal behavior) that one wonders whether it too has an evolutionary function. The same sociobiologists who think religion has a biological function likewise postulate that homosexuality may have had the biological role of strengthening the bonding instinct in males, necessary for warriors and hunters.

One may argue that skepticism also has an enviromental and evolutionary function, for, unless human beings can deal with the world as they find it—the reality principal—they cannot cope with threats in their environment. Those social systems that elevate mythology at the expense of technology or science have a lesser chance of surviving, and indeed have often been overcome by invading armies possessed of higher technologies and arts. Moreover, the priests and prophets who became celibate did not transmit their genetic propensities to future generations, whereas the soldiers and men of fortune who went out to conquer the world spread their seed everywhere. But one may still ask: Do human beings need some illusions, and are the mythmakers only feeding this need? Can the skeptics who reject the orthodox religions live and prosper without substituting their own system of illusions?

In one sense, this is true. Even "normal people" need some kind of myths and fantasies. The growth of Marxism in the twentieth century as a quasi-religious, ideological system to fill the void of collapsing religions perhaps illustrates this. Similarly, we have seen how newer forms of paranormal belief-systems have often replaced the orthodox systems of belief and how many children of liberal parents often become prey to newer cults. Aside from the historic mythic institutionalized systems of overbelief, there are other forms of illusion—individual illusions that we often need in order to get through the day.

There seems to be some psychological evidence that this is the case, and that some illusions are necessary for ordinary life. I am not talking about "delusions," but innocent illusions or "white myths." The process of manufacturing illusions seems to motivate people by giving them goals and purposes to attain. And this is part of the creative process that we have already discussed. Illusions may provide some basis for optimism and hope. One, for example, needs some illusions about one's own talents and abilities. If a person doesn't have any confidence that he will succeed, then he may give up too easily. If nothing else, an inflated ego and some self-love are essential for ambition and achievement. Some people do in fact give up early: they may have such an honest appraisal of their own limitations that this deflates any efforts on their part to overcome them and to excel.

Those who have illusions on the contrary are often willing to expend the arduous efforts needed to achieve great things, and though their illusions may in fact have been wrong, their wish is able to father the future fact. This is true not only about beliefs about ourselves, but our beliefs about others (which may be mistaken) or about our society (which may be undeserving) or about the future (which may be bleak). Surely, we need to be *realistic* about all of these things. But we need also to be dreamers; and some degree of illusion, alas, is the psychological soil out of which great adventures emerge and flourish.

The point is: there are illusions and illusions, and there are also delusional systems as well. Moreover, the long-surviving systems of religious belief that were introduced in the infancy of the race no longer can perform their original functions today. They clash rudely with reality. Judaism is over 3,500 years old, Christianity 2,000, and Islam 1,300. These "venerable systems" of belief are now contradicted by our scientific knowledge of nature; insofar as they are, they are dysfunctional. Although systems of belief and magic may have functioned to provide relief in the past, when primitive forms of science attributed causes to nature gods, they now have negative functions, especially when they are taken literally. In any case, it is clear that if the religious impulse—reverential piety toward the unknown—has biological and genetic roots within the primeval biology of the species, so does the reality-causality principle, which competes with it and seeks to understand nature by uncovering the actual causes at work, and in constructing technologies that are workable. The truth is that there are many biological impulses in the breast of man, which often compete, and we cannot be slaves to one at the expense of the others.

Indeed, there is a basic war between the saint and the practical man, the mystic and the doer, Jesus and Prometheus, and it is only the practical man, the doer who gives proper attention to the means of survival. This is particularly true for postmodern civilization. The language of instinct and genes is apt to be deceptive if it is used as a mask to read our moral biases into human nature. Konrad Lorenz has postulated an instinct for aggression, which serves the biological function of enabling the strongest males to breed and thereby contributes to the viability of the species. He finds this displayed in many different species, from goldfish and ducks to *homo sapiens.* If this is the case, then one should not conclude that we should satisfy every instinctive impulse, seek to root it out, or suppress it as evil. On the contrary, *if* there is an instinct for aggression, we need to discover constructive equivalents in order to channel it into wholesome forms of human conduct (such as competitive sports or cooperative efforts to fulfill social ideals). Even if it is impossible to overcome our biological roots—in this case our tendency to create mythic systems of overbelief in order to cope with adversity—it all depends on which systems we will accept and whether we can channel our primitive tendencies away from superstitious systems to alternative constructive substitutes. The historic religious systems evoke a passive-acquiescence model, in which human beings are led to submit to the universe and God in obedience and acceptance. It is based on the failure to understand nature and the tendency to read in magical causes and cures.

In its place is the active, heroic, Promethean model of nature. The options for men and women are first to understand nature and second to enter into the world to change it. Promethean men and women refuse to bend to nature, except at the moment of death—they seek to transform nature to suit their purposes. We may need substitute myths: these should be those that are appropriate to the future of humankind, and those that enable us to achieve. No doubt there should remain a residue of the primeval religious intuition: that the cosmic universe is a vast scene of impersonal forces, ultimately beyond our control. We can understand nature and adapt our behavior or modify the course of events by using knowledge, but only to a limited extent. There are things not within our power but beyond our control. Some of these must be accepted as inevitable, such as accidents, natural disasters, disease, and death.

And perhaps a religious response is warranted after all. Thus we need to cultivate some stoic sense of resignation. This does not mean that we should worship the transcendent or invest it with divine powers, but at the very least that we should have some sense of the majesty of the cosmos and some sense of our impotence within and before it. This is not mythological thinking, but is based upon a realistic appraisal of the human condition.

The moral function of religion

Still, there are, no doubt, other reasons for the persistence of systems of religious overbelief. Surely of great significance is their moral function. For even though the adoration of God the Father or the transcendental unseen has been shattered for modern man, religious institutions retain

another major function. They provide a system of rules and commandments guiding conduct. This is true of the thousands of sects and cults
that have been practiced by a wide variety of sociocultural groups in the
past. It is also true of the great historic religions, which provide us with
the Ten Commandments, the Sermon on the Mount, the Koranic virtues,
the Buddhist Path to Righteousness, etc. Does religion, in the last analysis, provide a necessary foundation for morality? Do moral rules also play
an important biogenetic role in the struggle for survival?

Sociobiologists have also suggested that if social groups are to cope
with adversity, then they need some internal rules to govern their behavior: moral systems, thus, have some adaptive value, and those rooted
in religion help sanctify their grounds and insure a sense of obligation and
obedience. Moreover, sacred systems of moral rules help to determine
who shall survive and who shall reproduce.

Religions thus are not simply systems of belief; they define a way of
life. They lay down norms of conduct and regulate various forms of
behavior. They may govern the relationship between the sexes and determine what sexual conduct is virtuous or sinful. This also applies to the
structure of the family, delineating the proper role of the father, mother,
and children. In a similar manner, this operates for other complex social
institutions, which introduce taboos and phobias. Some actions are
deemed the exemplars of virtue and nobility, and are rewarded both in
this life and next. Others are considered wicked and impermissible, and
these may be punished by death, excommunication, exile, physical
punishment, imprisonment, or general disfavor. The original meaning of
the term *morals* is related to *mores*, which refers to the normative traditions and habits of a group of people living and working together.

Presumably, those groups with a well-regulated system of conduct
passed down from generation to generation did not have to go about
inventing the wheel each time. They could instruct their children in the
moral faith of the fathers, and thus insure some cohesion and provide
some internal unity necessary for the perpetuation of the group. Those
individuals or groups who could not conform to the code would not
survive, and hence they could not transmit their genetic dispositions to
future generations. Morality, in this sense, thus is a method of adaptation;
and when tied to religion it is given divine sanction. There are divine
rewards and retributions, and the love or fear of God provides the
motives for moral obligations and duties. Morality is not simply manmade, but divinely inspired. Moses brought down from Mount Sinai
God's commandments to the children of Israel. This provided a sacred
basis for his patriarchal political system, and the divine right to rule. It is
the union of religion and morality that reinforces the moral system. Thus,

morality has a profound sociological function: it provides the framework and the integrative bonds that enable the group to preserve itself and function. Inasmuch as human beings are social animals unable to survive outside of some community, morality also has a biogenetic function.

All of this is essential if we are to understand the positive historical role that religions play. Even if a religious moral code is harsh, it nonetheless provides some basis for stability and order, and it guards against anarchic conduct. In obeying the traditional norms, some measure of social peace and harmony is maintained.

These are, however, dysfunctions that result from wedding morality too closely to religion, as human history vividly demonstrates. First, where there is social change, a fixed system may find it difficult to accommodate. It may be an intolerable strain to try to stretch the absolute commandments of religious morality and apply them to unique contexts where decision-making is required. Often in the moral sphere, it is not a question of choosing between good or bad and right or wrong, but between two or more rights or goods, both of which we cannot have, or between the lesser of two or more evils. Although a cohesive system may help to define rights and responsibilities, it may become repressive in not allowing for new interpretations or modifications of practices that may be necessary. It thus becomes an obstacle to progress. This failure is exacerbated when a system of traditional morality encounters alternative moral systems—as is inevitable in human history. The conflict of competing moral standards makes it difficult to negotiate compromises. The ancient Greeks characterized the moral values of other nations as "barbaric," but even among the Hellenic city-states there were widely varying customs, as seen in the differences between the moral codes of Sparta and Athens. The Christian crusaders clashed with Muslim moral ideals. Marco Polo later reported to Europe the strange cultural differences of Genghis Khan and the Chinese empire.

It is in this context that customary morality often becomes an obstacle to furthering human understanding and contact, especially if it is parochial, chauvinistic, ethnic, or nationalistic. The sword becomes the arbiter of right and wrong. Differences in morality are resolved by warfare and conquest, as one group seeks to impose its way of life on another, as Alexander the Great tried to do when he conquered Persia, only to have his successors subverted by Persian ways. A system of law must emerge eventually to adjudicate disputes between peoples to define certain forms of conduct as legally permissible and others as illegal and impermissible, and to enforce the code of life by the power of the state. In time, tribal religions themselves must become more universal, as when Christianity attempted to transform Judaism from a nationalistic, tribal religion into a universal message for all human beings, and as Islam still later sought to impose its moral faith and law on a wide stretch of geography that it conquered.

Returning to the biogenetic question in this context, we may ask: Do religious institutions influence the course of evolution? The answer is in the affirmative, for most religions, particularly the historic ones, determined who would reproduce and why. Interestingly, however, these reproductive rules varied. There are rules governing infanticide and abortion, so whether defective fetuses can survive depends on social rules. Ancient Greek societies widely practiced infanticide, whereas Christian and Muslim societies opposed it. Similarly, in Jewish and Muslim societies, which practice circumcision, some infants died under the rusty knife of the rabbi or imam. Some religions practice initiation rites or puberty ceremonies. Some extol celibacy, chastity, and virginity; others are more permissive in allowing sexual freedom. Christian and Jewish societies became monogamous; Muslims have practiced polygamy. Catholic and Buddhist priests were celibate, but Jewish rabbis, Muslim mullahs, and Protestant divines procreated and multiplied their kind. There are rules governing divorce and remarriage. Many practices were purely fortuitous; for example, Mohammed engaged in several wars where many of his soldiers were killed; hence it was thought all the more appropriate for a man to take more than one wife. The need to care for widows and orphans is constantly referred to in the Koran.

All of these commandments had an inevitable causal influence on the reproductive strategies and the kinds of people who would reproduce, much of it quite independent of adaptation to the environment. Similar profound influences on the biological strain are exerted in other ways: Hindus would wash their dead in the Ganges and also dispose of the bodies there, no doubt a cause of infectious diseases. Muslim pilgrims on the way to Mecca have suffered great hardship and exposure to diseases, such as cholera. Christians at mass would kiss the same crucifix and drink from the same wine chalice, thus transmitting infectious diseases. Some religions have adopted a fatalistic element toward disease and pain, which tended to prevent or weaken the efforts of medical science to discover cures. For a long time, there were religious prohibitions against autopsies. Thus there are powerful unconscious and largely irrational forces interfering with natural selection.

Whether this tended to develop genetic dispositions is difficult to say, since there has been such a wide range of practices. Ernest van den Haag has observed, however, that it might have had contrary effects on inbreeding, population, and gene pools: Catholic priests, who became scholars and studied in monasteries, were celibate, and hence could not bequeath their intellectual talents (insofar as they had them) to future generations, whereas Jewish rabbis and Talmudic scholars in the *shtetls* of Poland and Russia had large families. Similarly, Jewish religious prohibitions against marriage outside the group may have tended to exacerbate some negative genetic dispositions, such as the high incidence of Tay-Sachs disease.

There is another issue to be raised, and that is whether or not individuals can be released from the strict moral rules that govern the social group. Clearly, another positive development did occur in human history; this was the divorce of ethics from religion and customary morality, as well as independent efforts to establish ethics as an autonomous field of inquiry based upon reason, entirely free of religious sanctions.

It was the Greek Sophists who made this development possible. In traveling from city state to city state, they were impressed by the relativity of moral conduct. Men and women worshiped different idols and cherished different moral values, and the itinerant teachers noticed this diversity and came to reject them all. Thrasymachus, Callicles, Gorgias, Protagoras, and others saw that values were relative to the society in which they arose and that they were often maintained by social elites for their own self-interests. Hence the Sophists became skeptical and cynical about absolutes. Instead, they preached self-interest, the art of succeeding and getting ahead—essentially, how to make friends and influence people. Out of the challenge, Socrates and Plato sought to establish a new field of ethics, one that was not based simply on convention but which had some grounding in natural law. They also sought to debunk the Homeric myths—as do Socrates and Adaemantus in the *Republic*—as an inadequate and hypocritical basis for moral conduct. Were these moral ideals— good, virtue, beauty, truth, justice—in the nature of things, as Socrates thought, and could they serve as a guide for conduct?

Aristotle followed through on this inquiry in the *Nicomachean Ethics* by making it into an empirical inquiry, based upon practical reason. What is good, he asked, and how can we achieve the good life? He set forth a rational method for fulfilling human nature, actualizing our potentialities, achieving some measure of excellence and nobility, *eudaemonia* or happiness. Philosophers have demonstrated ever since that ethics can be based upon reason. It need not be anchored in customary morality or religious doctrine. Thus if moral codes had at their inception a social function, required by the nature of the human being if he is to live in community and survive, it need not have a divine sanction to be enforced or be enforceable. Modern men have ever since sought to establish the conditions of the good life on rational grounds and to judge moral conduct by their consequences in behavior for good or evil. An action is deemed good if it maximizes human happiness and minimizes suffering, and evil if the contrary. Moreover, there are standards of fairness, which can be justified quite independently of any theological grounds, as Kant has shown. The moral conscience has its own sources: empathy and altruism provide additional motives for moral behavior. Although it is no doubt the case that religious belief-systems may have had a moral and social function and these often served well, humankind has outgrown this need; from an earlier facilitator of moral conduct, religion has now become an obstacle. For morality based on entrenched social habits and religious authorities may be impediments to human progress, necessary moral reform, and the application of critical intelligence to the solution of human problems.

The quest for ethnicity

There is still another deeply rooted function of religion, so much so that it becomes at times virtually impossible to modify or change it, and any effort to do so must be within the preexisting framework. I am here referring to the sociological function of institutionalizing rules of behavior. Religion is not simply moral; it is a way of life. It is like the language that we have inherited and use almost unconsciously. This defines the rules of social grammar, which though nominally arbitrary, we all need to learn and use if we are to communicate and interrelate.

I do not mean simply the moral code, though that is an essential part of it, but also something far deeper: the best terms are *ethnicity* and *nationality*. For a religion, in time, actually defines one's being. One is a member of an identifiable, interbreeding religious ethnic or racial stock. Religion involves kinship and blood relations. It is not simply a cognitive system of beliefs that we can either accept or reject on intellectual terms; it structures the very essence of an individual's identity.

I am not referring to the first or second generations in which a new religious creed takes hold, or to the process of conversion and acceptance, where people come to accept a new faith. There are certain periods of religious convulsion when new messiahs or prophets spring forth, and they are able to convert an inner core, who are assured of their divine mission. In the final analysis it is not so much what a Moses, Jesus, Mohammed, Buddha, or Joseph Smith does or says that is important, but what the successive generations of believers do with the message (as later embellished and codified) that counts. How do they build the church, and what are the doctrines, rules, and regulations that they seek to impose?

Herein lies the secret of longevity of religious institutions and why some mythic systems survive the first generation after the death of their founders. Actually, aside from the early origins of a religion, and perhaps the periods of intense religious awakening or renewal that occur at various times, for most people the religion that they have is what they imbibe at their mother's knee and with the help of their father's rod. It is the religion of their forefathers that has been instilled in the young, the essential tribal lore. People are bred into a religion; it becomes part of them; it is usually not, except during periods of conversion, a matter of personal conviction. The best illustration is Judaism, a faith to which countless millions of Jews have tenaciously clung throughout the millennia out of love and respect for their forebears as much as for any other reason. The same is true, however, after a period of time, for most religions, which if they endure the first challenges to their claims, are passed down to the next generation by a set of strict rules which must be observed. Members of an orthodox religious group (1) indoctrinate the children in the beliefs and practices, (2) prohibit any intermarriage from those outside their clan, and (3) punish, expel, or excommunicate members of the religious tribe who are heretics. Moreover, (4) group sanctions are imposed to guarantee conformity: any dissent is viewed as a threat to the integrity of the group or to its preservation as a viable entity.

By this process of enculturation, religious institutions become enshrined, indeed entrenched in the very way of life of a people. Tevye in *Fiddler on the Roof,* living as an Ashkenazi Jew in Russia at the turn of the century, reveres and exalts tradition; so do Roman Catholics, Greek Orthodox, Muslims, and others. The point is that they *define* what a man or a woman is as a human being. This applies to dietary practices, speech patterns, the ways of expressing humor and coping with sorrow, and a wide range of practices and rituals. They are one's ethnic being. To oppose one's religion, thus, is to reject one's very roots, family and relatives, friends and neighbors.

Religion often involves geography: virtually everyone in Cambodia was a Buddhist; in India, a Hindu, Muslim, or Sikh; in Italy or Spain, a Roman Catholic. But it also involves *blood kinship,* and marriage within one's kind; all of one's relatives speak the same language, sing the same hymns, and worship in the same temple. Accordingly, maintaining one's fidelity to one's religion is equivalent to ethnic loyalty; to reject one's heritage is to be alienated from the original fountain of one's being.

There have been dramatic changes in this prognosis of religiosity in the postmodern world. For different individuals and groups now live side by side in pluralistic social environments, and one's tribal and ethnic allegiances are constantly being challenged. There are competing influences on the individual, hence conflicting loyalties. Even though we may share the same religion with others in our community, we may differ racially, or have come from different countries. Nationalities are being redefined. Americans, for example, are the product of a melting pot: new civic virtues, allegiances, and traditions have emerged. Thus one's religious faith is not as essential to a person's core of being in the postmodern era as in the past. In many societies it is now difficult to proscribe intermarriage or enforce narrow indoctrination on one's children, especially since they encounter so many other faiths in the schools and marketplace. In a free forum of ideas and values, different systems contend for our children's commitment, and individuals are enticed to adopt new codes of belief and practice. Free conscience competes with ethnicity, and the latter often find it difficult to hold its youth in a vise. The young often look outward, not homeward, for inspiration and fulfillment.

Celebrating the rites of passage

Nevertheless, there is another function of religious institutions that it is important to recognize, although it is increasingly difficult to maintain the hegemony of religion in postindustrial urban societies; and that is its traditional function as a support system within a community. Some sociologists and anthropologists have considered this function an essential feature of sociopsychological health, and they have deplored the fact that in modern and postmodern society religious institutions have been disintegrating. In their wake has emerged mass culture, with its depersonalized and rootless human beings, without structures, alienated from their origins, often living banal lives, without meaning or significance.

In a sense, it is not simply the loss of religion but the collapse of community that is the problem, and with this the abandonment of traditional rites and rituals associated with the local group. Historically, most human beings were conceived, nourished, educated, married, worked, lived, and died in one locality. The farm, rural village, or small town provided the parameters of one's life. In medieval Europe, the church played an integrative role: indeed it was the center of the community's activities, overwhelming in architectural grandeur all other buildings. One's forebears lived and died in the same area, and traditions became entrenched. This religious-social fabric, moreover, ministered to the common needs of human life at various stages, and these were celebrated by communal rites of passage, which the church consecrated. The birth of an infant, his or her baptism, circumcision, and first communion or confirmation was appropriately sanctified. Marriage was considered a sacred union, with vows openly proclaimed before the altar of God. And upon death, the last rites ensured one's entrance into heaven. The members of the group shared in common the joys and sorrows of life, and duly acknowledged them in public ceremonies. This provided a support system for the individual who, when faced for example with the death of a loved one, could be nourished by group solace and sympathy. Membership in a community—where face-to-face interactions were possible— gave a person a sense of identity and belonging.

Some humanists are opposed to these ancient tribal ceremonies; they want simple burials and no religious ceremonies for the passages of life. Yet they will gladly attend graduations, inaugurations, or retirement parties, grant *Festschriften* or celebrate wedding anniversaries. Why not? There can be deep psychic satisfaction at such rituals and rites, even if nontheistic, and their loss is often gravely felt.

This traditional sociological function of religion has been rudely interrupted time and again in human history. The most powerful countervailing influence to religious tradition is the intrusion of social change, which is as inevitable in human societies as death. One cannot remain fixated in eternal social structures. Calamities occur, people die, new generations emerge, economic and political change is ongoing. The disruption of the community is inevitable. Today a profound destabilizing force is mobility, the fact that individuals are now able to leave the soil of their youth and go out into the world. The urbanization and suburbanization of modern life has had a convulsive effect, for it brings into interaction and possible conflict individuals from different backgrounds, and it liberates them from traditions which were held to be sacrosanct. No doubt the most dramatic factor in the transformation of earlier life is the influence of technological innovation and the growth of industry, which has radically transformed economic and social conditions. Traditional religions die hard, however, and they persist in replaying their ancient roles.

North and South America, Australia, Africa, and other areas of the world have become new homes for countless millions of immigrants. Uncharted wildernesses have provided opportunities to settle in new frontier communities. Faced with unpredictable, strange new worlds and confronted with new challenges, the settlers invariably clung to their ethnic loyalties and religious allegiances. Or they created new sects and cults, which soon became established orthodoxies. Religions in these contexts attempted to provide cosmic explanations of the universe and man's place within it; often for unsophisticated folk, it offered certitudes, rules, and structures by which they could live.

The American experience is fascinating. Uprooted immigrants—English Anglicans, Scotch-Irish Presbyterians, Polish, Italian, Spanish, and Irish Catholics, Russian Jews, etc.—all brought their religious traditions with them and transplanted them in the new environment. Living on this continent were the aboriginal Indian tribes; transported to America in cargo ships were millions of black slaves. Eventually Chinese coolies, Japanese, Indian, Cuban, Vietnamese, and Mexican and Central American immigrants also came. Side by side with the proliferation of many diverse sects and cults, there emerged a new sense of nationality and a new set of civic virtues.

The United States is a melting pot, but mixed in the broth are plural ethnicities and religions, many of which have not dissolved. The issue we need to address is whether it is ever possible to transcend the limits of ethnic or racial differences and to move beyond the ancient parochial allegiances. I believe that it is, for new nationalities and ethnicities have emerged in various parts of the world and these provide individuals with a new sense of belonging and identity. Moreover, new rites of passage and holidays are celebrated to heighten the civic virtues, although ethnic and racial ties are maintained. The great battle in America is whether the "Judeo-Christian heritage" defines Americanhood, or whether a secular humanistic identity can or does. Given the existence of non-Judaic and non-Christian groups and large numbers of unbelievers and humanists, it is difficult to return to the old orthodoxies, especially given the fact that the Jeffersonian-Madisonian tradition, separation of church and state, and the First Amendment have tended to prevent the establishment of a church. Given the basic principle of religious liberty, any effort to re-establish the motto "Pro Ecclesia et Patria" is very difficult. With the state central in matters of education, a secularized republic emerges: concomitant methods of celebration and common values now fill the void left by the collapse of influence of the earlier dominant religions. With a plurality of competing sects, none dare dominate, and a humanistic society can better function.

This problem, however, is worldwide, for what is happening in the United States is a microcosm of what is happening worldwide. The technological revolution of the past three centuries, and the consequent social, political, and economic upheavals they have produced, have transformed the globe into one world. However, ancient religious loyalties still contend in daily battle. Most major urban centers contain people of various racial, religious, and ethnic backgrounds, many still clinging to ancient faiths, which are largely irrational, ill-founded, perfectly false, and increasingly difficult to maintain. Worldwide trade and commerce and rapid communication have made us part of one global village. What should be the moral foundation of such a world community? Should the ancient religions still play a strong role? Can we develop new convictions in which our commitment to humanity as a whole, not to one parochial group, commands our ethical loyalty?

The skeptic, agnostic, atheist, and secular humanist challenge Judaism, Christianity, Islam, Hinduism, and the cults of the paranormal. Is it possible to transform these and develop a new religion of humanity that is truly global in character? There is the danger of terrible secular gulags emerging; Fascism and Stalinism-Leninism have demonstrated the human capacity for inflicting unimaginable cruelties. How can we guarantee against terror and oppression by new and awesome totalitarian religious ideologies?

Can humankind live without the historic religions? If so, does something need to take their place? Most of the great religions came to be universal; they now battle on a worldwide scale for men's souls. Christian missionaries still seek to convert the heathen to the most primitive forms of Christianity, using the Bible as a weapon, impervious to any kind of criticism of its foundation, and with great success. Muslims carry the flag of Islam worldwide, and a few Indian gurus strive to indoctrinate Western youth with their ideals of spirituality. Paranormal faiths contend with the ancient religions. All of these give vent to the transcendental temptation lying deep within the human psyche.

Meanwhile, powerful secular forces are at work transforming the globe. Massive population increases, trade, imperialism, military and political conflicts between states, increased communication and travel, the growth of multinational conglomerates, and worldwide revolutionary forces in opposition all play dramatic roles. Two secular ideologies, Marxism and democratic humanism, compete for loyalties as alternative global philosophies. In this intense global struggle, one asks: Can there be a new scientific, moral, or philosophical outlook appropriate to the new age of space? Can it be skeptical, scientific, humanistic? These questions have large-scale implications that I do not hope to resolve here. Is it possible to transcend the transcendental temptation? What is the role of knowledge and critical intelligence in the process? Can a new ethical outlook emerge—with new civic virtues, rites, and rituals—that is more appropriate to the reality of the global village and the building of a genuine world community?

Can we transcend the transcendental temptation?

As we have seen, the transcendental temptation lurks deep within the human breast. It is ever-present, tempting humans by the lure of transcendental realities, subverting the power of their critical intelligence, enabling them to accept unproven and unfounded myth systems. Can we live without myths? Can we ovecome the defect, as it were, in our natures? Is it so rooted in our natures that it cannot be overcome, but will crop up in generation after generation, the forms and functions of the transcendental temptation the same, with only the content different?

Any impartial observer of human history must be duly impressed by the tenacious character of religious faiths and the persistence of transcendental myths in human history, as well as the ability of the most outrageous myths to sprout and take root, even in the graveyards of defeated and dying systems. Is this a perverse strain, like original sin, but beyond any hope of human redemption?

One can say that there is a *tendency,* yes, and that it affects all age groups, social strata, levels of education, occupation, or intelligence. Whether it is biogenetic and in what sense is unclear. For, as we have seen, it is absent in many individuals, and moreover, there are degrees of severity of the temptation or of domination by the prevailing mythologies. Even if there is a genetic disposition, perhaps it can be mitigated, sublimated, or deflected away from superstitious and otherworldly fixations into more constructive alternatives. Perhaps there are scientifically grounded moral equivalents for transcendental religiosity that we can discover and by which we can live.

The key factor in transforming consciousness and culture is the vital role that critical, scientific intelligence can play. The question is: To what extent have the increased knowledge and information made possible by science emancipated individuals and social groups from the tenacious hold of ancient religious dogmas and loyalties? To what degree can science moderate or liberalize, if not eliminate, these mythic systems, and avoid infection by new ones, even more virulent and irrational in intensity?

There is no guarantee that it can. For even human beings dedicated to the quest for knowledge and truth and committed to science and rationality will profess religious faith and ascribe to the truth of religious belief-systems, as the temptation of the transcendental gets hold of them. I can only claim something more modest, namely, that there is an important role that skepticism can play (as explained earlier in this book) in examining the pretentious claim of religious faiths or paranormal myths and in leading to their revision. This is not an overly ambitious program, and it makes no grandiose claims about what will ensue. It simply asserts that skeptical criticism is essential to the lifeblood of the mind, and that this can provide some therapy against the transcendental temptation gone wild, and some immunity against future infection. For that is what we are most troubled about: the possible transformation of religious or paranormal belief-systems into fanatic missionary cults that become the enemies of truth and the engines of repression. The transcendental temptation, if allowed free play without criticism, often develops authoritarian systems of moral absolutes and truths; the Holy Crusade syndrome is ever ready to impose its views on others, even to suppress dissent and to limit the free conscience.

The skeptic is not passionately intent on converting mankind to his or her point of view and surely is not interested in imposing it on others, though he may be deeply concerned with raising the level of education and critical inquiry in society. Still, if there are any lessons to be learned from history, it is that we should be skeptical of all points of view, including those of the skeptics. No one is infallible, and no one can claim a monopoly on truth or virtue. It would be contradictory for skepticism to seek to translate itself into a new faith. One must view with caution the promises of any new secular priest who might emerge promising a brave new world—if only *his* path to clarity and truth is followed. Perhaps the best we can hope for is to temper the intemperate and to tame the perverse temptation that lurks within.

Alas, skeptics are considered dangerous to humankind. They shake the foundations of the social order; they question and hence are considered a threat to the authority of the reigning mythology. That is why heretics and dissenters from Socrates on down through the ages have been condemned, forced to drink the hemlock, burned at the stake, and tortured on the rack by the Inquisition, in the dungeons of the gulag, or in the prisons of totalitarian states. All too often they have been excommunicated or rejected as pariahs. It is only in modern society, given the development of democratic values and the recognition of the power of critical scientific intelligence, that there has been an increased appreciation of the importance of the moral principle of tolerance. Where human beings differ about fundamentals, the best strategy is to engage in dialogue and debate, in order to negotiate differences and work out compromises.

A major problem we face is that the entrenched orthodoxies have resisted free inquiry about their basic premises, as if suspecting that any digging permitted around the foundations would lead to their toppling. If there can be free intellectual inquiry into the foundations of religion, without undue obstacles or restrictions, we have every reason to believe that there would be a loosening of the ancient ties. People might become more sensible and intelligent about their beliefs. There might be some lessening of self-righteous fanaticism and intransigence. The world might become a more humane place in which to live; at least others might be more wary about attempting to fill us with indigestible nonsense. Whether or when the dominant faiths of today will eventually disappear, as did the great Mithraic, Egyptian, Assyrian, and other religions of the past, is difficult to predict. No doubt they will be altered: instead of being taken as eternal dogma, they might be viewed as metaphorical and moral poetry, as is the case among liberal theologians, Protestant, Catholic, and Jewish. This can only have a civilizing effect. As the world moves from an era in which prayer-theology and metaphysical transcendence dominates to one in which science and rationality prevail, new forms of human creativity can flourish.

But it is not enough to weaken the hold of the ancient diseases of the soul, or even to seek to root them out, without asking: What will take their place? Perhaps it is dangerous to disabuse us of our traditional religious illusions about Moses, Jesus, or Mohammed, for once cleansed of our theological errors, we may be open to new infections far more virulent. The ancient faiths have a kind of venerability, at least when they are not taken too seriously. New ones may wreak havoc with our systems if they suddenly appear and are proclaimed as new orthodoxies.

This is why skepticism must be applied to the newer claims of the paranormal as well as the old. We need critical examination of all occult claims; for in rejecting the old myths, we must not be open to the call of new ones, equally nonsensical and perhaps even more dangerous. It is clear that high on the humanist agenda is the power of education; but this means that we must cultivate our intellectual talents as one of the highest virtues. We need to develop in society at large an appreciation for cognitive processes, for critical thinking, not simply deductive mathematical thinking or technological expertise in narrow specialties. A computer expert can be a barbarian of virtue, and a chemical engineer lack all wisdom outside his own field of technical expertise. We need the ability to evaluate hypotheses and ideas and to appraise values. Central to this agenda are the schools—the primary agents of social change—for the young need to be nourished from the start in the art of thinking. They need to be encouraged throughout the entire curriculum up through higher education.

The larger task, however, is to raise the levels of intelligence and knowledge of society at large. Unfortunately, this is increasingly difficult because the dominant media of communication are usually used for purposes other than providing information or enhancing knowledge. The media entertain, which is all well and good, but they also provide political and religious indoctrination and are operated for profit or power. Those who are able to control the media are in a position to determine the mind of the global village in the future. That is why keeping open the channels and media for criticism and for new ideas to be heard remains essential to the future of humankind. We need pluralistic centers of thought and inquiry to provide some immunity against dominant economic, political, ideological, or religious forces that seek to define and impose orthodoxies, whether old or new.

One question that has always puzzled rationalists is why Christianity, an error-filled mythology impervious to reason and thoroughly refuted in its own day, managed to survive its detractors and eventually succeeded in conquering the Roman empire and stamping out all opposition. It is often forgotten that the church was established after doctrinal disputes were ended by the Council in Nicaea in 325 C.E., and after Constantine declared it to be the official religion of the empire. It is estimated that only 5 percent of the people in the Roman empire at that time were Christians. Thus, it was the union of political power and belief and the crushing of dissent by the state that paved the way for the dark ages and the inordinate focus on salvation. Rome could not defend itself against the barbarians after the Christian virtues of submissive obedience and concern for the next life had suppressed the heroic pagan virtues that dominated in the Greek and Roman periods.

Similarly, Islam was forged out of the unity of the sword and the Koran, and large numbers of peoples and territories were enlisted in the cause of Mohammed only when political power compelled them to do so. Thus the great danger is that religious ideas will be imposed by political and military power, and that this will sweep human societies and crush dissent. Some skeptics have observed that revolutionary Marxism has all the earmarks of an ideological religion, and they fear that it will eventually sweep the world as a new religion—worshiping at the magic altar of the dialectical process. Thus it is of crucial importance that the separation of church and state be maintained and that efforts by theology or ideology to impose itself by political power be resisted.

There are the cynics who will demur that what I have proposed is not feasible. They insist that skepticism is the philosophical outlook of an elite, a dissident minority of intellectuals, forever confined to the fringes of the social fabric without touching the mainstream of humanity. The ordinary man, we are told, needs myths to survive, and he is more profoundly moved by his passions and his feelings than by his reason. Perhaps they are correct in this belief that critical thinking, fully applied, can only govern a small part of the human condition, and that even here great men and women acting as intellectuals are themselves not immune to the ravages of passion, instinct, habit, and unreason.

Surely this pessimistic appraisal has some merit, at least as viewed in historical perspective. For the transcendental temptation has held sway for millennia, and to hope to mitigate or obviate its continued power may be to engage in wishful thinking. The great era of classical Greco-Roman civilization, dedicated to the life of the mind, was overwhelmed by the mystery religions of the east, and Judaism, Christianity, and Islam have been able to dominate human consciousness for a long period of time. The Renaissance, Enlightenment, and the growth of modern science offered a challenge to this hegemony, but these are relatively recent developments and of short duration in human history. What guarantee do we have that science too will not be overwhelmed and superseded by new faiths of unreason commanding human imagination? Perhaps the escape from reason is the inevitable destiny of the human species, and perhaps a cyclical swing will again allow the transcendental temptation to seize our attention.

Prometheus was condemned by the gods for his audacity in the stealing of fire and bequeathing science and the arts of civilization to humankind. As his punishment, a vulture plucked at his liver. Perhaps the modern Promethean spirit will again be vanquished by acquiescence and passivity, and fear and trembling will outdistance courage and reason. The Promethean myth expresses the courage to seize destiny and master our fate, but it also marks a deep-seated fear of human independence.

One cannot predict the future course of human history with any degree of confidence. Regrettably, often the unthinkable becomes true. Will the unimaginable again overtake us, as we slip into a new dark age of unreason? The only option for us to prevent this is to continue to use the arts of intelligence and skeptical criticism against the blind faiths, old and new, forever sparked by the smouldering embers of the transcendental temptation. We need Occam's razor, honed by reason and science, to cut away new tentacles of the hydra-headed monster within ourselves.

But still this is not enough. Negative criticism, though necessary and vital for humankind, is not sufficient in itself. For men and women need to dream. We are not animals capable only of cerebral functions, logic and reason, but we are also creatures of passionate feelings; we are forever entranced and moved by great ideas, if they have dramatic intensity—and often even if they are false—as long as they satisfy our passionate needs, whether real or imagined.

Is there any hope that a scientific, secular, or humanist culture can develop and prevail, devoid of transcendental myths? This can only happen if in the place of the old myths of salvation there are new ideals and goals sufficient in grandeur and power so that they can inspire and move us to greatness. If the salvation myths are no longer tenable, what will take their place? The dilemma is always that new faiths and new myths may emerge, equally irrational.

Today we confront new powerful realities of the human condition: we are truly members of a world community, and parochial faiths and ancient loyalties no longer can cope with that situation. And we have embarked upon a voyage into space. Moreover, we have unleashed vast new powers in which we can immolate human civilization and possibly even destroy life on this planet. The urgent need for humankind is to develop a new philosophical, moral, and scientific outlook, appropriate to the emergence of the global situation and the age of space. Whether we will succeed in this venture, no one can say. But if we are to do so, we will need to enlist our highest talents and skills and to cultivate the two great humanistic virtues: reason and courage.

Secular humanism has a profound and urgent task in this challenge; and that is to help us develop, as best we can, moral ideals and values, which, though tested in the crucible of reason and evidence, are yet of sufficient imaginative power and intensity that they inspire new and meaningful goals in the emerging civilizations of the future.

Index